W0037861

Topics in Applied Physics   Volume 49

Topics in Applied Physics Founded by Helmut K. V. Lotsch

Volumes 1–56 are listed on the back inside cover

Laser Spectroscopy of Solids

Edited by W. M. Yen and P. M. Selzer

With Contributions by
A. H. Francis T. Holstein D. L. Huber
G. F. Imbusch R. Kopelman S. K. Lyo R. Orbach
P. M. Selzer M. J. Weber W. M. Yen

Second Edition

With 117 Figures

Springer-Verlag Berlin Heidelberg GmbH

Professor *William M. Yen*

Department of Physics, University of Wisconsin
Madison, WI 53706, USA

Peter M. Selzer, M.D., Ph.D.

Division of Diagnostic Radiology, Stanford University
Stanford, CA 94305, USA

ISBN 978-3-540-16709-9   ISBN 978-3-540-38605-6 (eBook)
DOI 10.1007/978-3-540-38605-6

Library of Congress Cataloging-in-Publication Data. Laser spectroscopy of solids. (Topics in applied physics; v. 49) Bibliography: p. Includes index. 1. Solids – Optical properties. 2. Laser spectroscopy. I. Yen, W. M. (William M.) II. Selzer, P. M. (Peter M.) III. Francis, A. H. (Anthony H.) IV. Series. QC176.8.06L37 1986   530.4'1   86-11880

© Springer-Verlag Berlin Heidelberg 1981 and 1986

Originally published by Springer-Verlag Berlin Heidelberg New York in 1986

2153/3150-543210

We wish to dedicate this reissue to the memory of

Theodore D. Holstein

(1915–1985)

A mentor, a colleague, and a friend

Preface to the Second Edition

My colleagues and I have been gratefully pleased by the reception accorded to this volume by the scientific community. This has resulted in the decision to reissue *Laser Spectroscopy of Solids* in a second edition. As we had predicted, the activity in this research area has enjoyed an explosive growth since 1981, thus some of the contents of the monograph have inevitably become dated. Fortunately, a great deal of the material has maintained its currency, specially those sections dealing with theoretical and methodological aspects. We feel consequently that this volume remains viable as an introduction and general reference to this area of spectroscopy.

As we have already noted, there has been an impressive amount of activity in the area of laser spectroscopy as applied to the condensed phases. The commercialization of many laser devices has been widespread and the general availability of these sources has contributed greatly to the dissemination of the experimental techniques advocated in this volume. At the same time, the needs of various advanced technologies for better and more efficient optical materials have continued to stimulate the study of optical properties of solids in both theoretical and experimental senses. We cannot see anything on the horizon which will diminish this activity and can safely predict continuing advances in spectroscopic methodologies, in fundamental improvements in optical materials and in applied optical technologies.

Since the volume appeared, all areas delved in the reviews have experienced considerable advances. Many experimental laser devices have since then been commercialized and hence have become more or less commonplace laboratory tools. Even as developments have begun which allow entry into the femtosecond temporal domain, picosecond pulsed lasers have become readily available and thus studies of dynamical processes in excited states have naturally expanded into this time regime. Simultaneously, by using nonlinear processes and by exploiting the optical properties of defects, the tunability of stimulated devices has been extended further into the vuv and to the near ir. Because of these developments, many more levels and materials have become amenable to laser-spectroscopic investigation.

In the course of the past five years, we have also witnessed the maturation of coherent spectroscopic techniques such as those described in this volume. These techniques have made it possible to obtain spectra with unprecedented resolution so that the intrinsic limits have been attained in many cases. A great deal of information on superhyperfine interactions affecting optically active electrons has been obtained through these means, resolving a number of the questions which had been raised in this volume.

Dynamical effects have also been pursued vigorously in the interim between the publication of the original volume and this revision. The theoretical developments reviewed in the monograph have by and large been placed on a very firm experimental foundation. As a consequence, we now possess a much more detailed understanding of both the microscopic and macroscopic aspects of processes which produce relaxation, quenching and transfer out of excited optical states in ordered and disordered solids.

The advances in the understanding have also received considerable assistance from new theoretical concepts such as those involved in fractals. Additionally, the exponential growth in general computing capabilities has made it feasible to engage in large scale modeling of structures and of dynamics in ways which had not been thought practical even a short five years ago. All of these developments have given us new insights and new opportunities for exploitation.

The present volume has been fairly criticized for having been restricted to the area of insulators. As it was noted in the preface to the first edition, this restriction was placed principally on account of space restrictions. It is our intention to amend this omission by the publication of a sequel volume to *Laser Spectroscopy of Solids* in the near future. The sequel will also include an extensive update of the various advances in the field which we have alluded to above.

We have been permanently saddened by the untimely passing of one of our colleagues in this endeavor. Professor Theodore Holstein was a valued colleague to all of us as well as a mentor and a friend. We wish then to dedicate this reissue volume to his memory as a small expression of our joint appreciation.

Finally, we are pleased that the reissue is appearing in a lower-price paperback form. This economy will allow a larger student audience to avail themselves of this review of ours. We are thankful to Dr. H. K. V. Lotsch and to Springer for their continued encouragement and support which has served to make this effort possible and a positive experience.

Madison, January 1986   *W. M. Yen*

Preface to the First Edition

In this volume we have attempted to present a concise survey of the spectroscopic properties of insulators as derived from the application of tunable laser spectroscopic techniques. As has been the case in gaseous atomic spectroscopy, the use of tunable lasers has allowed the extension and the refinement of optical measurements in the condensed phases to unprecedented resolutions in the frequency and temporal domains. In turn, this firmer base of empirical findings has led to a more sophisticated theoretical understanding of the spectroscopy of optically excited states with major modifications being apparent in the area of their dynamic behavior. Yet the revivalistic nature of these advances implies that additional advances are to be expected as the techniques and developments outlined in this volume are put to widespread use. Regardless, it is our hope and that of our distinguished colleagues in this venture that the reviews presented here will be useful to neophytes and veterans to this field alike — to the former as a laissez-passer into solid-state spectroscopy, to the latter as a useful synopsis and reference of recent developments.

We have also attempted to expose the reader to the concept that optically active materials, be they organic or inorganic, as universality would require, behave in a like manner and, though terminology may vary in detail, the outline and general features of all insulators remain constant.

The book is organized as follows: Chapter 1 surveys in general terms the field of spectroscopy of insulators and establishes the basic features the other chapters refer to. Chapters 2 and 3, respectively, deal with the microscopic and macroscopic aspects of the theory of dynamics of optically excited states with emphasis on ion-ion interactions which are responsible for optical energy transfer and diffusion in condensed phases. Chapter 4 details experimental techniques which are used in laser spectroscopy in solids. Finally, the last three chapters present surveys of the empirical status of these studies in ionic or crystalline, glassine or amorphous, and organic solids, respectively.

The other areas in the study of optical properties of condensed matter where lasers have played crucial roles and where considerable advances have been made, such as semiconductors and the various types of light scattering experiments, are not the principal focus of this volume and, hence, will not be reviewed here.

Finally, we wish to acknowledge the many people who have encouraged us and collaborated with us in various phases of compilation. We specially wish to thank Dr. H. K. V. Lotsch and the editorial staff of Springer for their help and patient guidance, and Dr. E. Strauss and Dr. S.T. Lai for their assistance during the preparation of this volume. A special note of thanks is due Ms. Karen M. Wick who did the majority of the type composition of this volume; her ability was only surpassed by her patience and understanding. One of us (W.M.Y.) has benefitted from support from the John Simon Guggenheim Memorial Foundation during the 1979-80 academic year. We also acknowledge support from the National Science Foundation and the Army Research Office for the preparation of the manuscripts.

January 1981   *W. M. Yen · P. M. Selzer*

Contents

Contributors

Francis, Anthony H.
   Department of Chemistry, University of Michigan, Ann Arbor, MI 48109, USA

Holstein, T. (deceased)

Huber, David L.
   Department of Physics, University of Wisconsin, Madison, WI 53706, USA

Imbusch, George F.
   Department of Physics, University College, Galway, Ireland

Kopelman, Raoul
   Department of Chemistry, University of Michigan, Ann Arbor, MI 48109, USA

Lyo, S. K.
   Sandia National Laboratory, Albuquerque, NM 87185, USA

Orbach, R.
   Department of Physics, University of California, Los Angeles, CA 90024, USA

Selzer, Peter M.
   Division of Diagnostic Radiology, Stanford University,
   Stanford, CA 94305, USA

Weber, Marvin J.
   Lawrence Livermore National Laboratory, University of California,
   Livermore, CA 94550, USA

Yen, William M.
   Department of Physics, University of Wisconsin, Madison, WI 53706, USA

1. Optical Spectroscopy of Electronic Centers in Solids

G. F. Imbusch and R. Kopelman[*]

With 20 Figures

1.1 Overview

In this chapter we present an outline of the optical spectroscopy of ions and molecules in solids which will serve as a background for the later chapters. We adopt a general approach which takes into account both organic and inorganic systems, and we attempt to elucidate the similarities and differences between the two systems. We start by considering the case of a low concentration of optically active centers in an otherwise optically inert crystalline host. The centers are considered to be too far apart from each other to interact. In this case, when all the centers are identical, we need only analyze the properties of one representative center and apply appropriate statistical considerations. Afterwards we consider the case where the concentration of optically active centers is large enough so that adjacent centers can interact; this interaction then leads to the dynamical processes whose investigation by laser techniques and subsequent theoretical analyses form the theme of this book.

1.2 Interaction of Electronic Center with Optical Radiation

Electromagnetic radiation interacts with an electronic center through the electric or magnetic fields of the radiation, the operators describing the interaction being $-\boldsymbol{p} \cdot \boldsymbol{E}$ or $-\boldsymbol{m} \cdot \boldsymbol{B}$ for these two fields respectively. The electric dipole moment is $\boldsymbol{p} = \Sigma_i\, e\boldsymbol{r}_i$, where the summation is over all the optically active electrons. Similarly, the magnetic dipole moment is $\boldsymbol{m} = \Sigma_i\, (e/2m)\, (\boldsymbol{l}_i + 2\boldsymbol{s}_i)$. The probability of a radiative transition, emission or absorption, between an initial level a and a final level b is proportional to the square of the matrix element between a and b of the appropriate operator, and this quantity is called the strength, S, of the transition.

$$S(ab) = \sum_{a,b} |\langle b\,|\,D\,|\,a \rangle|^2 \qquad (1.1)$$

where the general dipole operator, D, is replaced by $\boldsymbol{p}$ for an electric dipole transition and by $\boldsymbol{m}$ for a magnetic dipole transition. The summation in (1.1) is over the two manifolds of states which make up levels a and b.

This matrix element is zero if states a and b differ in total spin, hence we find the spin selection rule: $\Delta S = 0$. Because of spin-orbit coupling, some admixture of

[*] Supported by NSF Grant DMR 800679 and NIH Grant 2 R01 NS 08116-10A1

difference spin states occurs and as a result spin is not a valid quantum parameter and the spin selection rule is not rigorous. Nevertheless, it is found that, in general, spin-allowed ($\Delta S = 0$) transitions are stronger than spin-forbidden ($\Delta S \neq 0$) transitions. When we have discussed the orbital nature of the states we will derive orbital selection rules.

It is useful to define a dimensionless quantity called the *oscillator strength*, f [1.1, 1.2]:

$$f(ab) = \frac{1}{g_a} \sum_{a,b} \frac{8\pi^2 m\nu}{3he^2} |\langle b | D | a \rangle|^2$$

$$= \frac{1}{g_a} \frac{8\pi^2 m\nu}{3he^2} S(ab) \qquad (1.2)$$

where ν is the frequency of the transition and $h\nu = |E_b - E_a|$. D is the appropriate dipole operator, p or m, and g_a is the statistical weight of the initial state a. For an allowed electric dipole transition $f_{ED} \simeq 1$, while for an allowed magnetic dipole transition $f_{MD} \simeq 10^{-6}$, so the magnetic dipole process is very much the weaker. Higher multipole processes can also occur but they are negligibly small.

Macroscopically we measure the strength of an absorption transition by measuring the absorption coefficient $k(\nu)$ as the radiation passes through a material containing the optically active centers. $k(\nu)$ is defined by

$$I_\nu(d) = I_\nu(0) \exp[-k(\nu)d] \qquad (1.3)$$

where $I_\nu(d)$ is the intensity of the radiation of frequency ν after traversing a thickness d of the material. We will assume that $k(\nu)$ is the absorption coefficient found after averaging over all polarizations relative to the crystal axes of the host material. We define a normalized absorption coefficient, $\sigma(\nu) = N^{-1}k(\nu)$ where N is the number of centers per unit volume. If the absorption occurs by an electric dipole process, then $\sigma(\nu)$ is related to f_{ED} and S_{ED} by [1.1, 2]

$$\int \sigma(\nu)d\nu = N^{-1} \int k(\nu)d\nu = \frac{1}{4\pi\epsilon_0} \frac{\pi e^2}{mc} \left[\left(\frac{n^2+2}{3} \right)^2 \cdot \frac{1}{n} \right] f_{ED}(ab)$$

$$= \frac{1}{4\pi\epsilon_0} \frac{8\pi^3 \nu}{3hc} \left[\left(\frac{n^2+2}{3} \right)^2 \cdot \frac{1}{n} \right] \frac{1}{g_a} S_{ED}(ab) \qquad (1.4)$$

The term in the square brackets takes into account the polarization of the material by the electric field. n is the refractive index. In these formulas SI units are employed. The formulas for cgs units are identical to the above except that the $1/4\pi\epsilon_0$ term is omitted.

The Einstein spontaneous transition probability for a radiative transition between a and b is $A(ab)$. If the radiative process is electric dipole, then we have

$$A_{ED}(\nu) = \frac{1}{4\pi\epsilon_0} \frac{8\pi^2\nu^2 e^2}{mc^3} \left[\left(\frac{n^2+2}{3}\right)^2 \cdot n\right] f_{ED}(ab)$$

$$= \frac{1}{4\pi\epsilon_0} \frac{64\pi^4\nu^3}{3hc^3} \left[\left(\frac{n^2+2}{3}\right)^2 \cdot n\right] \frac{1}{g_a} S_{ED}(ab). \tag{1.5}$$

If $A_{ED}(ab)$ is the only radiative process from level a, then $A_{ED}(ab) = \tau_R^{-1}$, where τ_R is the radiative decay time. Then by putting numerical values into (1.5) we can write (SI units)

$$f_{ED}(ab)\tau_R(ab) = 1.5 \cdot 10^4 \lambda_0^2 \left[\left(\frac{n^2+2}{3}\right)^2 \cdot n\right]^{-1} \tag{1.6}$$

where λ_0 is the wavelength in vacuum. For $\lambda_0 = 5000$ Å and $n = 1$ the quantity on the right has the approximate value $4 \cdot 10^{-9}$ s. This means that an allowed electric dipole transition ($f_{ED} = 1$) will have a radiative decay time of around $4 \cdot 10^{-9}$ s. For the trivalent rare-earth ions and the transition metal ions we usually find τ_R between 10^{-6} s and 10^{-2} s. These then are either weak electric dipole transitions or magnetic dipole transitions. For molecular centers one find the whole range of τ_R values, up to 10 s and longer.

The above formulas for absorption coefficient and radiative decay rate also hold for magnetic dipole processes, but f_{MD} and S_{MD} are used, and the $(n^2 + 2/3)^2$ term is replaced by n^2.

Lastly we mention the Einstein stimulated transition probability, B. If ρ_ω is the energy density of the radiation field per unit volume per unit frequency range ($\Delta\omega = 1$), then the probability per second that a transition between levels a and b will be stimulated by the resonant radiation field is $B\rho_\omega$. B is related to the Einstein spontaneous transition probability between the same levels, A, through

$$A/B = (\hbar\omega^3/\pi^2 c^3)n^3 = (4h\nu^3/c^3)n^3. \tag{1.7}$$

1.3 Eigenstates for the Electronic Center in a Solid

The Hamiltonian describing the optically active electronic center in an optically inert crystalline solid can be written

$$\mathcal{H} = \mathcal{H}_{\text{isolated center}} + \mathcal{H}_{\text{elec-static lattice}} \tag{1.8}$$

$$+ \mathcal{H}_{\text{elec-dynamic lattice}} + \mathcal{H}_{\text{dynamic lattice}}.$$

$\mathcal{H}_{\text{isolated center}}$ describes the electronic center as if it were isolated from the remainder of the solid. For dopant ions in an inorganic solid this is $\mathcal{H}_{\text{free ion}}$ with eigenstates $^{2S+1}L_J$. For molecular centers this can be written $\mathcal{H}_{\text{isolated molecule}}$, and we shall consider this term in more detail later in this section. For an F center, which is an electron trapped at a cation vacancy, $\mathcal{H}_{\text{isolated center}}$ and $\mathcal{H}_{\text{elec-static lattice}}$ cannot be separated from each other.

$\mathcal{H}_{\text{elec-static lattice}}$ describes how the center is affected by the average static environment. For dopant ions this is the crystal field or ligand field energy, and it is larger for transition metal ions than for rare earth ions. This is not an important term for molecular centers which interact only weakly with their environments.

$\mathcal{H}_{\text{dynamic lattice}}$ is the vibrational energy of the crystalline lattice, which is described in terms of lattice phonon modes.

$\mathcal{H}_{\text{elec-dynamic lattice}}$ describes how the electronic center is affected by the lattice modes. For ions in a crystalline solid this can be visualized as a dynamic crystal field energy, and it can modify the shape and strength of the optical transitions. In addition, it can give rise to nonradiative processes on the ions. For molecular solids this is not an important term as the interaction between the optically active molecular center and the adjacent molecules of the host material is weak. When the concentration of electronic centers is high so that adjacent centers can interact with each other, $\mathcal{H}_{\text{elec-dynamic lattice}}$ can affect the transfer of optical excitation energy between adjacent centers. This transfer will be discussed in a later section. Since the molecular center is a much more complicated entity than the ionic center, we need to consider $\mathcal{H}_{\text{isolated molecule}}$ in more detail. We write it as

$$\mathcal{H}_{\text{isolated molecule}} = \mathcal{H}_{\text{static molecule}} + \mathcal{H}_{\text{mol. vibrations}}$$
$$+ \mathcal{H}_{\text{elec-mol. vibrations}}. \qquad (1.9)$$

The first term is the energy of the electrons in a rigid time-average molecular structure. The second term describes the allowed intra-molecular vibrations, while the third term describes how the electronic energy is affected by these intramolecular vibrations. This last term plays a similar role for molecular solids as $\mathcal{H}_{\text{elec-dynamic lattice}}$ plays for ionic solids. The three terms of (1.9) combined are often called $\mathcal{H}_{\text{vibronic}}$ but will be designated for simplicity as $\mathcal{H}_{\text{elec}}$.

In a spectroscopic treatment of ionic centers, one first works out the eigenstates of $\mathcal{H}_{\text{free ion}} + \mathcal{H}_{\text{elec-static lattice}}$. One gets a finite number of eigenstates with discrete energy levels, and one can calculate the strengths of the sharp optical transitions between these states. $\mathcal{H}_{\text{elec-dynamic lattice}}$ is then taken into account by perturbation theory. This treatment of ionic centers is described in Sect. 1.4.

In a spectroscopic treatment of molecular centers in solids, one first works out $\mathcal{H}_{\text{elec}}$ (actually $\mathcal{H}_{\text{vibronic}}$) from ab initio calculations or from *free molecule gas-phase spectroscopic data* (for which, if not available, liquid or solid solution data may be substituted). In the latter case one utilizes eigenvalues and optical transition strengths derived from spectroscopic analysis. The $\mathcal{H}_{\text{elec-static lattice}}$ term involves what is called the *site-shift* (energy shift) and *site-group splitting*. A group-theoretical consideration based on a correlation of the free molecule symmetry with that of the site gives the number of site-group components and predicts whether a transition, forbidden in the free molecule, becomes allowed in the solid. However, the magnitude of the shift and splitting are usually derived from spectroscopic measurements. The term $\mathcal{H}_{\text{elec-dynamic lattice}}$ is again treated in Sect. 1.4. The treatment of molecular centers is described in Sect. 1.5.

1.4 Energy Levels and Radiative Transitions in Ionic Centers

In this section we consider the optical spectroscopy of individual ionic centers in a
solid when the concentration is so low that the centers do not interact with each
other [1.3]. We treat rare earth ions and transition metal ions separately. Other
centers, such as color centers, are not considered as these are not the systems of
major interest in the later chapters.

1.4.1 Rare Earth Ions in a Static Environment

For the rare earth ions the crystal field energy, $\mathscr{H}_{\text{elec-static lattice}}$, is weak because of
the shielding of the outer $5s$ and $5p$ shells of electrons. So we consider $\mathscr{H}_{\text{free ion}}$ first
and take the other terms into account afterwards by perturbation theory.

    We write the free ion Hamiltonian as

$$\mathscr{H}_{\text{free ion}} = \mathscr{H}_0 + \mathscr{H}_C + \mathscr{H}_{\text{s.o.}} \tag{1.10}$$

$\mathscr{H}_0$ describes the interaction of each optically active electron with the spherically
symmetric ion core. This gives us the independent $4f$ atomic electron orbitals which
form the $(4f)^N$ electron *configuration*. $\mathscr{H}_C$ takes into account the Coulomb interaction of the $4f$ electrons with each other. $\mathscr{H}_C$ splits the $(4f)^N$ configuration state
into a number of states of different energy, each of which is characterized by values
of L and S. We call these the *LS terms*. Configuration mixing by $\mathscr{H}_C$ is small and
usually neglected. Particularly if the number (N) of $4f$ electrons is large, there can
be more than one state with the same values of L and S, and so an additional label
(γ) must be used. The states of different energy but with the same LS values are
mixed together by $\mathscr{H}_C$.

    Next spin-orbit coupling is taken into account. This interaction has the form

$$\mathscr{H}_{\text{s.o.}} = \sum_i \xi(r_i) l_i \cdot s_i \tag{1.11}$$

where the summation is over all the $4f$ electrons. Since $\mathscr{H}_{\text{s.o.}}$ commutes with J, the
eigenstates of $\mathscr{H}_{\text{s.o.}}$ are labelled by J, M_J. If we assume that the matrix elements of
$\mathscr{H}_{\text{s.o.}}$ are much smaller than the separation between terms, then we can neglect the
mixing of different LS terms. In this approximation (the Russell-Saunders approxi-
mation), $\mathscr{H}_{\text{s.o.}}$ can be written as $\zeta L \cdot S$, and it splits up each term into a number of
states each of which is characterized by the quantum numbers SLJ. These are the
J multiplets and the separation between them should be in accordance with the
Landé interval rule. Each of the LSJ levels is $(2J + 1)$-fold degenerate, these states
being characterized by the M_J values.

    In practice the different LS terms are too close together for the Russel-Saunders
approximation to be valid and so $\mathscr{H}_{\text{s.o.}}$ mixes up states of the same J (and M_J)

but from different *LS* terms. The resultant *intermediate coupling* eigenstates, $|f^N\{\gamma SL\}J\rangle$, can be written in terms of the Russel-Saunders states ($|f^N\gamma SLJ\rangle$)

$$|f^N\{\gamma'S'L'\}J\rangle = \sum_{\gamma SL} c(\gamma SL)\,|f^N\gamma SLJ\rangle. \tag{1.12}$$

The *c* coefficients are generally obtained by computer diagonalization of the $\mathcal{H}_C$ + $\mathcal{H}_{s.o.}$ matrix. The $\{\gamma'S'L'\}$ label for the intermediate coupling state in (1.12) is usually that which corresponds to the largest contributing Russell-Saunders state. Very elaborate and detailed methods have been worked out for obtaining fairly exact free ion states for the triply charged rare-earth ions. These methods are discussed in [1.4–7], in a recent review article by *Pappalardo* [1.8], and in some technical memoranda [1.9].

Figure 1.1 (the Dieke diagram) shows the energy levels of the trivalent rare earth ions in LaCl$_3$, and the states are labelled according to the $S'L'J$ values in the form $^{2S'+1}L'_J$. A measure of the deviation from Russell-Saunders coupling is the depature of the separation between different J levels in the same $^{2S+1}L_J$ group from the Lande interval rule.

The crystal field energy, $\mathcal{H}_{\text{elec-static lattice}}$, is now taken into account. This causes a splitting of the free ion levels, and the extent of the splitting in LaCl$_3$ is indicated by the width shown for the levels in the Dieke diagram. The crystal field for a particular site in a particular host crystal is parameterized by a number of parameters, B_q^k. Group theory tells us the number of parameters needed to characterize a given crystal field, and the values of these parameters are obtained by comparing calculated and observed splittings. The crystal field term can mix states with different J values (*J mixing*) which can affect radiative selection rules.

The positions of the levels shown in Fig. 1.1 were obtained by absorption studies. In addition, luminescence occurs from those levels in Fig. 1.1 which show a pendant semicircle. For a given rare-earth ion all these free ion levels are formed from the same $(4f)^N$ configuration and so transitions between them should occur only by a magnetic dipole process. Most transitions on rare earth ions, however, occur by an electric dipole process. Hence there must be some mixing of the opposite parity $(4f)^{N-1}5d$ configuration into the $(4f)^N$ configuration. This mixing can occur if the rare-earth ion experiences a crystal field which lacks inversion symmetry — which is the case in many crystals. But even when the rare earth ion appears to occupy a site with inversion symmetry, the transitions are often electric dipole. This may indicate a tendency for the substituting rare earth ion to distort its surroundings and remove the inversion symmetry.

We consider the transitions between free ion states, neglecting the splittings due to the crystal field but allowing for some admixture of opposite parity $(4f)^{N-1}5d$ states by the crystal field.

The strength of the transition $S_{\text{ED}}(ab)$ is given by

$$S_{\text{ED}}(ab) = \sum_{a,b} |\langle b|D|a\rangle|^2 \tag{1.1}$$

where a (or b) refers to the full set of states in the $f^N \{\gamma SL\} J$ multiplet. $S_{ED}(ab)$ is the sum of squares of a number of matrix elements. Now if an odd-parity crystal field term, $\mathcal{H}_{odd}$, mixes states of opposite parity and permits the electric dipole process, then the above matrix element has the form [1.10, 11]

$$\sum_\beta \frac{\langle b | \mathcal{H}_{odd} | \beta \rangle \langle \beta | D | a \rangle}{E_b - E_\beta} + \text{converse}$$

where β signifies all those opposite parity states mixed in by $\mathcal{H}_{odd}$. We do not have sufficient knowledge about the nature and positions of the β states to carry out an exact theoretical calculation of the above matrix element. *Judd* [1.11] and *Ofelt* [1.10], working on this problem independently, introduced a very helpful simplification by ascribing the same energy denominators to all β states. The constant energy denominator can now be removed from the summation and closure used. In this way one can regard

$$\sum_\beta \mathcal{H}_{odd} | \beta \rangle \langle \beta | D$$

as an even parity operator which operates between the free ion multiplets. S_{ED} can now be written

$$S_{ED}(ab) = e^2 \sum_{t=2,4,6} \Omega_t | \langle f^N \gamma_b S_b L_b J_b \| U^{(t)} \| f^N \gamma_a S_a L_a J_a \rangle |^2 \qquad (1.13)$$

Ω_t are the Judd-Ofelt intensity parameters and reflect the strength and nature of the odd-parity field. The reduced matrix elements of the tensor operator $U^{(t)}$ can be evaluated. Values have been tabulated for many ions in different hosts and the sources are listed by *Riseberg* and *Weber* [1.12]. *Carnall* et al. [1.9] have given a complete listing of values for transitions on trivalent rare earth ions in LaF_3.

The relationship between S_{ED} and f_{ED} and how they are related to the absorption coefficient are given in (1.2, 4, 5). In this case the statistical weight factor g_a is $(2J_a + 1)$.

By measuring the absorption strengths for a number of transitions from the ground multiplet one can determine the S_{ED} values and from these one can obtain values for the Judd-Ofelt parameters Ω_t. Once the Judd-Ofelt parameters have been obtained for a given rare earth-host combination they can be used to calculate absorption strengths and radiative decay rates between any two of the f^N levels of the system. The radiative decay rate from a particular state a is given by

$$1/\tau_R = \sum_b A(ab). \qquad (1.14)$$

If the observed decay time, τ, is shorter than τ_R the discrepancy is attributed to nonradiative processes. Hence one gets the nonradiative decay rate from state a:

$$W_{nr} = 1/\tau - \sum_b A(ab). \qquad (1.15)$$

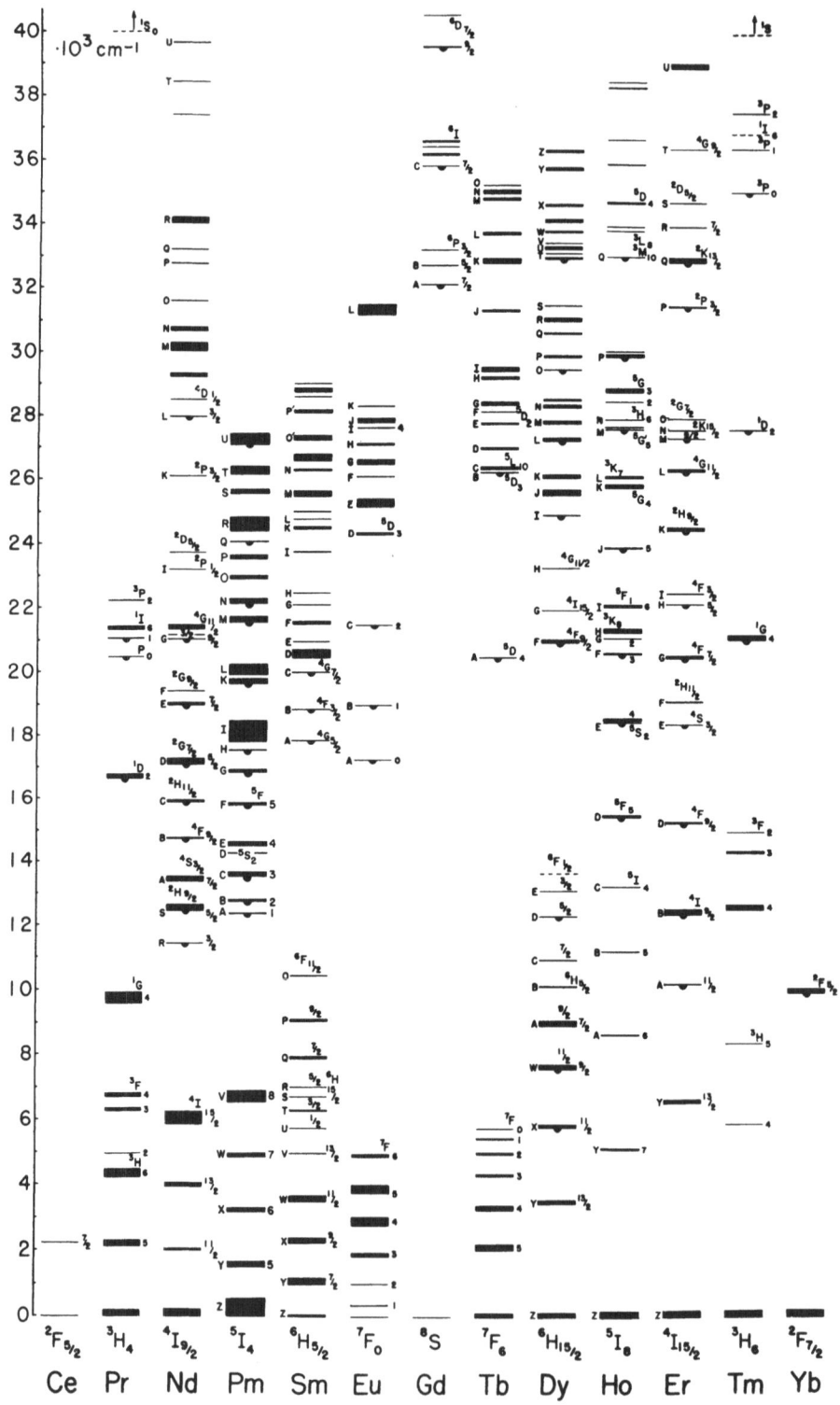

This has been the method employed by *Weber* [1.13] to obtain W_{nr} values for his analysis of nonradiative decay rates in trivalent rare-earth ions.

The Judd-Ofelt approach elicits some general selection rules for electric dipole transitions on the rare earths:

a) $\Delta J < 6$
b) For a rare earth ion with an even number of electrons −

   i) $J = 0 \rightarrow J' = 0$ is forbidden,
   ii) $J = 0 \rightarrow$ odd J' are weak,
   iii) $J = 0 \rightarrow J' = 2, 4, 6$ should be strong,
   iv) $J = 1 \rightarrow J' = 2$ should appear only in σ polarization.

We must recall that J mixing by the crystal field can relax these selection rules to some extent.

The analysis of magnetic dipole processes is easier. The magnetic dipole operator is $m = \Sigma_i \, e/2m(l_i + 2s_i)$ so we can similarly define the strength, S_{MD}, of the magnetic dipole transition as

$$S_{MD} = \sum_{a,b} \, | \langle f^N \{ \gamma_b S_b L_b \} J_b \, | \, m \, | f^N \{ \gamma_a S_a L_a \} J_a \rangle |^2 \qquad (1.16)$$

and the formulas for oscillator strength, absorption strength, and radiative decay rate (1.2, 4, 5) can again be used for magnetic dipole processes if S_{ED} is replaced by S_{MD} and the $(n^2 + 2/3)^2$ term is replaced by n^2. We also find the general selection rule $\Delta J = 0, \pm 1$ (with $0 \rightarrow 0$ forbidden) but J mixing can again relax this condition.

1.4.2 Transition Metal Ions in a Static Environment

The individual terms in the Hamiltonian being considered here are

$$\mathcal{H}_0 + \mathcal{H}_C + \mathcal{H}_{s.o.} + \mathcal{H}_{elec\text{-}static\ lattice}.$$

For the transition metal ions the lowest eigenstate of $\mathcal{H}_0$ is the $(3d)^N$ configuration state made up of independent one-electron $3d$ states. The $3d$ electrons are on the outside of the ion and are very sensitive to the effects of the crystalline environment, and as a result we find that $\mathcal{H}_{elec\text{-}static\ lattice}$ is comparable with $\mathcal{H}_C$, the Coulomb interaction of the $3d$ electrons with each other. It is customary now to adopt

◄ **Fig. 1.1.** Energy level diagram of the trivalent rare earth ions in LaCl₃. The width of a level represents the crystal field splitting in LaCl₃. Luminescence occurs from these levels with a pendant semicircle. This figure is a compilation of work carried out by the late Professor Dieke's group at Johns Hopkins, and is reproduced here through the courtesy of Dr. H. M. Crosswhite

the strong field scheme, in which one first considers how the d orbitals are affected by the electrostatic field of the surrounding ions which is approximately of either octahedral or tetrahedral symmetry. We shall only concern ourselves with the octahedral case. This octahedral field splits the fivefold degenerate d orbital into a twofold degenerate e_g state and a threefold degenerate t_{2g} state. The energy separation between e_g and t_{2g} is labelled $10\,Dq$, where Dq is a parameter which measures the strength of the octahedral field. This purely ionic model ignores covalency effects, and a more accurate theory — ligand field theory — can be developed which takes account of the electron transfer between the transition metal ion and the neighboring ligands. In ligand field theory the t_{2g} and e_g orbitals do not have d character exclusively. The symmetry properties of the system, however, remain unchanged, and as long as one does not expect quantitative accuracy the conceptually simpler crystal field theory picture can be maintained.

Starting with these one-electron t_{2g} and e_g orbitals one can now take $\mathcal{H}_C$, into account. The new eigenstates are characterized by the values of S and Γ, where Γ is a label which indicates the irreducible representation of the octahedral group to which the eigenstate belongs.

These $S\Gamma$ *terms* are analogous to the SL terms in the free ion case which were used in developing the rare earth eigenstates. Spin-orbit interaction is weaker for the transition metal ions than for the rare earth ions, and it is usual to treat $\mathcal{H}_{s.o.}$ as well as any deviations from perfect octahedral crystal symmetry by perturbation theory. These interactions split up the $S\Gamma$ terms into *multiplets*, analogous to the splitting of the SL terms into J multiplets in the case of the free ions. The text by *Sugano* et al. [1.14] gives the theoretical development of the transition metal ion eigenstates sketched here.

The Coulomb interaction between electrons is characterized by three Racah parameters, A, B, C, but since the A parameter does not enter into the expressions for the energy separation between $S\Gamma$ terms we only concern ourselves with B and C. And it is expected that the C/B ratio (which is labelled γ) will have almost the same value through the full transition metal ion group; this ratio is around 4.5. Hence for each ion the energy separations between the various $S\Gamma$ levels depend on two parameters, Dq and B, the value of γ being assumed constant. Energy level diagrams showing how energy E (in units of B) varies with the ratio of Dq/B have been given by *Tanabe* and *Sugano* [1.15] for all the transition metal ions and these are given in their text. Figure 1.2 shows the Sugano-Tanabe diagram for the $(3d)^3$ system (V^{2+}, Cr^{3+}, Mn^{4+}). The levels are labelled according to the $^{2S+1}\Gamma$ values and the strong field configuration from which the term is derived is given in parentheses. The energy is measured relative to that of the lowest $S\Gamma$ term, $^4A_{2g}$ in this case. (Sometimes the "g" subscript — indicating even-parity states — is omitted.)

The broken vertical line in Fig. 1.2 is drawn at the value of Dq/B appropriate to a number of Cr^{3+} systems (e.g., ruby). Figure 1.3 shows the energy level diagram for Cr^{3+} in ruby where the lower symmetry trigonal field and spin-orbit interaction are taken into account. The absorption and luminescence spectra taken at low temperatures are also shown.

Radiative transitions between purely $(3d)^N$ states should occur only by a magnetic dipole process. Just as in the case of the rare earth ions, however, any deviation from inversion symmetry in the site occupied by the ion can cause a slight mixing of $(3d)^{N-1}4p$ and $(3d)^{N-1}4f$ odd-parity configurations into the $(3d)^N$ configuration which will permit electric dipole transitions to occur. For example, the strong dichroism in the ruby spectrum can be satisfactorily explained by taking into account how the odd-parity trigonal crystal field at the Cr^{3+} site mixes the odd-parity configurations into the $(3d)^3$ states [1.16, 17]. In the absence of spin-orbit coupling we expect the selection rule, $\Delta S = 0$, to apply. The presence of the weak spin-orbit interaction weakens this selection rule but as Fig. 1.3 shows, spin-forbidden ($\Delta S \neq 0$) transitions are much weaker than spin-allowed ($\Delta S = 0$) transitions.

From Fig. 1.2 we notice than the energy separation between $^4T_{2g}(t_2^2 e)$ and $^4A_{2g}(t_2^3)$ depends strongly on Dq/B, as does the separation between $^4T_{1g}(t_2^2 e)$ and $^4A_{2g}(t_2^3)$. Because of lattice vibrations and the associated modulation of Dq the absorption transitions between these pairs of levels are broadened (Fig. 1.3). The shape of optical transitions will be discussed later in connection with the interaction of the electrons with the dynamic lattice. From Fig. 1.3 we also see that luminescence occurs only from the ^2E_g levels at low temperatures. If a Cr^{3+} ion is raised to a higher excited state, $^4T_{1g}(t_2^2 e)$ for example, it quickly decays nonradiatively through the small energy gaps to adjacent lower levels until the ^2E_g state is reached. Because of the large gap between ^2E_g and $^4A_{2g}$ nonradiative decay is unlikely and the ^2E_g state is de-excited by radiative decay. The radiative decay time is around 4 ms.

Figure 1.4 shows the energy level diagram for Mn^{2+} $(3d)^5$ in an octahedral field. When Mn^{2+} ions are excited luminescence occurs only from the lowest ex-

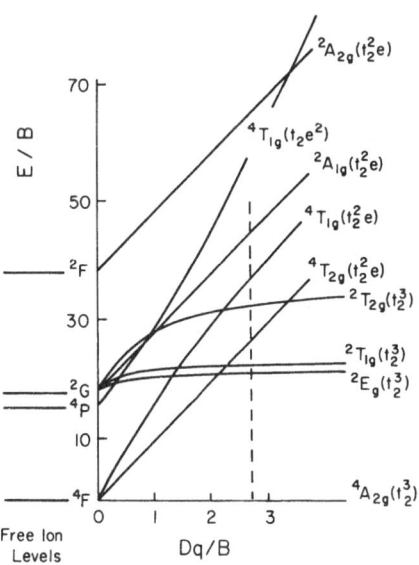

Fig. 1.2. Energy level diagram of the $(3d)^3$ electronic configuration in an octahedral field. The dependence of E/B on Dq/B varies slightly for different values of B and C. The above is plotted for $B = 640$ cm^{-1} and $C = 3300$ cm^{-1}. The vertical broken line is drawn at a value of Dq/B appropriate to Cr^{3+} is ruby

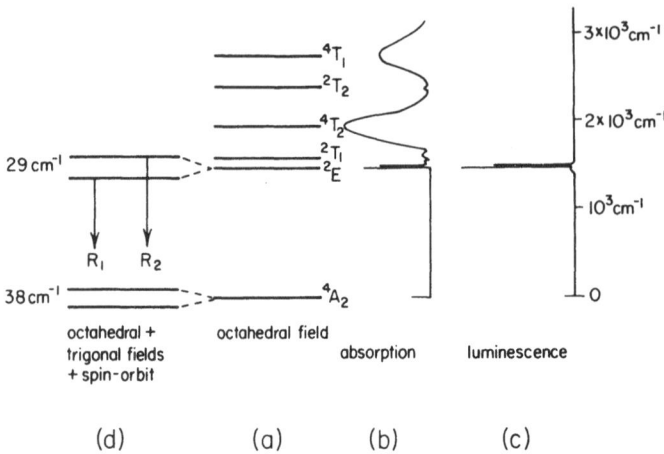

Fig. 1.3. **(a)** Energy level diagram of Cr^{3+} in an octahedral field; **(b)** Absorption spectrum of ruby at low temperatures. The spin allowed transitions are much stronger than the spin-forbidden transitions; **(c)** Luminescence spectrum of ruby at 77 K. The luminescence consists of two sharp lines and an associated vibrational sideband from the ^2E state. Luminescence does not occur at low temperatures from the higher excited states; **(d)** Fine structure in the ^2E and $^4A_2$ levels. The 29 cm^{-1} splitting in the ^2E state results in two sharp fluorescence lines. The 0.38 cm^{-1} splitting in the $^4A_2$ state can be resolved in good quality ruby crystals

cited $^4T_{1g}$ state. Again we note that this is the only excited state with a large gap ($> 10,000$ cm^{-1}) to the next lower level.

Nonradiative decay comes about because of the interaction of the optically active electrons with the dynamic lattice, and this nonradiative decay will be discussed in Sect. 1.7.

1.4.3 Interaction of the Optically Active Ion with Lattice Modes

Up to now we have considered the crystalline environment as merely supplying a static average field. Now we consider the effect of the dynamic lattice modes, $\mathcal{H}_{\text{elec-dynamic lattice}}$, on the radiative transitions on the rare earth and transition metal ions.

A very simple picture of the effect of the dynamic lattice is given by the *configurational coordinate model* [1.18, 19, 20] which is successful in explaining the shape of the broadband transitions. Rather than attempt to take into account the effect of all possible lattice modes this model assumes that we need only consider one representative lattice mode, which is taken to be the "breathing mode", in which the surrounding lattice pulsates in and out about the optically active ion. It also assumes that a harmonic oscillator model can describe the pulsations. We need only a single variable to describe this breathing mode — the distance between the

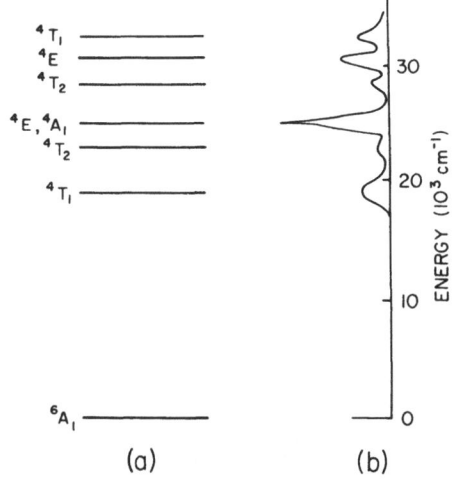

Fig. 1.4. (a) Energy level diagram of Mn^{2+} in an octahedral field; (b) Room temperature absorption spectrum of MnF_2

optically active ion and the neighboring environment — which is called the configurational coordinate Q. The full ion-plus-lattice state is described by $| \psi, \chi_n \rangle$, where ψ is the electronic state of the ion and χ_n is the vibrational state of the lattice. n is the occupancy of the vibrational mode.

The configurational coordinate diagram showing the ground and excited states of a hypothetical ion-plus-lattice system is shown in Fig. 1.5. When the ion is in the ground electronic state, g, the configurational coordinate has the average value Q_0. The ground state harmonic oscillator parabola, $\frac{1}{2} M\omega^2 (Q - Q_0)^2$, is shown in the figure. M is the equivalent ionic mass engaged in the vibration, and ω is the vibrational frequency. The nth vibrational state is described by the harmonic oscillator function, χ_n. When the ion is in the excited state, e, the average value of the configurational coordinate is Q_0'. The mth vibrational state is described by the harmonic oscillator function χ_m'.

We can assume, for simplicity, than the two parabolas have the same frequency, ω. The change in configurational coordinate, $Q_0 - Q_0'$, is a measure of the difference in sensitivity of the two electronic levels to changes in the environment. This difference in sensitivity is characterized by a dimensionless quantity called the *Huang-Rhys parameter*, S, defined by

$$\frac{1}{2} M\omega^2 (Q_0 - Q_0')^2 = S\hbar\omega. \tag{1.17}$$

The probability of an absorption transition from $| g, \chi_n \rangle$ to $| e, \chi_m' \rangle$ depends upon the square of the matrix element $\langle e | D | g \rangle \langle \chi_m' | \chi_n \rangle$ where D is the appropriate dipole operator. Particularly when $S > 1$ the overlap integrals, $\langle \chi_m' | \chi_n \rangle$, are non-zero for a large range of values of n and m, and so the radiative transition can be accompanied by the creation and annihilation of a number of phonons. Figure 1.6a

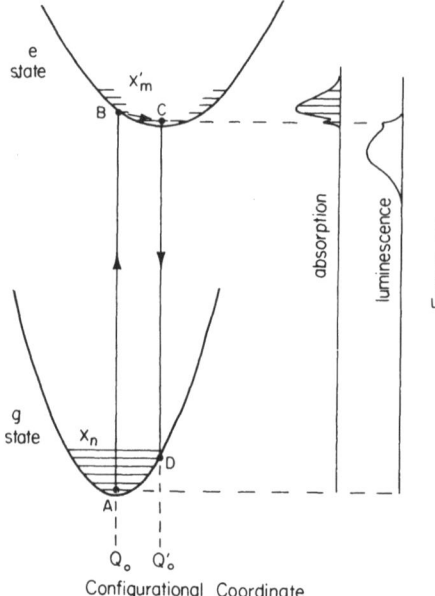

Fig. 1.5. Configurational coordinate diagram for the analysis of optical transitions between e and g. At low temperatures the absorption originates on the χ_0 level, and the peak of the absorption band corresponds to transition $A-B$. At low temperatures nonradiative $B-C$ relaxation is very rapid, hence the luminescence originates on the χ_0' level, and the peak of the luminescence band corresponds to transition $C-D$. The energy separation between the peaks is the Stokes' shift

shows the predictions of the model for a particular of S in the case of an absorption transition at low temperatures ($n = 0$ only). The vertical lines give the relative values of the $|\langle \chi_m' | \chi_0 \rangle|^2$ factors. The $|\langle \chi_0' | \chi_0 \rangle|^2$ factor, which has the values $\exp(-S)$, gives the relative size of the transition which occurs without the creation or annihilation of phonons (the *no-phonon line*). If we give a width to each photon-plus-phonon transition – remembering that there is not one but a range of values of phonon energies – we get the envelope shown in the figure. The no-phonon line also has some width, as will be discussed later, but at low temperatures it is usually distinguished by its relative sharpness compared with the remainder of the transition. To compare the shape predictions of the configurational coordinate model with actual broadband spectra we show in Fig. 1.6b the shape of the observed $^4A_{2g} \rightarrow {}^4T_{2g}$ absorption transition in ruby at 77 K.

If the initial and final states differ very little in their sensitivity to changes in the environment, as occurs for all rare earth ion transitions between $(4f)^N$ states as well as for some transition metal ion transitions, then $S \simeq 0$ and so $\exp(-S) \simeq 1$, and we only get a weak one-phonon sideband accompanying the no-phonon line. This case is best analyzed not by the configurational coordinate model but by considering how the ion interacts with each of the lattice modes. Each lattice mode contributes a one-phonon sideband so the full sideband gives us a clear picture of the density of lattice modes.

The success of the configurational coordinate model in predicting broadband line shapes should not cause us to lose sight of the very simplifying assumptions inherent in the model. For example, it does not take into account the odd-parity

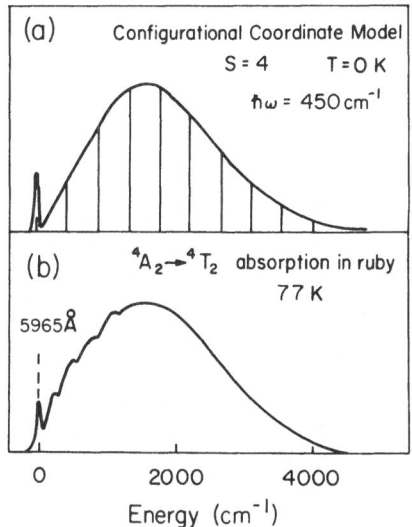

Figure 1.6. (a) Line shape predicted by the configurational coordinate model at $T = 0$ when $S = 4$ and $\hbar\omega = 450$ cm^{-1}; (b) observed line shape for the $^4A_2 \rightarrow {}^4T_2$ absorption transition in ruby at 77 K

component of the lattice modes. Because they introduce odd-parity crystal field terms these odd-parity distortions of the environment may allow the photon-plus-phonon sideband transition to occur by an electric dipole process, whereas the no-phonon line may occur only by the much weaker magnetic dipole process or by a weak electric dipole process. Such vibrationally induced electric dipole sideband processes are termed *vibronic*, and they can increase the intensity ratio of sideband to no-phonon line beyond that predicted by the configurational coordinate model.

Other types of modulation can also cause sidebands. For example, radiative transitions in magnetic crystals can be modulated by magnons, and magnon side-bands as well as phonon sidebands are found to accompany the no-phonon no-magnon line. These are discussed again in Sect. 1.9.

1.5 Energy Levels and Radiative Transitions in Isolated Molecular Centers

Molecular centers in mixed molecular crystals are known for the sharpness and rich-ness of their spectral features. As the basic "vibronic" structure differs little from that in the gas phase, and as it is *not* encumbered by the complex rotational enve-lopes which are typical of room-temperature gas phase spectra, such spectra were first studied for the sake of the information they provided on the free molecule [1.21]. The spectra are also essentially "transferable" from one host lattice to another.

As a typical example, we discuss here the aromatic molecule naphthalene ($C_{10}H_8$) with its typical π electron transitions. The π electrons are delocalized over the entire

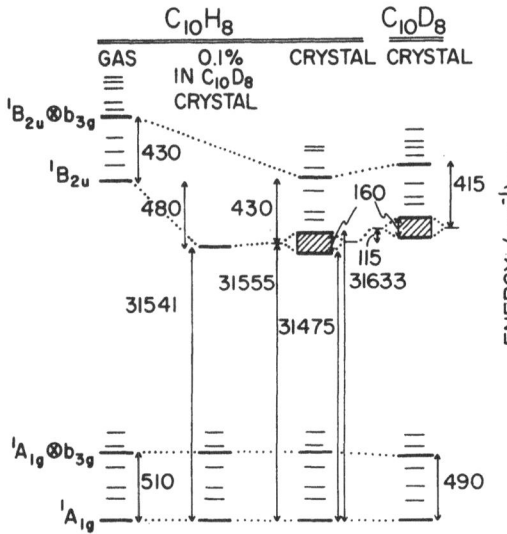

Fig. 1.7. Schematic energy level diagram of a molecular center in several environments: naphthalene electronic ground ($^1A_{1g}$) and first excited singlet ($^1B_{2u}$) states, with vibrational structure (representative ground state vibration b_{3g} at 510 cm^{-1}) and vibronic structure (same representative vibration in excited state at 430 cm^{-1}). The rotational (gas phase) and phonon (crystal lattice) fine structures are not shown. All numbers are in cm^{-1}

molecule (or at least its carbon ring skeleton) (Fig. 1.7). They are easy to excite and thus typically are responsible for the lower excited electronic states. The π electron states are further classified by their symmetry and multiplicity (electron spin is a very good quantum number in the organic molecules which are made of the light atoms: hydrogen, carbon, oxygen, and nitrogen). We can thus classify them by $S\Gamma$ terms, like the transition metal ions (Sect. 1.4.2). We note that Γ usually refers to the irreducible representation ("chemists' notation") of the *electronic* state in the *molecular* symmetry group, even though strictly speaking it should designate the *vibronic* state in the *site* group (an illustration is given in Fig. 1.7). We notice that the naphthalene molecule has $(3 \cdot 18 - 6) = 48$ nondegenerate internal molecular vibrations, not counting overtones and combination bands. Only schematic vibrations (or the most important ones) are usually shown on energy level diagrams. Vibrations are often designated by their molecular symmetry group irreducible representation, designated by a *small* letter (again using the "chemists' notation") and the overall vibronic state is often written as a direct product (such as $^3B_{1u} \cdot b_{2g}$). The relevant spin states are usually only singlets and triplets (except for free radicals and molecular ions) and the ground state is usually a *totally symmetric singlet* state (i.e., $^1A_{1g}$). The selection rules are again determined by group theory, but because of the vibrational states one again has overlap integrals or Franck-Condon factors $\langle \chi \mid \chi' \rangle$ due to vibrational state overlap, i.e., the transition strength is

$$S_{ED} = |\langle S\Gamma \mid D \mid S'\Gamma' \rangle \langle \chi \mid \chi' \rangle|^2 \tag{1.18}$$

where χ and χ' again refer to the vibrational states (and their irreducible representations). In addition there are *vibronically induced* (Herzberg-Teller) transitions, where the purely electronic matrix element vanishes [1.22].

The dynamical lattice Hamiltonians in molecular crystals are similar to those in ionic crystals, except that they consist of both translational and librational modes of the semi-rigid molecular centers and, except for special symmetry cases, of mixed translational-librational modes. As the space groups of most molecular crystals are nonsymmorphic, there is usually no clear distinction between "optical" and "acoustical" modes.

Molecular centers often create localized [1.23] or pseudolocalized [1.24] phonon modes (usually of librational nature). In this case the analysis of Sect. 1.4.3 is very appropriate, as there often exists only a *single* such localized mode which dominates the spectrum via sharp zero-phonon lines (Fig. 1.8). Strictly speaking, for librations one should use a restricted-rotor rather than an harmonic-oscillator model, but the difference is small for small amplitude librations.

In contrast to the centers with pseudolocalized phonons there are centers that barely perturb the phonon structure of the host (amalgamation limit) [1.25]. This is almost always true for centers that differ from the host molecules only by isotopic substitution. Amalgamation (also called the one-phonon case) is also typical for centers containing only "minor" chemical substitution compared to the host molecules (i.e., betamethylnaphthalene in naphthalene [1.26]). In such cases, as stated in Sect. 1.4.3, there is only one zero-phonon line, accompanied by a sideband representing a density of states of the host phonons weighted by electron-phonon coupling [1.27] (Fig. 1.9). The case (see Sect. 1.4.3) of vibronically induced (Herzberg-Teller analog) electric-dipole allowed sidebands, with an increased sideband-to-zero-phonon-line intensity ratio, is also common for molecular centers [1.28].

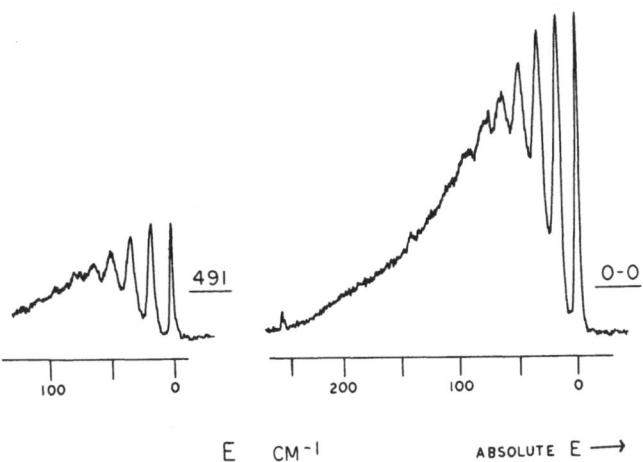

Fig. 1.8. Pseudolocalized phonon sideband of naphthalene in durene (0—0 electronic and 491 cm^{-1} vibronic phosphorescence bands). 1.1% (nominal) $C_{10}D_8$ in $C_{10}D_{14}$ at 2 K. The major phonon spacings are 17, 34, 51, 65 and 80 cm^{-1} [1.24]

1.6 Broadening of the No-Phonon Line

Let us now turn to a discussion of line-broadening mechanisms for sharp no-phonon lines in solids. Two types of process can contribute to the width of the no-phonon line: inhomogeneous and homogeneous broadening mechanisms. During crystal growth strains, dislocations, faults, unintentional impurities, etc., occur. Because of these the sites occupied by the optically active ions are not all identical; the ions are distributed among sites whose environments are perturbed to greater or lesser extent. Since the frequency of a transition on the ion is influenced by the environment we get a range of transition frequencies. Hence the observed line is *inhomogeneously broadened*, the profile being a composite of the transitions from ions in the different sites. Any randomness in the interaction of the optically active ions with their crystalline environments leads to inhomogeneous broadening. If we are dealing with transitions on ions in identical sites we would expect a much sharper line, showing only the *homogeneous width* of the transition (Fig. 1.10).

Variations in crystal field due to strains in the crystal are a major source of inhomogeneous broadening. We would expect transition whose initial and final states show a similar sensitivity to variations in crystal field to have least inhomogeneous broadening. In good quality crystal inhomogeneous broadening for some optical transitions may be only a small fraction of a cm^{-1}.

The use of broadband light sources (e.g., arc lamps) excites the ions in different sites with equal probability and the resultant luminescence exhibits the inhomogeneous broadening of the material. Narrow band tunable lasers on the other hand, can be used to excite ions in a small subset of sites, where the absorption frequencies

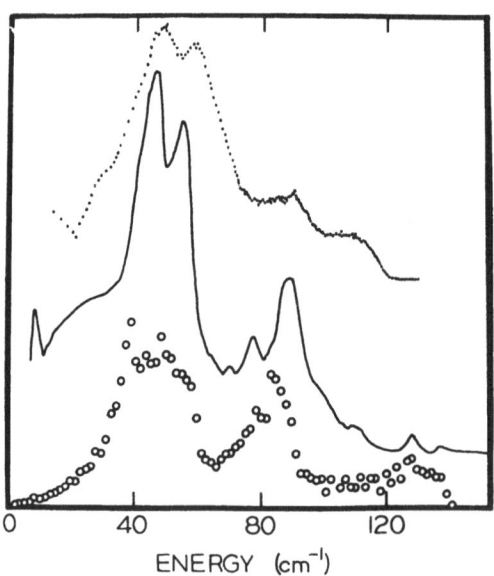

Fig. 1.9. Naphthalene crystal phonon density of states and weighted density of states. The top ($\cdots$) curve is derived from inelastic incoherent neutron scattering. The center (——) curve is a phosphorescence sideband of 0.25% $C_{10}H_8$ in $C_{10}D_8$ crystal at 2 K [1.27]. The bottom curve is from a semiempirical atom-atom calculation. (Fig. from [1.26])

ENERGY (cm^{-1})

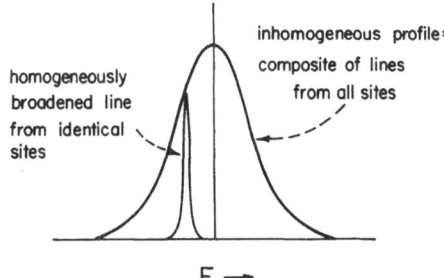

homogeneously broadened line from identical sites

inhomogeneous profile: composite of lines from all sites

E →

Fig. 1.10. Schematic representation of homogeneous and inhomogeneous broadening

are within the narrow laser linewidth. The luminescence re-emitted by these excited ions will be contained in the same narrow frequency band which is much narrower than the inhomogeneous linewidth of the material. This is termed *fluorescence line narrowing* (FLN). Using such narrow band excitation sources allows us to study effects which were hitherto masked by inhomogeneous broadening. If, in addition, the ions are excited by a narrow band pulse of very short duration the interesting time evolution of the narrow band excitation can be studied. Such experiments are described in the later chapters.

One can increase the inhomogeneous broadening in a material by incorporating sizeable amounts of optically inert impurity ions within the material. This results in a significant randomness in the crystal fields acting on the optically active centers incorporated in the material. For example, we see in Fig. 1.11 the inhomogeneous broadening of the optical transitions of Nd^{3+} in mixed materials [1.29].

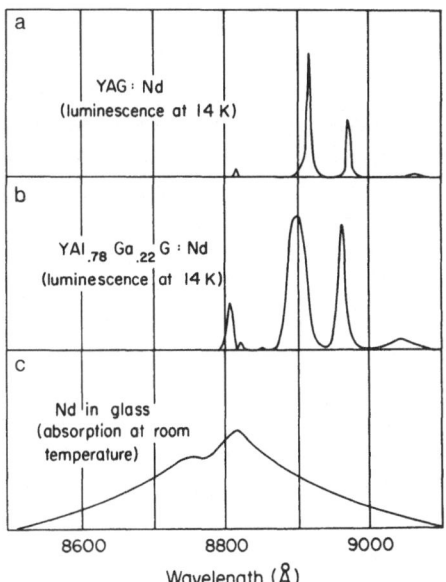

a

YAG : Nd
(luminescence at 14 K)

b

YAl$_{.78}$Ga$_{.22}$G : Nd
(luminescence at 14 K)

c

Nd in glass
(absorption at room temperature)

8600   8800   9000

Wavelength (Å)

Fig. 1.11 a–c. $^4F_{3/2} \leftrightarrow {^4I_{9/2}}$ transitions of Nd^{3+} in (a) $Y_3Al_5O_{12}$; (b) 78% $Y_3Al_5O_{12}$ and 22% $Y_3Ga_5O_{12}$ mixed crystal [1.29]; (c) glass

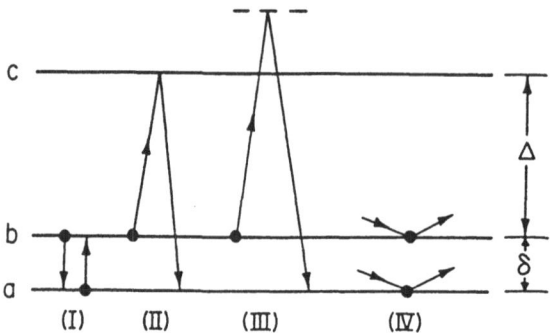

Fig. 1.12. Relaxation mechanisms leading to homogeneous line broadening

Homogeneous broadening is a consequence of the finite lifetime of the initial and final states involved in the transition. If τ is the lifetime of a state then there is an uncertainty in the energy of that state, ΔE, and $\Delta E\tau = \hbar$. This energy uncertainty shows up as a broadening (Γ) of the transition involving that state. A decay rate of $2 \cdot 10^{11}$ s^{-1} causes a broadening of 1 cm^{-1}. If a number of relaxation processes with relaxation rates W_l contribute to the lifetime then the broadening is $\Gamma = \hbar \sum_l W_l$. We now consider the relaxation mechanisms which can reduce the lifetime of a state.

First the radiative process itself. Most radiative decay times are longer than 1 μs, hence the broadening is negligible by comparison with other broadening mechanisms. Of more importance are relaxation processes involving lattice phonons. Some of these relaxation processes are shown in Fig. 1.12. Consider two states a and b, whose separation, δ, is within the range of phonon energies. We consider the mechanisms which reduce the lifetimes of these states and which will lead to a broadening of optical transitions involving one of these levels as an initial or final state.

Process (I) is the direct relaxation process. We can write $W_{ab}^{(I)} = W_0^{(I)}(\bar{n})$, $W_{ba}^{(I)} = W_0^{(I)}(1 + \bar{n})$, where $\bar{n}$ is the thermal occupancy of phonons of energy $\hbar\omega = \delta$. $W_0^{(I)}$ is the relaxation rate from b to a at absolute zero of temperature, and it depends on the difference in sensitivity to lattice distortion between states a and b. In the long wavelength Debye approximation it also varies as ω^3/v^5, where v is the velocity of sound. Because of the ω^3 factor which comes principally from the density of phonon states this direct process relaxation rate becomes much larger as the energy gap between a and b increases — as long as the gap is within the range of phonon energies. Figure 1.13 shows the additional homogeneous broading attributed to downward direct process relaxation from the upper split level of the luminescent ^2E state of Cr^{3+} in different crystals. The deviation from a strict δ^3 law is attributable to the inadequacy of the Debye approximation, to possible variation of v with phonon energy, and to the difference in sensitivity to lattice distortion in the different host materials. These rates are much faster than the radiative decay rate from either level. And since $W_{ab}/W_{ba} = \exp(-\delta/kT)$ this direct process will maintain the populations of these excited levels in a Boltzmann ratio.

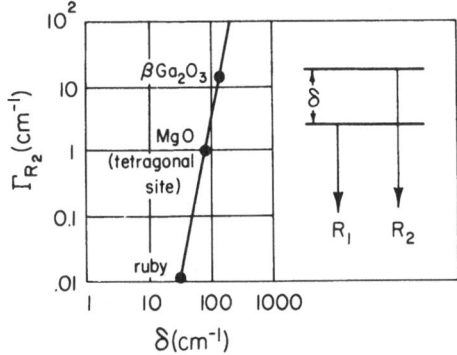

Fig. 1.13. Additional broadening of the upper luminescent level of Cr^{3+} (from which R_2 originates) attributable to direct process relaxation from the upper level. The value for ruby is taken from [1.32], the values for the tetragonal site in MgO and for the trigonal site in β-Ga_2O_3 were taken from the observed additional broadening of the R_2 lines at low temperatures

When the separation between a and b is beyond the range of phonon energies a direct process can still occur by a multiphonon process. If the number of phonons involved is not too large this can be an efficient relaxation process for the upper level, removing ions from the state rapidly enough that luminescence cannot occur. This important nonradiative relaxation process will be considered again in Sect. 1.7.

Process (II) is an Orbach process [1.30]. This is a resonant two-phonon process involving the absorption of a phonon, k_1, and the subsequent emission of another phonon, k_2. This process can occur whenever there is an electronic state, c, within the range of phonon energies. In discussing this process it is helpful to assume that $\Delta \gg \delta$. The process depends upon the availability of thermal k_1 phonons, and the relaxation rate varies as

$$W_{ba}^{(II)} = W_{ab}^{(II)} = W_0^{(II)}\bar{n} \simeq W_0^{(II)} \exp\left(-\Delta/kT\right) \tag{1.19}$$

when $\Delta \gg kT$. In this case $\bar{n}$ is the thermal occupancy of phonons. $W_0^{(II)}$ is related to the sensitivities of levels a, b, c to lattice distortions.

(III) is a Raman relaxation process which is a two-phonon process proceeding through a virtual intermediate state (broken level in Fig. 1.12) [1.3]. In the Debye approximation and where δ is much less than the maximum phonon energy the rate for the Raman relaxation is given by

$$W^{(III)} = A\left(\frac{kT}{h}\right)^7 \int_0^{x_0} x^6 e^x (e^x - 1)^{-2}\, dx \tag{1.20}$$

where A is a constant related to the sensitivitiy of the electronic levels to lattice distortion. $x_0 = \hbar\omega_0/kT$, where ω_0 is the Debye cut-off frequency. At low temperatures the rate goes as T^7. Just as in the Orbach relaxation process this Raman process will operate to broaden both levels a and b by equal amounts.

In addition to these processes which cause transitions out of electronic levels a or b there is also an intrinsic Raman broadening mechanism which is due to a relaxation from one phonon state to another of equal energy while the electronic state stays the same. Since the description of the state of the ion must specify both the electronic state and the phonon state ($| \psi, n >$) this intrinsic Raman process reduces the lifetime of the state and can lead to broadening. The temperature dependence of this rate is the same as that of (1.20).

McCumber and *Sturge* [1.31] demonstrated that the observed temperature-dependent broadening of the ruby R lines above 77 K is explained by the Raman process, while from 4.2 K to 77 K *Muramoto* et al. [1.32] have shown that the observed broadening is explained by a combination of Raman and direct processes. Some more recent laser spectroscopy results are discussed in Chap. 5.

The interaction with the phonons which causes the broadening mechanisms can also shift the electronic energy levels, and this can lead to a temperature dependent frequency shift [1.33]. It can be shown that this line shift varies as the lattice "total heat," and its magnitude depends upon the difference in sensitivity to lattice vibrations of the initial and final states. In the case of the ruby R lines both lines shift to lower energy by about 20 cm^{-1} from 0 K to room temperature, and the additional homogeneous broadening at room temperatures is about 10 cm^{-1} [1.31].

A discussion of homogeneous and inhomogeneous broadening and line shapes of molecular centers is given in Chap. 7.

1.7 Nonradiative Relaxation

If we study the Dieke diagram which shows the levels of the trivalent rare earth ions we notice that only some of the states emit luminescence when excited – these with the pendant semicircle. When the other states are excited they lose their energy nonradiatively. Further, we observe that the larger the energy gap between two levels the greater the probability of radiative decay from the upper level; for small gaps the upper level decays nonradiatively. Similar effects are found for the transition metal ions.

For each of the rare earth ions all the states shown in the diagram are derived from the same $(4f)^N$ configuration and they have approximately the same coupling with the lattice. Therefore the lattice can be described by the same phonon states irrespective of the electronic state of the rare earth ion. We can therefore speak of multiphonon emission between the different electronic states.

Consider the competition between radiative (W_r) and nonradiative (W_{nr}) rates from a level situated at energy E above the next lowest level (Fig. 1.14). The larger E the greater the number of phonons involved in the multiphonon process. For the rare earth ions the electron-dynamic lattice interaction can be handled by perturbation theory, and the larger the number of phonons involved the higher the order of perturbation theory required, and consequently the weaker the probability of nonradiative decay. Thus the larger the gap the less likely is the probability of nonradiative decay. On the other hand the probability of a radiative process increases as the gap is increased [as E^3 or ν^3, see (1.5)].

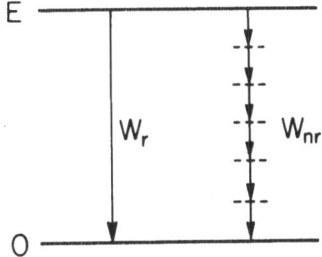

Fig. 1.14. Schematic representation of the radiative and nonradiative processes. The radiative process occurs by the emission of a single photon, the nonradiative process involves the emission of a number of phonons

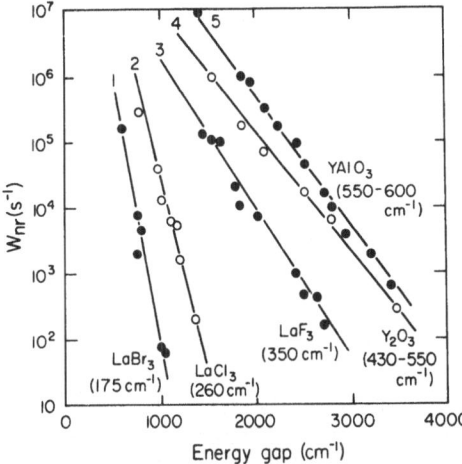

Fig. 1.15. A plot of nonradiative decay rates against energy gap to the next lowest level for the excited states of various trivalent rare earth ions in different host crystals. For a given host, the nonradiative rate varies exponentially with energy gap; the solid lines give the best fit to this exponential dependence. The solid lines are labelled 1–5 for later reference [1.12]

A systematic study of nonradiative decay rates in rare earth ions was undertaken by *Moos* [1.34] and the Johns Hopkins group as well as by *Weber* [1.13]. Figure 1.15 shows the values of W_{nr} for various levels of *different* rare earth ions in a number of crystals plotted as a function of the energy gap to the next lowest state, E. For each crystal the particular lattice phonon expected to be most effective is given. For each material one finds an exponential dependence upon energy gap, and this is consistent with simple perturbation theory calculations.

If for each of these materials one plots the exponential dependences of the W_{nr} values against the number of effective phonons ($p = E/\hbar\omega_{\text{effective}}$) (Fig. 1.16) the results for the different materials are more similar. The shaded area in the figure indicates the range of typical values for radiative decay in the trivalent rare earth ions. This plot shows that the breakdown point for ions in LaBr$_3$ and LaCl$_3$ is around five phonons; the nonradiative process involving more than five phonons will be too weak and the radiative process should dominate, while the nonradiative process should dominate for these host materials if the process can occur by the release of less than five phonons. For the other materials the breakeven occurs at around 6–8 phonons.

The pth order multiphonon process in the rare earth ions is expected to vary with temperature through a $(1 + \bar{n})^p$ factor, where $\bar{n}$ is the thermal occupancy of

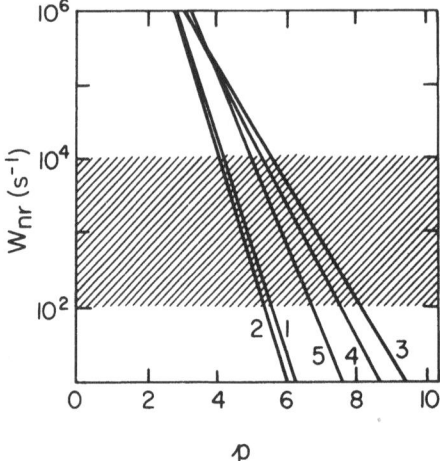

Fig. 1.16. A plot of the nonradiative relaxation rates of rare earth ions in different host crystals (the solid lines of Fig. 1.15) against the number of effective phonons emitted in the relaxation process. The shaded area covers the range of radiative relaxation rates normally found in the trivalent rare earth ions

effective phonons. *Weber* [1.35] has verified this temperature dependence in the case of the $^4I_{11/2} \rightarrow {}^4I_{13/2}$ transition on Er^{3+} in $YAlO_3$.

Nonradiative processes in transition metal ions are more difficult to analyze because of the much stronger coupling of the transition metal ions to the lattice and because the strength of the coupling can vary so much level from to level in the same ion. If one adopts a configurational coordinate model in which one type of phonon is assumed to be effective and calculates the nonradiative relaxation rate between a and b one gets a formula whose leading term is [1.36, 37]

$$W_{nr} = \frac{2\pi}{\hbar} \, |\langle a \, | H' | \, b \rangle|^2 \, \frac{e^{-S(1+2\bar{n})} S^p (1 + \bar{n})^p}{p!} \tag{1.21}$$

where H' is an interaction which couples the two electronic states, S is the Huang-Rhys parameter, and p is the number of phonons produced in the relaxation process. When S is small, as for rare earth ions, and when p is unchanged the temperature dependence is all contained in the $(1 + \bar{n})^p$ term, which is the result we expected for rare earth ions.

Sturge [1.37] in his analysis of the nonradiative relaxation between the $^4T_2$ and $^4T_1$ states of Co^{2+} in $KMgF_3$ showed that, although the temperature dependence of W_{nr} is well described by the configurational coordinate model, the numerical values of W_{nr} are incorrect by two orders of magnitude. He showed how anharmonic effects may have a profound effect upon the overlap integrals which determine the theoretical relaxation rate, and he argued that inclusion of anharmonic effects may explain the discrepancy between theory and experiment for this and other transition metal ion systems.

1.8 Increased Concentration and Excitation Transfer

In this section we consider the effect on the spectroscopic properties of increasing the concentration of optically active centers.

1.8.1 Increased Concentration – Ionic Solids

When the concentration of optically active ions in increased the possibility of clustering arises. Two ions may be near enough to interact strongly and form a distinct resolvable center – in this case a dimer (pair). Trimers (triads) can also be formed as well as more complex centers. For example, at very low concentrations the only sharp luminescent features seen in ruby are the two R lines which originate on isolated single ions. At concentrations of around 0.1% additional sharp lines, which originate on exchanged-coupled pair of chromium ions are found (Fig. 1.17). These pairs act as distinct luminescence centers. Above 1% a broad band appears farther to the red which has been attributed to luminescence from centers involving three coupled ions.

In addition, at high concentrations the "isolated" ions may be near enough to each other so that a weak interaction can occur between them. This interaction will be too small to affect the energy levels but it might allow excitation on one of the ions to be transferred nonradiatively to another unexcited ion nearby. If this transfer is efficient enough, energy can migrate nonradiatively through the crystal and it can travel to the vicinity of sinks – impurities, defects, etc. The energy may then be transferred to the sinks from which it is lost nonradiatively. This can explain the common phenomenon of *concentrations quenching*, in which the luminescence efficiency of a material decreases when the concentration is increased significantly.

In addition to energy transfer to such quenching centers, energy may also be transferred to secondary radiating centers which act as sinks for the energy and which in turn emit their own characteristic luminescence very strongly. This is the phenomenon of *sensitized luminescence*. The occurrence of trace amounts of rare earth ions in crystals of MnF_2 results in very strong rare earth luminescence from these crystals at low temperatures. This is caused by the efficient migration of Mn excitation to the rare-earth ions. Other examples of energy transfer leading to enhanced luminescence from specific dopant ions are discussed by *Blasse* [1.38].

The microscopic picture of the transfer of excitation between two ions, D (donor) and A (acceptor), separated by a distance R and between which some interaction exists, was first treated by *Förster* [1.39] and *Dexter* [1.40]. They showed that multipolar interactions and exchange can cause transfer, the transfer rate varying as R^{-n}, where $n = 6, 8, 10$ for dipole-dipole, dipole-quadrupole, and quadrupole-quadrupole interactions, respectively. The transfer rate due to exchange falls off exponentially with increasing R. *Förster* and *Dexter* considered that the absorption bandwidth on A overlapped the emission bandwidth on D to some extent and

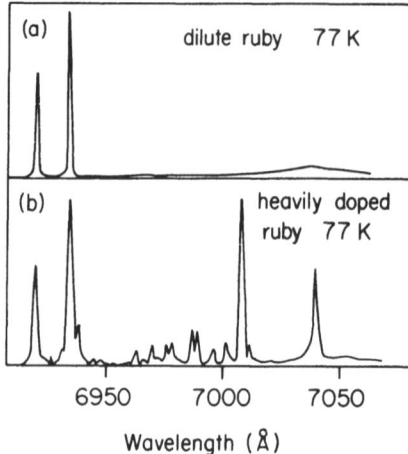

Fig. 1.17a, b. A comparison of the spectrum of (a) dilute ruby, and (b) heavily doped ruby, taken at 77 K. The additional lines in (b) originate on exchange-coupled pairs of Cr ions. The strong pair line at 7009 A (the N_2 line) acquires part of its excitation by energy transfer from single Cr ions. The transfer can be studied by examining the decay pattern of the N_2 line, as is discussed in Chap. 5

this transfer was regarded as a resonant process. The distinction between spectral overlap and true resonance will be considered in Chaps. 2 and 3.

If the energy mismatch between the relevant levels is large the transfer is accompanied by a multiphonon emission process. In view of the exponential dependence on energy gap of the nonradiative (multiphonon) relaxation rate, discussed in the last section, it is not surprising to find that the multiphonon-assisted excitation transfer probability also has an exponential dependence upon the energy mismatch [1.36].

In carrying out experiments on energy transfer one measures decay rates and luminescence efficiencies from donors and acceptors in a macroscopic section of the crystal. Hence one must statistically average the microscopic formulas in order to calculate macroscopically measurable quantities. In addition, the efficiency of transfer from the donor species to the acceptor species can be affected by energy transfer among the donor ions themselves (D-D transfer). We consider the case where a number of donors is excited by some excitation pulse. If $P_n(t)$ is the probability that the nth donor is excited at time t then $P_n(t)$ varies as [1.41]

$$\frac{dP_n}{dt} = -(\gamma_R + X_n + \sum_{n''} W_{nn''})P_n(t) + \sum_{n'} W_{n'n}P_{n'}(t) \qquad (1.22)$$

where γ_R is the radiative decay rate of donors, $W_{nn'}$ is the D-D transfer rate from n to n, and X_n is the D-A transfer at ion n. X_n will depend upon the distribution of traps in the vicinity of the nth donor. Back transfer from acceptors to donors is assumed to be absent. This condition is satisfied in all systems when the acceptor ions have a rapid decay probability. The solution of this equation under various experimentally realizeable conditions is the subject of Chap. 3.

We now look a little more closely at D-D transfer and in particular we consider what is involved in a resonant transfer process. In the original Förster-Dexter treatment, the appellation "resonant" was applied to the situation where the emission linewidth of the donor species overlapped the absorption linewidth of the acceptor species. This definition probably reflects an attempt to compare the dipole-dipole transfer process with a simultaneous emission-plus-absorption process. A more rigorous criterion for a resonant process would require that the electronic levels of the two ions be sufficiently close that transfer can take place without the need for a phonon or phonons to make up any energy difference between the two ions. If we consider the solid material cooled to zero K and if the intrinsic homogeneous widths of the ionic levels in question are very small, then we might expect a true resonant transfer process to occur if the interaction strength (J) between the two ions is greater than the energy separation (ΔE) between the excited states of the ions arising due to inhomogeneous broadening. The precise criterion for a resonant process will be discussed more fully in Chap. 2.

We recall that strains in the crystal are the major source of inhomogeneous broadening. If two adjacent optically active ions are in sites of similar strain the energy separation between their levels may be small enough to satisfy the criterion for resonant transfer. In that case a rapid *spatial migration of excitation* may occur in the solid. Excited ions may also transfer excitation to other ions in sites of significantly different strain by a process in which phonons are employed to make up the energy mismatch between the ionic levels. This process, if rapid enough, can be detected by FLN experiments. By means of a narrow band fast laser pulse one can excite ions in sites of almost identical strain — whose fluorescence frequencies lie in the narrow component of Fig. 1.10, for example. The occurrence of nonresonant transfer will then show up as a gradual spreading out in time of the luminescence from the narrow component into the full inhomogeneous line. This is *spectral transfer* and is expected to be much slower than the resonant transfer process. The observation of the resonant spatial transfer process is a more indirect procedure. If this transfer results in the feeding of luminescent traps (which emit at their own characteristic frequency), the relative intensity of trap luminescence to donor luminescence will be a measure of the efficiency of the spatial transfer process. Such experiments are described in the later chapters.

The efficiency of the resonant spatial transfer depends on the average spatial separation of the nearest *resonant* ions, and this depends on the nature of the strains in the material. Two extreme points of view can be considered. In one we visualize a random distribution of microscopic strains throughout the crystal, and in the second we regard the strains as being of macroscopic size, with adjacent parts of the crystal experiencing almost equal strain. If strain is macroscopic adjacent ions are essentially in resonance, and resonant spatial transfer is efficient — the transfer can take place along channels of equal strain in the crystal. We then have a delocalized extended state in the crystal. On the other hand, if the crystal experiences a random distribution of microscopic strains then the average spatial separation between resonant ions can be large. In addition, if the interaction between the ions is short range (exchange or quadrupole-quadrupole) then at a sufficiently low concentration transfer may not be able to occur and the excitation may remain localized. According to the ideas of *Anderson* applied to optical excitation transfer [1.42], a random distribution of strains and a short range interaction will lead to the disappearance of delocalized states at some critical concentration. Above this critical concentration the inhomogeneous line will have *mobility edges* marking the boundaries of regions containing only localized states. As the concentration decreases, the regions of the line where there are only localized states expand filling the entire line at the critical concentration. This is the phenomenon of *Anderson localization*. The question as to whether or not Anderson localization is showing up in some experiments on energy transfer in inorganic solids is being hotly debated at present and this is discussed in Chaps. 2 and 5. The application of these ideas to energy transfer in molecular crystals is discussed in Chap. 7.

1.8.2 Increased Concentration – Molecular Solids

It is possible to achieve the whole range of donor-acceptor concentrations in many binary molecular crystals. This is almost always true for isotopic mixed crystals (we use the term "isotopic" rather than "isotopically" to distinguish between a random substitution of isotopic *molecular* species and a random substitution of atoms, for example, a random "isotopic" mixed hydrogen crystal may consist of a random mixture of H_2 and D_2 molecules, *but no HD molecules*, while a random "isotopically" mixed hydrogen crystal will always contain HD molecules). Thus one can produce larger and larger clusters with increased concentration until one reaches the "percolation concentration." The latter is mathematically defined [1.25] as the concentration at which there is a finite probability of producing an infinite guest (donor) cluster (assuming an inifinite host crystal). For finite crystals more practical definitions have been given [1.43], but the result (i.e., the critical concentration) is identical for all practical purposes. In Table 1.1, we list some examples of critical concentrations for simple lattices. We note that the connectivity ("bond") among sites, which *defines* the *clusters* is of utmost importance.

Well above the percolation concentration, most of the "guest" sites are included in the "infinite" cluster ("supercluster"). Under such circumstances it is often useful to use the various theoretical approaches utilizing "average" (i.e., "smeared") Hamiltonians (and Green's functions) such as effective medium theories, T-matrix approach, coherent potential approximation (CPA), etc. [1.25]. However, such symmetric Hamiltonians, which retain the translational symmetry and the k-state classification (see below), are not useful for the description of the localized small-cluster states [1.25].

In molecular crystals, small clusters, such as "dimers" (pairs), have much in common with their counterparts in ionic crystals (see above). However, there are aspects which make them much more interesting, at least in the case of isotopic mixed crystals. This is because the zeroth-order pairwise interaction in such a "resonance pair" (dimer) is the *same* as in a pure and perfect crystal. It is thus possible to design experiments in which one can effectively *measure* the separate pairwise

Table 1.1

Lattice	Bond definition	Number of neighbors	Percolation Concentration
Square	nearest neighbor (n.n.)	4	0.5927[a]
Square	2nd n.n.	8	0.4073[a]
Square	3rd n.n.	12	0.29[a]
Triangular	n.n.	6	0.5000 (exact)[b]
Simple cubic (sc)	n.n.	6	0.3135[a]
bcc	n.n.	8	0.243[b]
fcc	n.n.	12	0.195[b]
hcp	n.n.	12	0.204[b]

[a] J. Hoshen, R. Kopelman, E. M. Monberg: J. Stat. Phys. **19**, 219 (1978)
[b] V. K. S. Shante, S. Kirkpatrick: Adv. Phys. **20**, 325 (1971)

terms in the pure, perfect crystal Hamiltonian [1.25]. On the other hand, knowledge of the pure crystal Hamiltonian can be used for calculations of the very concentrated crystal properties, via the above-mentioned averaged Hamiltonians (Green's functions, etc., see above) [1.25]. There is thus an interesting complimentarity between medium-low concentration studies and medium-high concentration studies. Analytical techniques, such as Green's functions, are sufficient to treat both cases theoretically.

As the size of the cluster grows, the analytical solutions become quite complex [1.44]. However, it is possible to follow a combined experimental and theoretical program in which experiments on very dilute crystals, on medium dilute crystals and on "fully concentrated" crystals (see below) are combined in an iterative manner [1.25]. The underlying principle is that the site shift and splitting (crystal-field type "static" effects) do not change with isotopic substitution while, on the other hand, excitation exchange type interactions, which are analogous to "resonance" interactions, do depend on an exact energy matching. Because the real excitations are "vibronic" and not electronic, the excitation energies depend significantly on isotopic constitution. (The $C_{10}D_8$ first singlet excitation is 115 cm^{-1} above that of $C_{10}H_8$, the first triplet excitation is 99 cm^{-1} higher, etc.) (Fig. 1.7). This replaces the resonance with "quasiresonance" interactions. As a result, an *isolated* $C_{10}H_8$ center energy-level in a $C_{10}D_8$ host is "pushed down" (15 cm^{-1} in the first singlet excitation) [1.25] by the host exciton band. The analogous effects on $C_{10}H_8$ *cluster* states are more complex. Thus the ordinary *cluster* energy splitting due to direct excitation exchange ("resonance") interactions are further affected by "super-exchange" (formally guest-host-guest interactions [1.25]). An example where such a "superexchange" formalism is used to calculate eigenvalues and spectral transition intensities and polarizations for all possible monomer, dimer, trimer, and tetramer states has been gives by *Hoshen* and *Kopelman* [1.44] and has been used to assign experimentally observed cluster states by *LeSar* and *Kopelman* [1.45]. Such assignments (derivable from simple absorption spectroscopy) provide information on the excitation exchange and superexchange terms. The latter are of much interest for energy transfer and transport in mixed crystals (see Chap. 7).

For larger clusters (> 4), analytical solutions become impractical. There are about seventy topologically different pentamers in a simple binary square lattice (with nearest neighbor only bonds)! Just to derive their relative statistical weights is a serious undertaking, not to mention complete eigenvalue calculations. *Hong* and *Kopelman* [1.46] resorted to various numerical techniques in order to calculate approximate eigenvalue distributions (and polarized spectral distributions) for concentrations high enough that large clusters abound. *Hoshen* et al. [1.47] calculated cluster distributions via simulations on lattices with up to $64 \cdot 10^6$ sites. Such simulations give the percolation concentrations (Table 1.1), the "average cluster size" (at a given concentration) and other information of potential interest. There still is a basic question left open. Small cluster states are obviously localized; "infinite" cluster states appear to be extended. How sharp a transition from "localized" to extended states will there be at the percolation concentration? With artificial models, e.g., with only nearest neighbor interactions, it is easy to show that no extended states are possible below the percolation concentration. However, they are most likely to appear above it, thus producing an "analog" Anderson-Mott transition, as suggested by *Monberg* and *Kopelman* [1.48]. However, it is not obvious whether such a model is valid for a real binary crystal at a practical temperature (say above 1 K). For further discussion of percolation and localization the reader is referred to the very recent review by *Thouless* [1.49].

If the guest-host energy separation is small compared to the excitation bandwidth (of the fully concentrated crystal — see below), then one does *not* expect to have discrete guest cluster states. The latter become "virtual" states which are "immersed" in the host band. This is the *amalgamation* case discussed by *Hoshen* and *Jortner* [1.50] and also by *Prasad* and *Kopelman* [1.51] (who demonstrated 'amalgamation" experimentally for librational excitations in binary naphthalene crystals — [$C_{10}H_8/C_{10}D_8$)]. In such a case, the energy dynamics in the mixed crystal will resemble that in a fully concentrated crystal. We note that practially all of the discussion on ionic and molecular crystals implied a *separated band case* [1.25, 50], with excitations confined essentially to guest "centers" or guest quasilattices (above the percolation concentration). In other words, the optical spectroscopy of the electronic center has been assumed to occur in a spectral region where the host is effectively transparent. This fact is also of prime importance to the energy transfer, which is essentially a center-to-center transfer (see also Chap. 7).

1.9 Fully Concentrated Materials

We now consider the limiting case where we have 100% concentration of active centers, e.g., $PrCl_3$, Cr_2O_3, MnF_2, C_6H_6, $C_{10}H_8$. Once again, in principle, all centers are in identical sites, which is similar to the situation found at low concentrations where the centers were all isolated. But in the fully concentrated case the adjacent centers are close together and may interact strongly with each other. Particularly in the case of the transition metal ions, this interaction can be quite strong. Cooperative magnetic phenomena in fully concentrated transition metal salts are an indication of the strength of this interaction. The interaction between neighboring centers may strongly affect the properties of these materials. For example, whereas Al_2O_3 : Cr (ruby) emits strong luminescence, no luminescence occurs from Cr_2O_3. At low concentrations MnF_2 emits strong luminescence, but it originates mainly on traps in the material. The spectroscopic properties of PrF_3 are very similar to those of LaF_3 : Pr. On the other hand, whereas $LaCl_3$: Nd emits strong luminescence, no luminescence is observed from $NdCl_3$. In molecular crystals with 100% concentration the strength of the effect ranges from very weak (triplet states) to very strong (intense singlet states). The effect of 100% concentration is strongest in the transition metal ion systems because of the stronger coupling between adjacent ions.

1.9.1 Fully Concentrated Material – Ionic Solids

The coupling between adjacent ions can lead to efficient excitation transfer among the ions resulting in concentration quenching in some cases and in sensitized luminescence in others. Because of this coupling between ions the absorption of light by the material occurs not through single ions acting in isolation, but the light interacts with all the ions acting collectively. Such collective states are *exciton states* and are characterized by a k vector [1.52]. The range of allowed k values covers the first Brillouin zone, and the size of the exciton dispersion is a measure of the coupling between ions in adjacent lattice points. If at each lattice point there is a basis of more than one ion an additional splitting (Davydov splitting) occurs in the excited state. The size of the splitting is a measure of the strength of the coupling between the ions making up the basis at one lattice point. The situation in a hypothetical concentrated material is shown in Fig. 1.18. The ground state, G, where no ion is excited is clearly a $k = 0$ state.

In the process of absorption of a photon conservation of k must be satisfied. Since the k value of a photon is very small the change in the k value of the exciton is, essentially, $\Delta k = 0$. Hence the pure electronic absorption from the ground state, G, creates only a $k = 0$ excited state exciton. This is represented as transition 1 in Fig. 1.18. One cannot measure the exciton dispersion by pure electronic absorption transitions. Excitons with $k \neq 0$ can be created by the absorption of a photon if this is accompanied by the absorption or emission of a phonon or a magnon which can help to satisfy the $\Delta k = 0$ selection rule. For example, transition 2 in Fig. 1.18 showing the creation of an exciton of wave vector k can occur if it is accompanied by transition 3, the creation of a magnon of wave vector $-k$. This combined exciton-magnon transition appears as a magnon sideband accompanying the sharp $\Delta k = 0$ pure exciton line.

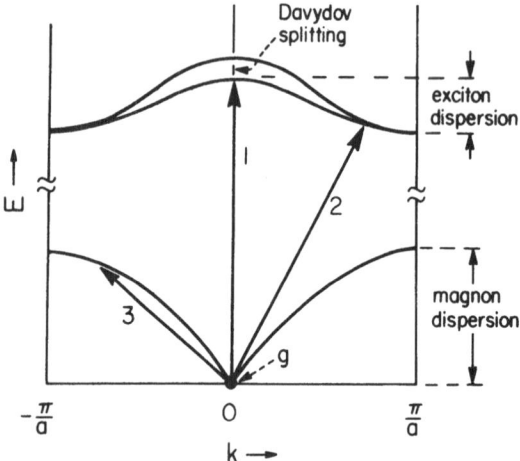

Fig. 1.18. Plot of exciton energy against k for a hypothetical fully concentrated material. The magnon and exciton dispersions as well as any Davydov splitting are much smaller than the energy separation between the ground and optically excited states

Many of the phenomena discussed in this and in previous sections are exhibited by MnF_2 [1.53]. We first consider the energy levels of the single Mn ions which are shown in Fig. 1.4. Because the energy gaps between adjacent Mn levels are small, each being less than about 5000 cm^{-1}, rapid nonradiative transitions occur between them. Hence Mn ions raised to the higher excited levels by absorption will quickly decay nonradiatively, cascading down through the upper excited levels until the lowest excited $^4T_1$ level is reached. The gap of around 18,000 cm^{-1} to the ground state is too large for an efficient multiphonon nonradiative process, so the ions in this $^4T_1$ level decay radiatively to the $^6A_1$ ground state. Since this is a spin-forbidden transition it has a long decay time — about 30 ms.

The interaction of the Mn ion with the crystalline environment is different in the $^4T_1$ and $^6A_1$ states, hence we expect that the transition between these two levels will be characterized by a large value of the Huang-Rhys parameter and so should show a weak sharp no-phonon line accompanied by a broad band with the shape shown in Fig. 1.6.

We now consider the Mn ions in MnF_2. This is an orange colored crystal, reflecting the strong absorptions in the green and blue shown in Fig. 1.4. Because of the exchange coupling between nearby Mn ions MnF_2 is an antiferromagnet with a Neel temperature of 68 K. And this coupling leads to rapid energy transfer among the ions, so that during the 30 ms lifetime of the $^4T_1$ state the excitation in this state will migrate among the Mn ions and "sample" a large portion of the crystal. This leads to efficient feeding of any traps in the material.

The $^6A_1 \rightarrow {}^4T_1$ absorption transition at 2 K is shown in Fig. 1.19. The broadband absorption stretching from 5500 Å to 4700 Å is shown in the lower trace, and it has the shape predicted by the configurational coordinate model (Fig. 1.6). The sharp features on the low energy side are shown in greater detail in the upper trace [1.54]. Because of fine structure in the $^4T_1$ level (caused by lower symmetry crystal

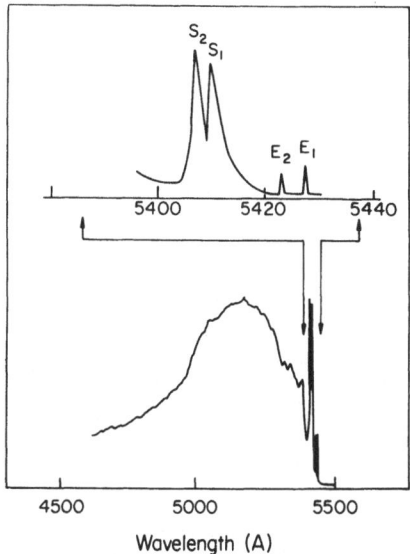

Fig. 1.19. Lower trace shows the $^6A_1 \rightarrow {}^4T_1$ absorption transition in MnF_2 at 2 K. The upper trace shows in greater detail the sharp features on the low energy side of the broad band

field, exchange, and spin-orbit coupling) two no-phonon lines, E_1 and E_2, occur. The two features at 5410 Å (S_1) and 5408 Å (S_2) are magnon sidebands accompanying the E_1 and E_2 lines, respectively. Whereas absorption into E_1 corresponds to transition 1 in Fig. 1.18 and creates a $k = 0$ exciton, absorption into the peak of the S_1 sideband creates a zone boundary exciton and a zone boundary magnon; this is transition 2 + 3 of Fig. 1.18. Interesting experiments on the exciton dynamics in MnF_2 are described by Selzer and Yen in Chap. 5. The basic optical spectroscopy of magnetic insulators is treated in [1.55, 56]. More recent experiments dealing with exciton dynamics in this type of material are described in Chap. 5.

From the absorption data shown here and from the fact that the luminescence should occur only from the $^4T_1$ level, one might expect that the luminescence from MnF_2 at low temperatures will occur from the lowest energy level in the $^4T_1$ state and will consist of the E_1 no-phonon line accompanied by sidebands at longer wavelength. This picture. however, is greatly modified by the rapid energy transfer and subsequent feeding of traps. Even in very pure carefully grown crystals most of the luminescence found at low temperatures originates on traps, and only a very weak E_1 luminescence line is observed. The traps in this case are Mn ions adjacent to impurities — usually Mg, Zn, Ca ions which occur at concentrations of a few parts per million [1.53]. The Mn ions in the vicinity of such impurities are perturbed by the impurities and as a result their $^4T_1$ levels are slightly lower in energy than that of the intrinsic Mn ions. At low temperatures the excitation transferred down to these Mn ions cannot return back up to the intrinsic E_1 level, so the excitation stays trapped there, ultimately being emitted as luminescence characteristic of the traps. Although they are present in only very small amounts in pure crystals, the efficiency of the transfer is such that essentially all the excitation on the regular Mn ions end up as traps. (Fe and Ni ions which are present in trace amounts also act as traps, but since they do not emit visible luminescence these are quenching traps.) At higher temperatures the perturbed Mn ions cease to be effective as sinks, all the excitation is transferred to the quenching traps and no luminescence occurs.

If MnF_2 is grown with trace amounts of rare earth ions these rare earth ions also act as traps. They are excited by transfer from the regular Mn ions and emit their own characteristic luminescence, and because of the efficiency of the transfer, only trace amounts are needed for bright rare earth luminescence to occur. In the case of MnF_2 doped with Eu, the $^5D_0$ level of Eu^{3+} is the only excited level below the $^4T_1$ level of Mn^{2+}. Hence the Eu luminescence occurs only from this level.

Figure 1.20 shows the luminescence from MnF_2:Eu at 2 K [1.57, 58]. The lower trace shows the expected broadband luminescence from Mn traps stretching from around 5500 Å to around 6400 Å. In addition, the sharp $^5D_0 \rightarrow {}^7F_J$ Eu^{3+} luminescence is also shown. Each of the ^7F_J states is split by the crystal field, hence the structure in the transitions. The $^5D_0 \rightarrow {}^7F_1$ transition is magnetic dipole — which is in accordance with the selection rules listed in Sect. 1.4.1 where only a $J = 0 \rightarrow J = 1$ transition is allowed from a $J = 0$ state — while the others are electric dipole. The $^5D_0 \rightarrow {}^7F_2$ and $^5D_0 \rightarrow {}^7F_4$ are the strongest electric dipole transitions, again in accordance with the electric dipole selection rules listed in Sect. 1.4.1. The $^5D_0 \rightarrow {}^7F_0$ should be forbidden but occurs weakly because of J mixing.

The fine structure in the low wavelength region is shown in greater detail in the upper trace. The arrow indicates where the E_1 line would occur; none, however, is found in this material. Instead two no-phonon lines are seen at lower energy, and

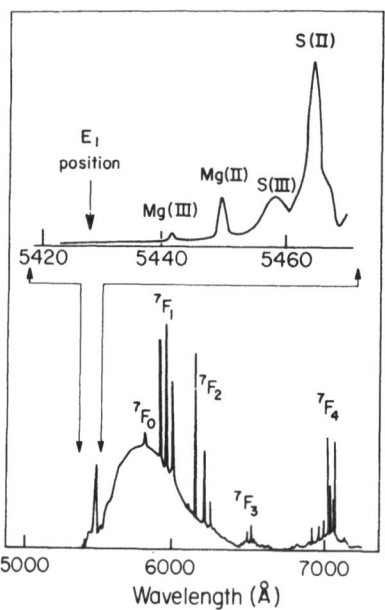

Fig. 1.20. Lower trace shows the luminescence spectrum of MnF_2:Eu at 2 K. The fine structure in the high energy side of the broad band is shown in greater detail in the upper trace

these are labelled Mg(II) and Mg(III). These are the no-phonon lines from Mn traps, which occupy sites which are, respectively, second nearest neighbors and third nearest neighbors to Mg impurity ions. The more intense features at S(II) and S(III) are magnon sidebands of these no-phonon lines, respectively. The multiphonon sideband shown in the lower trace is associated with all these sharp transitions.

Pure MnF_2 and MnF_2 : Eu exhibit the phenomena of rapid excitation transfer, concentration quenching, sensitized luminescence, and cooperative magnetization with its resultant effect upon the optical spectroscopy. This behavior is representative of that found in all concentrated Mn systems.

1.9.2 Fully Concentrated Materials – Molecular Solids

Historically the study of molecular solids started on "fully concentrated" (also called "neat" or undoped) organic crystals, such as benzene, naphthalene and anthracene, and only later moved towards mixed solids [1.21, 59]. Obviously the early luminescence studies on such systems were plagued by unknown impurity center lines (see Chap. 7). However, spectroscopic absorption studies usually revealed Davydov [1.59] splittings which are due to the *bulk* crystal states (Frenkel excitons) [1.59]. The number of Davydov components ($k = 0$ states) is equal to the number of interchange equivalent [1.60] molecules per primitive unit cell (which is the number of symmetry equivalent molecules in the lattice-point basis). Depending on the method of excitation, some or all of the Davydov components may be spectroscopically "forbidden" (zero strength). The "restricted-Frenkel limit" [1.25] is one in which the excitation exchange interactions fall off sharply with distance so that only a small number of nearest neighbor (n.n.) and next nearest neighbor pairwise interactions are relevant. This enables one to use "resonance pair" (dimer) spectra (Sect. 1.8.2) to predict dispersion relations, densities of states and Davydov splittings for perfect and pure ("fully concentrated") crystals [1.25]. The exciton density of states can be derived experimentally from phonon sidebands (Sect. 1.9.1) but here these are "phonons" that are actually vibrational excitons (based on intramolecular vibrations), many of which have a very narrow k-band (order of 1 cm^{-1}). This has been achieved for benzene [1.61], naphthalene [1.61, 62] and hexamethylbenzene [1.63] and has given excellent agreement with the exciton-densities of states as derived from resonance pair spectra (mediumly dilute binary mixtures). In isotopic mixed organic (molecular) crystals the pure crystal (e.g., $C_{10}H_8$) exciton dispersion relation is necessary for a full characterization of cluster states in medium-dilute crystals [1.25, 44] (say $C_{10}H_8$ in $C_{10}D_8$). However, for monomers, only the density-of-states function is required [1.25, 64]. The much studied case of the naphthalene first singlet excitation has been reviewed in [1.25]. We note that we find in molecular crystals a gamut of excitation exchange interactions, from the very strong transition-dipole-transition-dipole couplings in the anthracene singlet excitations (reviewed by *Kepler* [1.65]) to the very weak "excitation plus electron exchange only" interactions typical of most organic triplet excitations [1.66]. Only where transition-dipole type interactions play a minor role can practical and simple relationships be made between the localized cluster states in dilute systems and the extended bulk states of fully concentrated systems [1.25]. Excitation dynamics studies on fully concentrated molecular solids, including ultrapure crystals, are described in Chap. 7.

One of the authors (RK) would like to thank Laurel Harmon for a critical reading of his part of the text and for Fig. 1.7.

References

1.1   D. L. Dexter: In *Solid State Physics*, Vol. 6, ed. by F. Seitz, D. Turnball (Academic Press, New York 1958)

1.2   G. F. Imbusch: In *Luminescence Spectroscopy*, ed. by M. D. Lumb (Academic Press, London 1978)

1.3   A more detailed treatment is found, for example, in B. Di Bartolo: *Optical Interactions in Solids* (Wiley, New York 1968)

1.4   G. H. Dieke: *Spectra and Energy Levels of Rare Earth Ions in Crystals* (Wiley-Interscience, New York 1968)

1.5   B. G. Wybourne: *Spectroscopic Properties of Rare Earths* (Wiley-Interscience, New York 1965)

1.6   B. R. Judd: *Operator Techniques in Atomic Spectroscopy* (McGraw-Hill, New York 1963)

1.7   S. Hüfner: *Optical Spectra of Transparent Rare-Earth Compounds* (Academic Press, New York 1978)

1.8   R. C. Pappalardo: In *Luminescence of Inorganic Solids*, ed. by B. Di Bartolo (Plenum Press, New York 1978)

1.9   W. T. Carnall, H. Crosswhite, H. M. Crosswhite: "Energy Level Structure and Transition Probabilities of the Trivalent Lanthanides in LaF_3," Argonne National Laboratory Report No. 60439, Argonne, Illinois

1.10   G. S. Ofelt: J. Chem. Phys. 37, 511 (1962)

1.11   B. R. Judd: Phys. Rev 127, 750 (1962)

1.12   L. A. Riseberg, M. J. Weber: Prog. Opt. 14 (1976), p. 91.

1.13   M. J. Weber: Phys. Rev. 8B, 54 (1973)

1.14   S. Sugano, Y. Tanabe, H. Kamimura: *Multiplets of Transition Metal Ions in Crystals* (Academic Press, New York 1970)

1.15   Y. Tanabe, S. Sugano: J. Phys. Soc. Jpn. 9, 766 (1954)

1.16   S. Sugano, Y. Tanabe: J. Phys. Soc. Jpn. 13, 880 (1958)

1.17   G. Klauminzer: Ph.D. Thesis, Stanford University (1970)

1.18   J. J. Markham: *F-Centers in Alkali Halides*; Supplement 8 to Solid State Physics, ed. by F. Seitz, D. Turnbull (Academic Press, New York 1966)

1.19   T. H. Keil: Phys. Rev. A140, 601 (1965)

1.20   B. Di Bartolo, R. C. Powell: *Phonons and Resonances in Solids* (Wiley-Interscience, New York 1976)

1.21   D. S. McClure: *Electronic Spectra of Molecules and Ions in Crystals* (Solid State Reprints), (Academic Press, New York 1959)

1.22   G. Herzberg: *Molecular Spectra and Molecular Structure III* (Van Nostrand, Princeton 1966)

1.23   K. Rebane: *Impurity Spectra of Solids* (Plenum Press, New York 1970)

1.24   P. H. Chereson, P. S. Friedman, R. Kopelman: J. Chem. Phys. 56, 3716 (1972)

1.25   R. Kopelman: In *Excited States II*, ed. by E. C. Lim (Academic Press, New York 1975) p. 33

1.26   D. C. Ahlgren: Ph.D. Thesis, University of Michigan (1979)

1.27 R. Kopelman, F. W. Ochs, P. N. Prasad: J. Chem. Phys. **57**, 5409 (1972)
1.28 P. S. Friedman: Ph.D. Thesis, University of Michigan (1974)
1.29 M. Zokai, R. C. Powell, G. F. Imbusch, B. Di Bartolo: J. Appl. Phys. **50**, 5930 (1979)
1.30 E. A. Harris, K. S. Yngvesson: J. Phys. C **1**, 990 (1968)
1.31 D. E. McCumber, M. D. Sturge: J. Appl. Phys. **34**, 1682 (1963)
1.32 T. Muramoto, Y. Fukuda, T. Hashi: Phys. Lett. **48A**, 181 (1974)
1.33 G. F. Imbusch, W. M. Yen, A. L. Schawlow, D. E. McCumber, M. D. Sturge: Phys. Rev. **133**, A1029 (1964)
1.34 H. W. Moos: J. Lumin. **1, 2**, 106 (1970)
1.35 M. J. Weber: Phys. Rev. **B8**, 54 (1973)
1.36 F. Auzel: In *Luminescence of Inorganic Solids*, ed. by B. Di Bartolo (Plenum Press, New York 1978)
1.37 M. D. Sturge: Phys. Rev. **B8**, 6 (1973)
1.38 G. Blasse: In *Luminescence of Inorganic Solids*, ed. by B. Di Bartolo (Plenum Press, New York 1978)
1.39 T. Forster: Ann. Physik **2**, 55 (1948)
1.40 D. L. Dexter: J. Chem. Phys. **21**, 836 (1953)
1.41 D. L. Huber: Phys. Rev. (to be published)
1.42 S. K. Lyo: Phys. Rev. **B3**, 3331 (1971)
1.43 R. Kopelman: "Exciton Percolation in Molecular Alloys and Aggregates," in *Radiationless Processes in Molecules and Condensed Phases*, ed. by F. K. Fong, Topics in Applied Physics, Vol. 15 (Springer, Berlin, Heidelberg, New York 1976), p. 297
1.44 J. Hoshen, R. Kopelman: Phys. Status Solidi **B81**, 479 (1977)
1.45 R. LeSar, R. Kopelman: Chem. Phys. **29**, 289 (1978)
1.46 H.-K. Hong, R. Kopelman: J. Chem. Phys. **55**, 3491 (1971)
1.47 J. Hoshen, R. Kopelman, E. M. Monberg: J. Stat. Phys. **19**, 219 (1978)
1.48 E. M. Monberg, R. Kopelman: V. L. Broude Memorial Issue, Mol. Cryst. Liq. Cryst. **57**, 271 (1980)
1.49 D. J. Thouless: *Les Houches Conference on Ill-Condensed Matter*, ed. by R. Balin (North-Holland, Amsterdam 1979), p. 1
1.50 J. Hoshen , J. Jortner: J. Chem. Phys. **56**, 933 (1972)
1.51 P. N. Prasad, R. Kopelman: J. Chem. Phys. **57**, 863 (1972)
1.52 D. S. McClure: In *Optical Properties of Ions in Solids*, ed. by B. Di Bartolo (Plenum Press, New York 1975)
1.53 R. L. Greene, D. D. Sell, R. S. Feigelson, G. F. Imbusch, H. J. Guggenheim: Phys. Rev. **171**, 600 (1968)
1.54 R. L. Greene, D. D. Sell, W. M. Yen, A. L. Schawlow, R. M. White: Phys. Rev. Lett. **15**, 656 (1965)
1.55 R. Loudon: Adv. Phys. **17**, 243 (1968)
1.56 V. V. Eremenko, E. G. Petrov: Adv. Phys. **26**, 31 (1977)
1.57 J. Hegarty, G. F. Imbusch: Colloq. Int. CNRS **255**, 199 (1977)
1.58 B. A. Wilson, M. W. Yen, J. Hegarty, G. F. Imbusch: Phys. Rev. **B19**, 4238 (1979)
1.59 A. S. Davydov: *Theory of Molecular Excitons* (Plenum Press, New York 1971)
1.60 R. Kopelman: J. Chem. Phys. **47**, 2631 (1967)
1.61 S. D. Colson, D. M. Hanson, R. Kopelman, G. W. Robinson: J. Chem. Phys. **48**, 2215 (1968)
1.62 H.-K. Hong, R. Kopelman: J. Chem. Phys. **55**, 724 (1971)
1.63 S. D. Woodruff, R. Kopelman: Chem. Phys. **22**, 1 (1977)
1.64 V. L. Broude, E. I. Rashba: Pure Appl. Chem. **37**, 21 (1974)
1.65 R. G. Kepler: *Treatise on Solid State Chemistry III* (Plenum Press, New York 1976), p. 615
1.66 G. W. Robinson: Ann. Rev. Phys. Chem. **21**, 429 (1970)

2. Excitation Transfer in Disordered Systems

T. Holstein[1], S. K. Lyo, and R. Orbach[2]

With 7 Figures

Disorder profoundly affects excitation transport in solids. Plane wave solutions lose relevancy, and marked departures from conventional band transport can occur. In particular, in the limit of strong disorder, the wave functions of an otherwise pure crystal become localized. At zero temperature, transport becomes impossible. As the temperature is raised, inelastic processes can make up the energy mismatch between sites introduced by the disorder, allowing spatial transfer to occur. The purpose of this chapter is to examine those mechanisms which depend on thermal motion of the lattice. This "phonon-assisted transport" is responsible for a variety of conduction processes, from electrical conductivity to exciton diffusion. We shall treat the extreme limit of localization in this chapter, namely an excitation localized on an optical site. The reasons for the localization and the character of localized states will be briefly outlined in Sect. 2.1; the nature of phonon-assisted excitation transfer is discussed in Sect. 2.2; and explicit calculations developed for the transfer rates for multipolar site–site coupling in Sect. 2.3 and for phonon-assisted radiative transport in Sect. 2.4. Section 2.5 presents a short discussion of the reverse: resonantly trapped phonons which can escape the trapping volume because of spectral energy transport. Our results are summarized in Sect. 2.6.

2.1 Excitation Localization

Consider a regular array of sites, each of which possesses an identical set of energy levels. Assume that each site is coupled to the rest of the sites in an identical manner through some sort of interaction which falls off sensibly with distance, and that the site–site coupling strength is large compared to single site energy differences. Then the correct physical description of the system is a set of delocalized exciton eigenstates extending over the entire crystal, each related to a particular excitation level of a site. If the system is prepared in such a manner that the excitation is initially localized at a given site, the excitation will propagate away from that site as a superposition of plane waves, with a velocity given roughly by the exciton bandwidth appropriate to that particular site energy level.

---

1 Deceased.
2 Supported by the U.S. Office of Naval Research, Contract no. N00014-75-C-0245.

Introduction of disorder can profoundly affect the character of the energy eigenstates. Disorder in the energy eigenvalues of the individual sites is one type of disorder (usually termed diagonal disorder); disorder in the interaction energy between sites is another type of disorder (usually termed off-diagonal disorder). For both small compared to the interaction energy between sites, the proper description of the system is one of weak scattered plane waves, the differences in local energy eigenvalues and/or inter-site coupling strengths acting as weak scatterers. This description is appropriate to three-dimensional systems; in one dimension even weak disorder can be much more potent.

As the disorder strength increases, a remarkable change in the character of the states for short-range site–site interactions (falling off more rapidly than the third power of the separation) occurs. *Anderson* [2.1], in a seminal paper, argued that, for sufficiently large diagonal disorder, the spatial extent of the energy eigenfunctions underwent a sharp transition from extended to localized. The precise conditions are still open to some question, but his criterion for localization can be stated simply: the magnitude of the diagonal disorder is comparable to the site–site coupling strength. The concept of off-diagonal disorder, and the calculation of the condition for localization, was developed extensively by *Antoniou* and *Economou* [2.2]. They found that states near the center of the band were delocalized in the presence of off-diagonal disorder alone, but became localized sharply when the energy of the state fell outside a "mobility edge." They went on to calculate the position of the mobility edge when both off-diagonal and diagonal disorder were present.

Optical spectroscopy is uniquely suited to determine the position of a mobility edge in random systems. Fluorescent line narrowing (FLN) experiments [2.3] are capable of exciting eigenstates with energy resolutions approaching the MHz range. Altering the excitation frequency of the exciting laser light enables one to probe the existence of a mobility edge by monitoring the fluorescence of very dilute traps. An example of this approach is that of *Koo* et al. [2.4a] on dilute ruby. They probed the inhomogeneously broadened Cr^{3+} single ion R_1-fluorescence line with a laser-narrowed excitation, observing the optical emission (N-line) from Cr^{3+} pairs. The latter are very few in number (varying as c^2, where c is the chromium ion concentration). Monitoring the pair fluorescence as a function of the excitation energy of the laser within the Cr^{3+} single ion emission line can provide evidence for long range (spatially extended) eigenstates of the Cr^{3+} ions. A sharp drop of the pair line fluorescence was observed as the laser excitation light moved towards the wings of the single Cr^{3+} line. This was interpreted as the existence of a sharp mobility edge. The position of the edge was in relatively good agreement with the critical concentration calculated for the extended-localized transition, as reported by *Lyo* [2.5].

Very recent experiments of S. *Chu* et al. [2.48] call into question the existence of a mobility edge in ruby. Their experiments were performed with a dye laser, which in contrast to the ruby laser of *Koo* et al. [2.4] could probe both sides of the single ion Cr^{3+} R_1 emission line. They reported nearly universal ratios of pair/single ion emission as a function of laser excitation energy for concentrations spanning those of *Koo* et al. It has been established by *Chu* and others [2.48, 49] that the results of *Koo* [2.4] were spurious and that resonant transfer in ruby systems is generally slow. For details see [2.50].

2.2 Phonon-Assisted Excitation Transfer

The localized state amplitude, as introduced above, falls off exponentially with distance, though the characteristic length may be much greater than a lattice constant. The question addressed in this chapter is how excitation is transferred from one position in the crystal to another, when the eigenstates are localized in the *Anderson* sense [2.1]. We argue that because disordered eigenstates differ from one another energetically, spatial transport of excitation requires that the energy mismatch between localized states be made up from a reservoir external to the interacting spin system itself. Our approach is to use lattice vibrations (phonons) to make up the energy mismatch, and therefore to allow excitations to move through the crystal. The transfer process is energy conserving, of course, with either the phonon energy itself (for one-phonon processes), or the difference in phonon energies (for two-phonon processes) being equal to the energy mismatch of the participating sites.

The spatial transport of excitation under phonon assistance can be most easily observed through FLN experiments. Excitation at one energy eigenstate leads to emission from other energy eigenstates because of phonon-assisted transfer from the initially excited localized eigenstate to other eigenstates with differing energy. Concomitantly, temperature dependent pair-line emission and other consequences of spatial transport would be direct measures of phonon-assisted transport under conditions of spatially localized eigenstates. Chapter 5 of this volume discusses these classes of experiments in detail.

It should be clear that phonon-assisted transfer requires the presence of ion-ion couplings to effect spatial transport of excitations. Dilute systems, or amorphous hosts, possess a spread in strength of the ion—ion couplings, leading to a probability density for their magnitude. This in turn affects the time development of the spatial transfer (and, in FLN experiments, the spectral transfer). We shall not develop this aspect of phonon-assisted transfer because it is covered in detail in Chap. 3. We do wish to enumerate the interactions in a solid which play a role in ion—ion excitation transfer. The more obvious are the multipolar couplings, developed, for example, by *Dexter* [2.6] in his classic extension of the fundamental work of *Förster* [2.7] on energy transfer. One usually encounters dipole—dipole, dipole—quadrupole, and quadrupole—quadrupole couplings, each with their characteristic range and angular dependence. However, as realized by *Pryce* [2.8] and by *Birgeneau* [2.9], the exchange coupling between ions is also a potent transfer mechanism. For example, differences in the one-electron exchange couplings between different Cr^{3+} orbitals allow excitation transfer via exchange coupling between states of different total spin multiplicity. The rapid fallof of exchange with increasing number of cation—anion superexchange bonds leads to the prediction of excitiation localization [2.5] at moderately dilute Cr^{3+} concentrations in ruby, for example. Yet other couplings can transfer excitations spatially. As shown explicitly by *Sugihara* [2.10], and then amplified by others [2.11], phonons can be emitted virtually at one site and re-absorbed at another, leading to a phonon-induced site–site coupling. This coupling is of quite long range for transitions between states whose energy falls within the phonon spectrum, but falls off exponentially with distance for transition with larger energies. Finally, radiative transfer can be a potent mechanism at low ion concentrations where direct multipolar interactions are weak. This process requires photon trapping to be effective. As shown Sect. 2.4, energy transfer via phonon-assisted photon transfer is independent of ion concentration under conditions of extreme photon trapping. The process is therefore very efficient (essentially every fluorescent photon emitted takes part in the energy transfer process).

2.3 Nonradiative Excitation Transfer

As discussed above, spatial transfer between localized excitations with different energies requires that the energy mismatch be made up by the phonons. We shall work in the weak coupling limit and consider optical transitions which are no-phonon in their primary intensity. The strong coupling case has been treated by a number of authors. The reader is referred to [2.12, 13] for details.

The historical approach to this problem is that of *Förster* [2.7]. He arrived at a transfer rate given by the square of the site–site coupling, times the convolution of the emission profile of one site with the absorption profile of the other. For the weak coupling case considered in this chapter, the Förster result for a single phonon-assisted transfer process would be proportional to the spectral overlap of the one-phonon sideband of the optical transition. For higher order phonon processes (e.g. inelastic two-phonon scattering), the Förster result would be given by the direct spectral overlap of the phonon-broadened optical transitions. This approach is of great utility when the detailed character of the ion–phonon coupling is unknown. However, important transfer processes are omitted with spectral overlap treatments (see Sect. 2.3.3a). The reason for the omission lies with the character of the ion–phonon coupling. If the ion–phonon interaction takes place at only a single site, then the Förster approach, and the one to be presented in this chapter [2.14], are coincident (with certain qualifications). However, if phonon interactions take place at both sites in the transfer process, there is no obvious connection with the optical linewidth. As a consequence, the energy transfer rate cannot be obtained from spectral overlap considerations. It is for this reason that we shall choose an alternate formalism, treating the ion–phonon interaction on the same footing as the ion–ion spatial coupling [2.14]. This method is similar to those of [2.12, 13].

Our results can be categorized in the following way. If excitation transfer is to occur between different ions, or between different states of similar ions, one-phonon-assisted energy transfer dominates if the phonon density of states at the mismatch energy is not too small. For transfer between similar ions, a cancellation will take place which strongly inhibits one-phonon-assisted transfer. This "interference" will be discussed in detail in Sect. 2.3.2. Under these conditions, or for energy mismatches greater than the maximum phonon energy in the host crystal, two-phonon-assisted energy transfer is most important. Under conditions of extreme energy mismatch, multiphonon transitions dominate, leading to a transfer rate which diminishes exponentially with increasing energy mismatch [2.12]. In the two-phonon regime, a number of distinct processes must be considered. If, for example, another ionic level lies close to those participating in the energy transfer, resonant phonon assistance dominates. This typically depends exponentially on the temperature, with the activation energy the energy difference to the nearby level. This will be discussed at length in Sect. 2.3.3c. If no nearby level exists, then a Raman-like phonon absorption on one site, coupled with phonon emission on the other, is the dominant process. This will be described in Sect. 2.3.3a. Each of these processes leads to a transfer rate with a characteristic dependence on energy mismatch and temperature. Both of these processes have been isolated recently (see Chap. 5).

2.3.1 Formulation of the Problem

We construct the formal expression for the elemental two-site jump rate in this section. The expression will be the basis for the formulation of the transport equatin for phonon-assisted energy transfer (see Chap. 3).

Consider excitation transfer between two ions with energy levels as exhibited in Fig. 2.1. We label the sites as 1 and 2, and the electronic ground and excited state by $|j>$ and $|j^*>$, respectively, with $j = 1, 2$. We seek the rate for the transition from $|1^*, 2>$ to $|1, 2^*>$, where these states are simple products of the single-site states. We assume that sites 1 and 2 are coupled non-radiatively (i.e., by multipolar or exchange interactions), and we denote the matrix element of the site–site coupling Hamiltonian by

$$J = \langle 1, 2^* | H_{s-s} | 1^*, 2 \rangle. \tag{2.1}$$

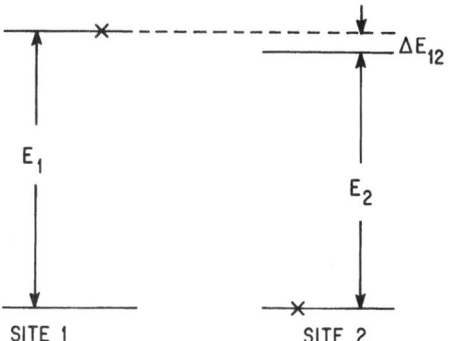

Fig. 2.1. The electronic energy levels of two sites between which excitation is to be transferred. The crosses symbolize the initial occupations, and the quantity ΔE_{12} the energy mismatch between excitations on sites 1 and 2

As shown in Fig. 2.1, the electronic energy mismatch for such a transition, ΔE_{12}, requires that a phonon (or phonons) be emitted or absorbed in order to conserve overall energy in the transfer process. We denote the coupling Hamiltonian between the ion and the phonons by $H_{ph}(j)$, noting that phonon emission or absorption can take place on either (or both) of the sites $j = 1,2$.

The matrix element of the phonon Hamiltonian will depend on the eigenstate of the ion. For the jth ion in its ground state, we write

$$\langle j, n_{s,q} \pm 1 \mid H_{ph}(j) \mid j, n_{s,q} \rangle = f(j) \langle n_{s,q} \pm 1 \mid \epsilon(j) \mid n_{s,q} \rangle \tag{2.2}$$

where $f(j)$ is the ground-state coupling strength for the ion at the jth site, and $\epsilon(j)$ the strain operator at the jth site. Similarly, for the excited state of the jth ion, we write

$$\langle j^*, n_{s,q} \pm 1 \mid H_{ph}(j) \mid j^*, n_{s,q} \rangle = g(j) \langle n_{s,q} \pm 1 \mid \epsilon(j) \mid n_{s,q} \rangle. \tag{2.3}$$

The phonon occupation number $n_{s,q}$ is labeled by the polarization index s, and the phonon wave vector q. Explicit evaluation of spatial part of the strain operator yields,

$$\langle n_{s,q} \pm 1 \mid \epsilon(j) \mid n_{s,q} \rangle = \langle n_{s,q} \pm 1 \mid \epsilon \mid n_{s,q} \rangle \exp[\mp iq \cdot r_j] \tag{2.4}$$

where r_j is the vector position of the jth site. The strain operator is usually written as a tensor,

$$\epsilon_{\alpha\beta} = \sum_{s,q} \left[\frac{1}{2} (\hbar/2MN\omega_{s,q})^{1/2} (a^\dagger_{s,q} + a_{s,-q})(e_{s,\alpha}q_\beta + e_{s,\beta}q_\alpha) \right] \tag{2.5}$$

where M, N are the mass of the ion and the total number of lattice sites, respectively, $e_{s,\alpha}$ is αth component of the polarization vector for the s polarization, and the $a^\dagger_{s,q}$ ($a_{s,q}$) the creation (destruction) operator for a phonon of polarization index s and wave vector q. Our treatment will not include crystal anisotropy, for simplicity, so that we shall be concerned only with averages of the strain tensor $\epsilon_{\alpha\beta}$ over the solid angle Ω. We write

$$\frac{1}{4} \langle (e_{s,\alpha}q_\beta + e_{s,\beta}q_\alpha)^2 \rangle_\Omega = \alpha_s q^2 \tag{2.6}$$

where α_s will be of order unity.

The transition probability per unit time for excitation transfer to take place is given by the "golden rule"

$$W_{2\leftarrow1} = (2\pi/\hbar) \sum_{s_1,q_1,s_2,q_2,\dots} |t_{f\leftarrow i}|^2 \delta[\sum(\Delta n_{s,q}\,\hbar\omega_{s,q}) \pm \Delta E_{12}] \tag{2.7}$$

where $t_{f\leftarrow i}$ is the "t" matrix, and the sum in the delta function goes over all phonon occupation numbers which change between the initial and final states, the latter labeled by i and f respectively. $\Delta n_{s,q}$ is the change in occupation number of the phonon of polarization index s and wave vector q going between i and f.

The t matrix is given by the usual expression

$$t_{f \leftarrow i} = \langle f|H|i \rangle + \sum_{m_1} \frac{\langle f|H|m_1 \rangle \langle m_1|H|i \rangle}{E_i - E_{m_1} + is}$$

$$+ \sum_{m_1, m_2} \frac{\langle f|H|m_2 \rangle \langle m_2|H|m_1 \rangle \langle m_1|H|i \rangle}{(E_i - E_{m_1} + is)(E_i - E_{m_2} + is)} + \dots \tag{2.8}$$

where s is a positive infinitesimal. The Hamiltonian appearing in (2.8) is the sum of the site–site coupling Hamiltonian, H_{s-s}, and the sum of the single-site phonon Hamiltonians, $\sum_j H_{ph}(j)$.

This formulation treats the site–site coupling on the same footing as the interaction with the phonons. One-phonon-assisted transfer processes involve only the second term in (2.8). Two-phonon-assisted transfer processes for nonradiative coupling will involve both the third term of (2.8) using (2.2) and the second term of (2.8) using the second-order expansion of H_{ph} (see Sect. 2.2.3b). In Sect. 2.4 we calculate phonon-assisted radiative transfer. In this instance, we need the third-order term in (2.8) for one-phonon-assisted transfer, while the fourth-order term is necessary for two-phonon-assisted transfer. This is because nonradiative (e.g., multipolar) site–site coupling enters the perturbation chain directly as H_{s-s}, while for radiative transfer one must introduce the ion–radiation field coupling twice to account for excitation transfer.

With these preliminaries completed, we are now in a position to calculate phonon-assisted transfer rates.

2.3.2 One-Phonon-Assisted Process

We consider the case where the emission or absorption of a single phonon can make up the energy mismatch between sites. This process was first calculated by *Orbach* [2.15] using the method described above. It was expanded upon in some detail by *Birgeneau* [2.9] and by *Soules* and *Duke* [2.13].

The initial state of the system is described by the ket

$$|i> = |1^*, 2, n_{s,q}> \tag{2.9}$$

while the final state is described by

$$|f> = |1, 2^*, n_{s,q} \pm 1>. \tag{2.10}$$

The upper and lower signs designate phonon emission and absorption, respectively. The two states differ in the phonon occupation number and the site excitations. Modulation of H_{s-s} will connect (2.9, 10), and the transfer rate can be calculated using the first term in (2.8) inserted into (2.7). We have found such terms smaller in general than direct modulation of the single-site energies as a consequence of $H_{ph}(j)$. The latter require the second-order t matrix with intermediate states $|m_1'> = |1^*, 2, n_{s,q} \pm 1>$ and $|m_1''> = |1, 2^*, n_{s,q}>$. The energy denominators in the second term of (2.8) are $E_i = E_1$; $E_{m'_1} = E_1 \pm \hbar\omega_{s,q}$; $E_{m''_1} = E_2$. Here, E_1 and E_2 are the unperturbed electronic excitation energies on sites 1 and 2, respectively. Using these expressions in the t-matrix expression, (2.9), one has

$$t_{f \leftarrow i} = \sum_{j=1,2} \frac{\langle 1, 2^*, n_{s,q} \pm 1 | H_{s-s} | 1^*, 2, n_{s,q} \pm 1 \rangle}{E_1 - (E_1 \pm \hbar\omega_{s,q})}$$

$$\times \langle 1^*, 2, n_{s,q} \pm 1 | H_{ph}(j) | 1^*, 2, n_{s,q} \rangle$$

$$+ \sum_{j=1,2} \frac{\langle 1, 2^*, n_{s,q} \pm 1 | H_{ph}(j) | 1, 2^*, n_{s,q} \rangle}{E_1 - E_2}$$

$$\times \langle 1, 2^*, n_{s,q} | H_{s-s} | 1^*, 2, n_{s,q} \rangle \tag{2.11}$$

where we have dropped the unnecessary infinitesimal. Using (2.1–4), (2.11) reduces to

$$t_{f \leftarrow i} = \frac{J \langle n_{s,q} \pm 1 | \epsilon | n_{s,q} \rangle \exp [\mp i q \cdot r_2]}{- \Delta E_{12}}$$

$$\{[f(1) - g(1)] \exp [\pm i q \cdot r] - [f(2) - g(2)]\} \tag{2.12}$$

where r_2 is the position of site 2, and $r = r_2 - r_1$.

The structure of (2.12) is interesting. Note first of all that the strength of the t matrix depends on the difference in phonon coupling constants between the ground and excited levels on *each* site. Further, for small phonon energies (long phonon wavelengths) the product of qr is small. This means that if ions 1 and 2 are of a similar type, a cancellation occurs in (2.12). This is an interference and has its physical origins in the following argument. The phonon coupling modulates the excited state energy, relative to the ground state, of each site. If the static state energies of the two sites are not equal, phonon modulation must instantaneously make the site energies equal so that excitation transfer can take place. This can occur only if the couplings are different on the two sites, or if the phonon wavelength is less than the distance between the two sites. We therefore arrive at the following conclusions. If the energy mismatch between sites 1 and 2 is large (i.e., a sensible fraction of the Debye energy), then the phonon wavelength $\lambda = 2\pi/q$ can be less than the effective interaction distance between sites. Excitation transfer can then occur even between sites with identical phonon coupling strengths. If the energy mismatch between sites 1 and 2 is small, then the phonon wavelength λ (equal to $2\pi/q = 2\pi\hbar v_s/\Delta E_{12}$ where v_s is the sound velocity for the sth polarization) will be much larger than the effective interaction distance between sites, so that $qr \ll 1$. The site energies can be made instantaneously equal only by virtue of very different phonon couplings on the two sites. In summary, for sites with similar phonon couplings, the distance must be greater than $\lambda = 2\pi\hbar v_s/\Delta E_{12}$ for excitation transfer to occur. However, at such distances, the site–site coupling strength may be small.

The sum of all these effects is the strong diminution of the importance of the one-phonon-assisted process between similar sites with small energy mismatch. Chapter 5 discusses recent excitation transfer experiments in optical systems. They show that one seldom observes a one-phonon-assisted transfer process for narrow optical spectra (i.e., small energy mismatch in FLN experiments). The remarkable exception is that of ruby, where FLM experiments at low temperatures in the moderate concentration range exhibit a behavior indicative of a one-phonon-assisted transfer process [2.16]. It is conceivable that this could be caused by excitations between weakly coupled pairs of Cr^{3+} ions whose spectra overlap the single Cr^{3+} ions, and Cr^{3+} single ions. The pairs are known [2.17] to be much more sensitive to strain than single Cr^{3+} ions, so that the coupling strengths for sites 1 (the pairs) and 2 (the single ions) are quite different, reducing the interference even for small energy mismatch.

For the purposes of the remainder of this section, we take the ions to be identical, apart from the difference in their energy. Then, inserting (2.12) into the golden rule (2.7), we obtain

$$W_{2\leftarrow1} = \frac{2\pi J^2 (f-g)^2}{\hbar(\Delta E_{12})^2} \sum_{s,q} |\langle n_{s,q} \pm 1 | \epsilon | n_{s,q}\rangle|^2 \, h^I(q,r) \, \delta(\hbar\omega_{s,q} \mp \Delta E_{12}), \tag{2.13}$$

where the coherence factor $h^I(q,r)$ is given by

$$h^I(q,r) = |\exp(iq \cdot r) - 1|^2. \tag{2.14}$$

We proceed to evaluate (2.13) using the Debye approximation. For any function of s and q,

$$\sum_{s,q} F(s,q) = \frac{V}{2\pi^2} \sum_s \frac{1}{v_s^3} \int_0^{\omega_{s,D}} \langle F(s,q)\rangle_\Omega \, \omega^2 \, d\omega, \tag{2.15}$$

where $\omega = q v_s$, and V and $\omega_{s,D}$ are the sample volume and Debye frequency for the sth polarization mode, respectively.

We now analyze (2.13) in the limits of large and small energy mismatch.

a) Large Energy Mismatch

The energy mismatch between even nearby sites can be of the order of hundreds of wave numbers for amorphous hosts. Under this condition, the energy conserving phonon has a wavelength of the order of the inter-site separation. The interference factor in the expression for the coherence factor, (2.14), then averages out, and one is left with $\langle h^I(q, r)\rangle_\Omega \simeq 2$. Using (2.15) to evaluate (2.13), and making use of our previous expressions for the phonon averages, (2.5, 6), we find

$$W_{2\leftarrow 1} = \frac{J^2(f-g)^2 \,|\,\Delta E_{12}\,|}{\pi \hbar^4 \rho} \left(\sum_s \frac{\alpha_s}{v_s^5}\right) \begin{Bmatrix} n(|\,\Delta E_{12}\,|) + 1 \\ n(|\,\Delta E_{12}\,|) \end{Bmatrix} \tag{2.16}$$

for $\begin{Bmatrix} \text{emission} \\ \text{absorption} \end{Bmatrix}$ of a phonon of energy ΔE_{12}. Here, $n(x) = [\exp(\beta x) - 1]^{-1}$ is the Bose factor ($\beta \equiv 1/k_B T$), ρ is the mass density, and α_s is of order unity [defined by (2.6)]. An interesting aspect of (2.16) is that when $\beta\,|\,\Delta E_{12}\,| < 1$, (2.16) reduces to

$$W_{2\leftarrow 1} = \frac{J^2(f-g)^2}{\pi \hbar^4 \rho} \left(\sum_s \frac{\alpha_s}{v_s^5}\right) k_B T. \tag{2.17}$$

It is seen that the one-phonon-assisted transfer rate is independent of energy mismatch, and linearly dependent on temperature. The former aspect is rather extraordinary. It can be tested directly using FLN experiments. The time development of the fluorescent emission profile after a short laser pulse exhibits transfer to spatial sites which form the remainder of the emission line. This buildup of the fluorescence of other sites would mirror the shape of the equilibrium emission line for a transfer rate independent of energy mismatch. This appears to be the case in the low temperature region for moderately concentrated ruby [2.16].

b) Small Energy Mismatch

When the energy mismatch is small, the quantity qr becomes small, as does the coherence factor (2.14). Under this condition,

$$\langle h^I(q,r)\rangle_\Omega \simeq (qr_{12})^2/3. \tag{2.18}$$

This alters the transfer rate from our previous expression (2.17) to

$$W_{2\leftarrow 1} = \frac{J^2(f-g)^2 |\Delta E_{12}|^2 \, r_{12}^2}{6\pi \hbar^6 \rho} \left(\sum_s \frac{\alpha_s}{v_s^7}\right) k_B T. \tag{2.19}$$

In general, (2.19) is smaller than the corresponding expression for large energy mismatch by the factor $(\Delta E_{12} r_{12}/v_s \hbar)^2/6$. Under the condition of energy transport between similar ions at small energy mismatch, (2.19) is in general too small to be observable, and higher order mechanisms predominate.

We should like to point out that a spectral overlap theoretical approach would not yield the detailed results presented in this section. Interference effects would be absent from such treatments, and the real reason for the impotence of one-phonon-assisted transfer processes would be lost.

We now examine processes which are important either at high temperatures, or even at low temperatures when a small energy mismatch would render the one-phonon-assisted transfer processes unimportant.

2.3.3 Two-Phonon-Assisted Processes

The smallness of the coherence factor for small energy mismatch, and/or the small phonon density of states when $|E_{12}| \ll k_B T$, leads to the consideration of higher-order processes. We consider the third-order term in the t matrix, (2.8), which allows for two-phonon-assisted excitation transfer. One power of H_{s-s}, and two of $\Sigma_j H_{ph}(j)$, will appear in (2.8) at third order. This means that two phonons, of energy $\hbar\omega_{s',q'}$ and $\hbar\omega_{s,q}$, will inelastically scatter off the two ion sites, such that overall system energy is conserved

$$\hbar\omega_{s',q'} - \hbar\omega_{s,q} = \Delta E_{12}.$$

Such processes are higher order than one-phonon-assisted processes, but they can dominate for two reasons:
a) limitation of the strength of the one-phonon-assisted process because of small phonon density of states at $\Delta E_{12}/\hbar$; and/or
b) lack of interference in the two-phonon-assisted transfer coherence factor. The phonons involved in higher-order transfer processes have short wavelengths $\lambda \simeq 2\pi\hbar v_s/k_B T$, and as a consequence cancellation in the interference factor is avoided.

There are a number of two-phonon-assisted transfer processes to consider. We group them according to their character.

The first we consider is that formed from the one-phonon Hamiltonian acting at the first site, transfer of the excitation to the second site via H_{s-s}, and then the one-phonon Hamiltonian acting at the second site; and those obtained by rearrangement of the above perturbation sequence. This is termed the "two-site nonresonance process" (Sect. 2.3.3a), with a transfer rate denoted by $W_{2\leftarrow 1}^{\text{non res} I}$. The second process involves the second-order phonon Hamiltonian acting at a single site. This results in a lower-order perturbation expression, with inelastic phonon scattering at one site, and excitation transfer to the other site via H_{s-s}. This is termed the "one-site Raman process" (Sect. 2.3.3b), with a transfer rate denoted by $W_{2\leftarrow 1}^{\text{non res} II}$. The third process we consider is important when a third level (or levels) is (are) near either the electronic excited state $|j^* >$ or ground state $|j >$. The one-phonon Hamiltonian acts at the first site, carrying the excitation to the immediate vicinity of the third level. The excitation is transferred by virtue of H_{s-s} from that level to the same level on the second site, and then a phonon is emitted or absorbed at the correct energy to conserve overall system energy. The near resonance in the denominator for the first step in the perturbation chain causes a great enhancement of this process. Other processes generated by rearrangement of the above perturbation sequence are also important. The totality of these processes is termed the "one-site resonance process" (Sect. 2.3.3c), with a transfer rate denoted by $W_{2\leftarrow1}^{res}$. The last process we consider is the nonresonant component of the third process. This is termed the "one-site nonresonant process" (Sect. 2.3.3d), with a transfer rate denoted by $W_{2\leftarrow1}^{non\,res\,III}$.

The relative importance of each of these four classes of two-phonon-assisted excitation transfer processes depends on the particular structure of the electronic energy levels of the physical system. In general, if no nearby levels are present, the first, or two-site nonresonant process, dominates. Its "signature" is a rate proportional to T^3. If a nearby level is present, lying near the states involved in the excitation transfer, the third process generally dominates. Its characteristic signature is a rate proportional to $\exp(-\Delta/k_B T)$, where Δ is the splitting to the third level.

In all four cases we work in the limit that $k_B T \gg \Delta E_{12}$. If the opposite limit pertains, then

$$\hbar\omega_{s',q'} + \hbar\omega_{s,q} = \Delta E_{12}.$$

That is, instead of the phonons inelastically scattering off the electronic sites, one absorbs both phonons in the two-phonon-assisted transfer process.

Under the former conditions (when $k_B T \gg \Delta E_{12}$), the initial and final states are, respectively,

$$|i> = |1^*, 2, n_{s,q}, n_{s',q'} >;$$

$$|f> = |1, 2^*, n_{s,q} - 1, n_{s',q'} + 1 >.$$

The energies of these states are, respectively,

$$E_i = E_1;$$

$$E_f = E_2 - \hbar\omega_{s,q} + \hbar\omega_{s',q'}.$$

The transition rate is given by

$$W_{2\leftarrow1} = \frac{2\pi}{\hbar} \sum_{s,q,s',q'} |t_{f\leftarrow i}|^2 \, \delta(\hbar\omega_{s',q'} - \hbar\omega_{s,q} - \Delta E_{12}). \tag{2.20}$$

With these general expressions in hand, we now calculate the phonon-assisted excitation transfer rates.

a) Two-Site Nonresonant Process

We introduce $\sum_j H_{\text{ph}}(j)$ twice in this process, once at each site or twice at one site. To make our presentation tractable, we shall temporarily omit the matrix element of $H_{\text{ph}}(j)$ in the ground state. These terms simply subtract off the strength of the ion–phonon coupling of the ground level, g, from that of the excited level in the final result [e.g., see (2.12)].

We represent the physical process we are considering in Fig. 2.2. Initially, site 1 is the excited state, and site 2 is in the ground state. The first step in the perturbation chain consists of a phonon (represented in the figure by an incoming wiggly line) of mode s, q being absorbed at site 1, leading to a first intermediate state denoted as $\lceil m_1 >$. The electronic state remains unchanged. The transfer Hamiltonian, $H_{\text{s-s}}$, then acts to put site 1 in its ground electronic state and site 2 in its excited electronic state. This configuration is denoted as $| m_2 >$. Finally, the second phonon (represented in the figure by the outgoing wiggly line) of mode s', q' is emitted at site 2. Note that phonon emission and absorption takes place at different sites. There is no obvious connection with phonon-induced linewidth processes, so that at the least it is not obvious how one could compute this excitation transfer rate using a spectral overlap approach.

The two intermediate states are

$$| m_1 > = | 1^*, 2, n_{s,q} - 1, n_{s',q'} >$$

$$| m_2 > = | 1, 2^*, n_{s,q} - 1, n_{s',q'} >$$

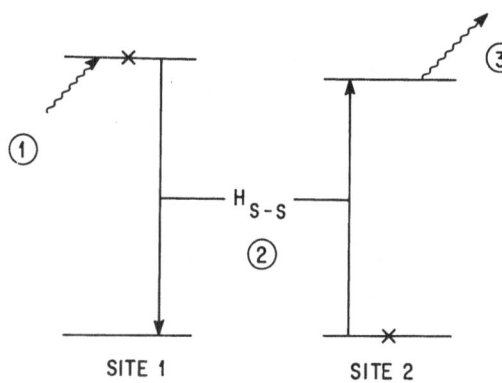

Fig. 2.2. Two-site nonresonant process. The ion–phonon interaction acts twice, the site–site coupling Hamiltonian once. The circled numbers indicate the order of transitions. The arrows denote the sense of transition. There are a total of sixteen processes of this kind

with respective energies

$$E_{m_1} = E_1 - \hbar\omega_{s,q}$$

$$E_{m_2} = E_2 - \hbar\omega_{s,q}.$$

Following the same procedures as before, the t matrix for this process becomes,

$$t_{f \leftarrow i} = \frac{Jf(1)f(2)}{\hbar\omega_{s,q}\,\hbar\omega_{s',q'}} \langle n_{s,q} - 1 \mid \epsilon \mid n_{s,q} \rangle \langle n_{s',q'} + 1 \mid \epsilon \mid n_{s',q'} \rangle$$

$$\times \exp(i q \cdot r_1 - i q' \cdot r_2). \tag{2.21}$$

A total of six distinct graphs of this type are obtained by applying the three steps exhibited in Fig. 2.2 in different orders. Examples are: a phonon is emitted at site 1, excitation exchange takes place between sites 1 and 2, and a phonon is absorbed at site 2; or, a phonon is absorbed at site 1, another phonon is emitted at site 1, and excitation exchange takes place between sites 1 and 2. Summing these graphs, the t matrix becomes (making use of the energy conservation requirement that $\hbar\omega_{s',q'} - \hbar\omega_{s,q} = \Delta E_{12}$),

$$t_{f \leftarrow i} = \frac{J}{\hbar\omega_{s,q}\,\hbar\omega_{s',q'}} \langle n_{s,q} - 1 \mid \epsilon \mid n_{s,q} \rangle \langle n_{s',q'} + 1 \mid \epsilon \mid n_{s',q'} \rangle$$

$$\times \{ f(1)f(2) \left[\exp(i q \cdot r_1 - i q' \cdot r_2) + \exp(-i q' \cdot r_1 + i q \cdot r_2) \right]$$

$$- f^2(1) \exp[i(q - q') \cdot r_1] - f^2(2) \exp[i(q - q') \cdot r_2]. \tag{2.22}$$

As can be seen from (2.22), all phonon interactions have been calculated with the ions in their excited state. We must also include graphs which involve phonon couplings to ions in their ground states. There are six which simply correspond to (2.22) with f replaced by g. There are twelve which contain phonon interactions in both the excited and ground states. Summing all twenty-four graphs is equivalent to replacing $f(j)$ in (2.22) by $f(j) - g(j)$. For simplicity, we take all ions to be similar, whence we can remove the dependence of $f(j)$ and $g(j)$ on the site index j. Writing $f(j) = f$ and $g(j) = g$, and inserting into the golden rule (2.7), we find the excitation transfer rate

$$W_{2 \leftarrow 1}^{\text{non-res }I} = \frac{2\pi}{\hbar} \sum_{s,q,s',q'} \left[J^2(f - g)^4 / \hbar^2 (\omega_{s,q}\,\omega_{s',q'})^2 \right]$$

$$\times \mid \langle n_{s',q'} + 1 \mid \epsilon \mid n_{s',q'} \rangle \langle n_{s,q} - 1 \mid \epsilon \mid n_{s,q} \rangle \mid^2$$

$$\times h^{II}(q, q'; r) \, \delta(\hbar\omega_{s',q'} - \hbar\omega_{s,q} - \Delta E_{12}) \tag{2.23}$$

where the coherence factor is given by

$$h^{II}(q, q'; r) = |\exp(-iq' \cdot r) + \exp(iq \cdot r) - 1 - \exp[i(q - q') \cdot r]|^2. \quad (2.24)$$

It is interesting to note that the energy transfer rate (2.23) is independent of energy mismatch in the very high temperature regime $(\beta |\Delta E_{12}| \ll 1)$. In this limit, the phonon wavelength is assumed sufficiently short that $\langle h^{II}(q, q'; r)\rangle_{\Omega} \simeq 4$. Equation (2.23), in the Debye approximation, reduces to

$$W_{2 \leftarrow 1}^{\text{non-res } I} = \frac{f^2(f-g)^4}{2\pi^3 \hbar^7 \rho^2} \left(\sum_s \frac{\alpha_s}{v_s^5}\right)^2 (k_B T)^3 I_2 \qquad (2.25)$$

where the integral I_n is defined by

$$I_n = \int_0^{\theta_D/T} \frac{x^n \, dx}{(e^x - 1)(1 - e^{-x})}; \quad x = \hbar\omega_{s,q}/k_B T. \qquad (2.26)$$

For temperatures low compared to the Debye temperature, $I_2 \simeq 2$. In the opposite limit, $I_2 \simeq \theta_D/T$, and the transfer rate (2.25) is reduced to a quadratic dependence on temperature.

The short wavelength condition which led to (2.25) is rather stringent. *Hamilton et al.* [2.18] noted that an expansion in qr and $q'r$ of (2.24) led to higher powers of the temperature than T^3 for $W_{2 \leftarrow 1}^{\text{non-res } I}$. At the temperatures of measurement for PrF$_3$ (between 6 and 20 K), the transfer rate was observed to vary as $T^{4.3}$, while over a larger temperature range between 5 and 35 K), the rates for LaF$_3$:Pr varied as T^3. *Hamilton* calculated $W_{2 \leftarrow 1}^{\text{non-res } I}$ from qr and $q'r \gg 2\pi$, where the T^3 dependence is obtained, to qr and $q'r \lesssim \pi/2$ where a T^7 dependence is found. [2.18] reports a rather more detailed calculation for PrF$_3$, yielding a $T^{6.1}$ dependence. The calculated temperature dependence is very sensitive to the precise shape of the phonon spectrum, so that at present these authors feel the disagreement between exponents (experiment, 4.3; theory 6.1) is not significant. In any case, at higher temperatures in the dilute case (LaF$_3$:Pr, 5% and 20%), the T^3 dependence is observed, as one would predict in the short wavelength regime.

We now go on to the calculation of the two-phonon-assisted excitation transfer process. Interaction with the phonons takes place at a single site, arising from the next higher order in the ion—phonon interaction, quadratic in the strain tensor ϵ.

b) One-Site Raman Process

The ion—phonon Hamiltonian (2.3) was written in terms of a single phonon interaction [linear in the strain tensor $\epsilon(j)$]. The effect of the next higher order term in the expansion of $H_{ph}(j)$, quadratic in $\epsilon(j)$, will be considered in this section. The physical process which results from these terms is pictured in Fig. 2.3. One phonon

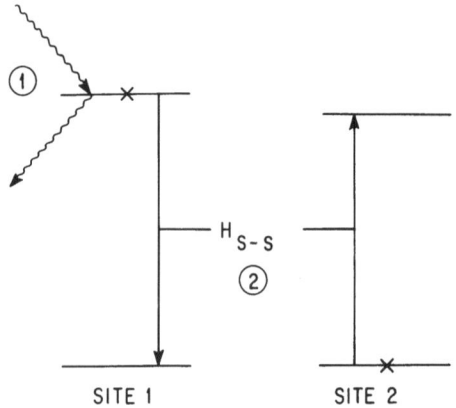

Fig. 2.3. One-site Raman process. The two-phonon term in the ion—phonon interaction acts once, the site—site coupling Hamiltonian acts once. The circled numbers indicate the order of transitions; the arrows the sense of the transitions. There are a total of four processes of this kind

is absorbed and another emitted simultaneously at site 1 in the first step of the perturbation chain. The second step is the excitation exchange between sites 1 and 2 by virtue of H_{s-s}. The energy mismatch between sites is taken up by the difference in phonon energies. The low order of this process requires only the second term in the t matrix expansion (2.8). A single intermediate state is involved,

$$| m > = | 1^*, 2, n_{s,q} - 1, n_{s',q'} + 1 >$$

with an energy

$$E_m = E_1 - \hbar \omega_{s,q} + \hbar \omega_{s',q'}.$$

We denote the phonon Hamiltonian by the approximate form

$$H_{\text{ph}} = \sum_j f_2(j) \, \epsilon^2(j) \tag{2.27}$$

for the excited level [with $g_2(j)$ replacing $f_2(j)$ for the ground level]. Equation (2.27) ignores the detailed properties of the second-order ion—phonon coupling. These can be found in, for example, [2.19], but will only change the results of this section by factors of order unity.

The contribution to the t matrix (2.8) for this process is,

$$t_{f \leftarrow i} = \frac{J f_2(1)}{-\Delta E_{12}} \langle n_{s,q} - 1, n_{s',q'} + 1 \mid \epsilon^2 \mid n_{s,q}, n_{s',q'} \rangle \exp\left[-(q - q') \cdot r_1 \right]. \tag{2.28}$$

There are three more graphs of the class exhibited in Fig. 2.3 (including those where H_{ph} acts in the ground level). Adding all four together,

$$t_{f \leftarrow i} = \frac{J \langle n_{s,q} - 1, n_{s',q'} + 1 \mid \epsilon^2 \mid n_{s,q}, n_{s',q'} \rangle}{\Delta E_{12}} \{[f_2(2) - g_2(2)] \exp [i(q - q') \cdot r_2]$$

$$- [f_2(1) - g_2(1)] \exp [i(q - q') \cdot r_1]\}. \tag{2.29}$$

We see that the t matrix is inversely proportional to the energy mismatch between sites 1 and 2. This is an example of the energy dependence for excitation transfer that one would predict using a spectral overlap model [2.6, 7]. As discussed earlier, the two-phonon part of H_{ph} can act to produce a homogeneously broadened optical transition at either site. The convolution of the resulting Lorentzian line shapes from both sites yields an excitation transfer amplitude inversely proportional to the energy mismatch ΔE_{12}, for ΔE_{12} much greater than the homogeneous broadening. The result (2.29) is identical to that which one would obtain from such an approach, apart from the interference factor (term in the curly brackets). Similar results will be obtained whenever H_{ph} acts at a single site in the excitation transfer process.

We assume that all ions are similar, as before, so that insertion of (2.29) into (2.7) yields

$$W_{2 \leftarrow 1}^{\text{non-res }II} = \frac{2\pi}{\hbar} \sum_{s,q,s',q'} (J/\Delta E_{12})^2 (f_2 - g_2)^2 \mid \langle n_{s,q} - 1, n_{s',q'} + 1 \mid \epsilon^2$$

$$\times \mid n_{s,q}, n_{s',q'} \rangle \mid^2 \; h^I(q - q', r) \delta(\hbar \omega_{s',q'} - \hbar \omega_{s,q} - \Delta E_{12}), \tag{2.30}$$

where h^I is given by (2.14). The rate (2.30) is proportional to the inverse square of the energy mismatch ΔE_{12} in the high temperature limit ($\beta \mid \Delta E_{12} \mid \ll 1$). If, as before, we assume the phonons are of sufficiently short wavelength that $qr, q'r \gg 1$, then $\langle h^I \rangle_\Omega \simeq 2$. Equation (2.30) becomes, in the Debye approximation,

$$W_{2 \leftarrow 1}^{\text{non-res }II} = \frac{J^2 (f_2 - g_2)^2}{4\pi^3 \hbar^7 \rho^2 (\Delta E_{12})^2} \left(\sum_s \frac{\alpha_s}{v_s^5} \right)^2 I_6 (k_B T)^7. \tag{2.31}$$

The dominant contribution to I_6 arises from phonons of energy $\hbar \omega_{s,q} \simeq 6 k_B T$, so that the T^7 power law holds for temperatures $T < \theta_D/7$. This limits the applicable temperature regime for (2.31). At temperatures greater than $\theta_D/7$, one must use the precise form of I_6, (2.26). At very high temperatures, (2.31) is proportional to T^2. At rather low temperatures where $qr, q'r \ll 1$, the interference term h^I becomes important. One obtains a temperature dependence for $W_{2 \leftarrow 1}^{\text{non-res }II}$ proportional to T^9, and a magnitude much smaller than (2.31).

It is interesting to compare the rates (2.25) and (2.31) for nonresonant phonon-assisted excitation transfer. We approximate $I_n \simeq n!$. In the high temperature limit (but below $\theta_D/7$), we find

$$\frac{W_{2 \leftarrow 1}^{\text{non-res }I}}{W_{2 \leftarrow 1}^{\text{non-res }II}} = \frac{(f - g)^4 (\Delta E_{12})^2}{180 (f_2 - g_2)^2 (k_B T)^4}. \tag{2.32}$$

As a general rule, the coupling constants in H_{ph} are all roughly equal (of order of the static crystal field coupling constants [2.20]). This means the ratio of the rates is very roughly $f^2(\Delta E_{12})^2/100(k_B T)^4$. Rare earth optical transitions are characterized by $f \sim 1,000$ cm^{-1}, $\Delta E_{12} \sim 1$ cm^{-1}. Hence, at a temperature of 10 K, the ratio (2.32) is of order unity. Nearly all recent excitation transfer experiments (see Chap. 5) exhibit $W_{2\leftarrow 1}^{\text{non-res }I}$ temperature dependences. This dominance over $W_{2\leftarrow 1}^{\text{non-res }II}$ may reflect an overestimate of I_6 from the Debye approximation, and rather greater ion–phonon coupling constants than those adopted here.

This section, and the preceding one, have considered phonon-assisted excitation transfer using only those levels actually involved in the optical excitation process. In many instances (e.g., Cr^{3+} in Al_2O_3 or $LaAlO_3$) a third level may be nearby (i.e., within the energy range of an acoustic phonon) which can effect excitation transfer. The remainder of Sect. 2.3.3 is concerned with this possibility. Section 2.3.3c treats direct resonant excitation of the third level by one of the phonons assisting transfer, while Sect. 2.3.3d treats the nonresonant component of the phonon transition to the third level.

c) One-Site Resonant Process

The availability of a third electronic level allows the single-phonon term in H_{ph} to act in a resonant transition. This results in a resonance fluorescence type of enhancement, conceptually the same as in the resonant spin-lattice relaxation process [2.20]. We calculate its contribution to excitation transfer in this section.

We designate the third level by $|j^{**}>$, where j labels the spatial position of the site. Although we shall discuss the situation where the third level lies near the upper level, as in ruby, the same approach applies to the case where it lies near the ground level, as in $LaF_3 : Pr^{3+}$. The excitation transfer process A is pictured in Fig. 2.4. First, a phonon of energy $\hbar\omega_{s,q}$ is absorbed at site 1, causing the electronic state to change from the initial level $|1^*>$ to the first intermediate level $|1^{**}>$. The electronic energy difference between these two levels is denoted by Δ. Resonance means that $\hbar\omega_{s,q} \sim \Delta$. Second, a phonon of energy $\hbar\omega_{s',q'}$ is emitted and the electronic level returns to $|1^*>$ (remember that this step is virtual). Finally, H_{s-s} transfers the excitation from site 1 to site 2. The primary feature of this excitation transfer process is that the first phonon's energy is fixed by the splitting to the third electronic level, Δ. The energy of the second phonon is determined at the last step. Overall energy conservation requires the difference between the first and second phonon energies to equal the electronic energy mismatch between sites 1 and 2. This is pictured in process A in Fig. 2.4 by drawing the length of the energy arrow for the second phonon transition equal to $\Delta + \Delta E_{12}$.

The two intermediate states are

$$|m_1> = |1^{**}, 2, n_{s,q} - 1, n_{s',q'}>$$

$$|m_2> = |1^*, 2, n_{s,q} - 1, n_{s',q'} + 1>$$

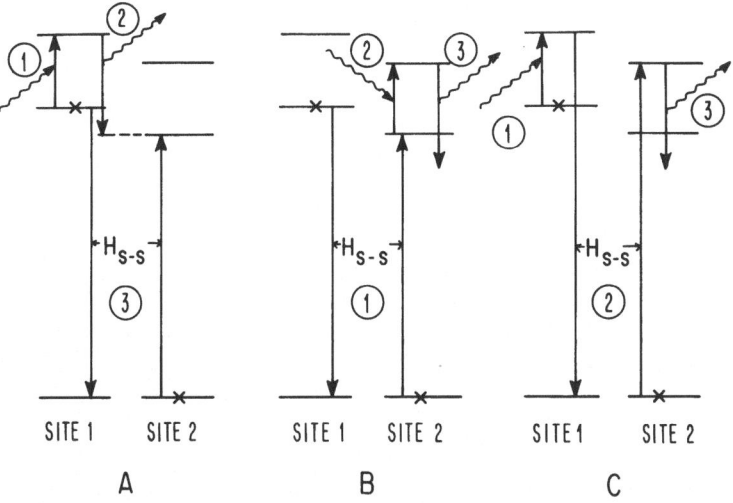

Fig. 2.4. One-site resonant process. The ion–phonon interaction acts twice, the site–site coupling Hamiltonian once. The circled numbers indicate the order of transitions; the arrows the sense of the transitions. There are three more processes of a similar type, corresponding to inversion of the phonon lines (these are important for the one-site nonresonant process)

with respective energies

$$E_{m_1} = E_1 + \Delta - \hbar\omega_{s,q}$$

$$E_{m_2} = E_1 - \hbar\omega_{s,q} + \hbar\omega_{s',q'}.$$

We define the matrix element of H_{ph} between the electronic states $|1^*>$ and $|1^{**}>$ by,

$$\langle 1^{**} | H_{ph}(1) | 1^* \rangle = A\epsilon(1). \tag{2.33}$$

The overall process is third order in H. The third term in (2.8) yields the following expression for the excitation transfer process, A, exhibited in Fig. 2.4:

$$t_{f\leftarrow i}^A = \frac{J}{\Delta E_{12}} \left(\frac{|A|^2}{\Delta - \hbar\omega_{s,q} + i\Gamma_2} \right) \langle n_{s,q} - 1, n_{s',q'} + 1 | \epsilon^2(1) | n_{s,q}, n_{s',q'} \rangle. \tag{2.34}$$

We have utilized the well-known procedure [2.21] of inserting the imaginary quantity $i\Gamma_2$ into the denominator of (2.34) to avoid divergence at resonance (when $\hbar\omega_{s,q} = \Delta$). Here, Γ_2 is the phonon-induced halfwidth of the electronic level $|j^{**}>$ by virtue of resonant phonon emission to $|j^*>$

$$\Gamma_2 = \pi \sum_{s,q} |A \langle n_{s,q} + 1 | \epsilon | n_{s,q} \rangle|^2 \, \delta(\Delta - \hbar\omega_{s,q}). \tag{2.35}$$

There are two further processes (B, C) of the resonant type, exhibited in Fig. 2.4. These give rise to the contributions, respectively,

$$t^B_{f \leftarrow i} = - \frac{J|A|^2}{\Delta E_{12}} \frac{\langle n_{s,q} - 1, n_{s',q'} + 1 \mid \epsilon^2(2) \mid n_{s,q}, n_{s',q'} \rangle}{\Delta - \hbar \omega_{s',q'} + i\Gamma_2}, \tag{2.36}$$

$$t^C_{f \leftarrow i} = \frac{J_2 |A|^2 \langle n_{s,q} - 1 \mid \epsilon(1) \mid n_{s,q} \rangle \langle n_{s',q'} + 1 \mid \epsilon(2) \mid n_{s',q'} \rangle}{(\Delta - \hbar \omega_{s,q} + i\Gamma_2)(\Delta - \hbar \omega_{s',q'} + i\Gamma_2)}, \tag{2.37}$$

where the matrix element of H_{s-s} between the ground level $|j\rangle$ and the third level $|j^{**}\rangle$ has been denoted by J_2 [compare (2.1)]

$$J_2 = \langle 1, 2^{**} \mid H_{s-s} \mid 1^{**}, 2 \rangle. \tag{2.38}$$

The three matrices, (2.34, 36, 37), contain not only resonant contributions, but also nonresonant contributions for the phonon frequencies away from the equalities $\hbar \omega_{s,w}$, $\hbar \omega_{s',q'} = \Delta$. In addition, further nonresonant contributions to the t matrices corresponding to the reversal of sign of the phonon energies in (2.34, 36, 37) (emission instead of absorption) must be added to these terms. These will be considered in Sect. 2.3.3d. For the present, we isolate the contribution to the t matrices near resonance. Combining (2.34, 36, 37), and using (2.4) for the phonon matrix element,

$$t_{f \leftarrow i} = |A|^2 \langle n_{s,q} - 1, n_{s',q'} + 1 \mid \epsilon^2 \mid n_{s,q}, n_{s',q'} \rangle \exp[i(q - q') \cdot r_1]$$

$$\times \left\{ \frac{J}{\Delta E_{12}(\Delta - \hbar \omega_{s,q} + i\Gamma_2)} - \frac{J \exp[i(q - q') \cdot r}{\Delta E_{12}(\Delta - \hbar \omega_{s',q'} + i\Gamma_2)} \right.$$

$$\left. + \frac{J_2 \exp(-iq' \cdot r)}{(\Delta - \hbar \omega_{s,q} + i\Gamma_2)(\Delta - \hbar \omega_{s',q'} + i\Gamma_2)} \right\}. \tag{2.39}$$

Insertion into (2.7) gives the desired rate. Before carrying out the algebra, we note that a few simplifications in (2.39) can sometimes be made. Should Δ be sufficiently large that the resonant phonon wavelength

$$\lambda = 2\pi \hbar v_s / \Delta \tag{2.40}$$

is less than the spatial separation between sites 1 and 2 ($\lambda \ll r$), one can neglect the interference between the first and second terms in the curly brackets of (2.39). If the additional relationship

$$|\Delta E_{12}| \ll \Delta \tag{2.41}$$

is also obeyed, one can neglect the interference between the first two terms and the third term in the curly brackets of (2.39) as well.

These conditions are fulfilled for Cr^{3+} in Al_2O_3 (ruby). The levels $|j^*>$ and $|j^{**}>$ correspond to $|\bar{E}>$ and $|2\bar{A}>$, respectively, the two subvectors of the ^2E level split by the trigonal field where $\Delta = 29$ cm^{-1}. The site–site energy mismatch $\Delta E_{12} \sim 1$ cm^{-1}, and (2.41) is obeyed. We estimate from $\Delta = 29$ cm^{-1} that $\lambda \sim 50$ Å. This means $qr = 1$ at a separation of approximately 9 Å. Hence, for excitation transfer beyond this range, condition (2.40) is also satisfied.

Using these inequalities, we obtain from (2.30),

$$
W^{res}_{2 \leftarrow 1} = \frac{2\pi}{\hbar} \sum_{s,q,s',q'} |A^2 \langle n_{s,q} - 1, n_{s',q'} + 1 | \epsilon^2 | n_{s,q}, n_{s',q'} \rangle |^2
$$

$$
\times \left\{ \frac{J^2}{(\Delta E_{12})^2} \left[\frac{1}{(\Delta - \hbar\omega_{s,q})^2 + \Gamma_2^2} + \frac{1}{(\Delta - \hbar\omega_{s',q'})^2 + \Gamma_2^2} \right] \right.
$$

$$
\left. + \frac{J_2^2}{[(\Delta - \hbar\omega_{s,q})^2 + \Gamma_2^2][(\Delta - \hbar\omega_{s',q'})^2 + \Gamma_2^2]} \right\} . \tag{2.42}
$$

The form of (2.42) makes it easy to separate out the resonant portion of the summation for the first term in the curly brackets using the approximate relationship,

$$
\delta(\Delta - \hbar\omega_{s,q}) \simeq \frac{1}{\pi} \frac{\Gamma_2}{(\Delta - \hbar\omega_{s,q})^2 + \Gamma_2^2}. \tag{2.43}
$$

The second term in (2.42) can be simplified by noting that

$$
\frac{1}{[(\Delta - \hbar\omega_{s,q})^2 + \Gamma_2^2][(\Delta - \hbar\omega_{s',q'})^2 + \Gamma_2^2]}
$$

$$
\simeq \frac{2}{(\Delta E_{12})^2 + [2(\sqrt{2})\Gamma_2]^2} \left[\frac{1}{(\Delta - \hbar\omega_{s,q})^2 + \Gamma_2^2} + \frac{1}{(\Delta - \hbar\omega_{s',q'})^2 + \Gamma_2^2} \right] . \tag{2.44}
$$

Both these approximations require that the Bose factors do not vary appreciably over the range Γ_2, viz., $\Gamma_2 \ll k_B T$.

Inserting (2.43, 44) into (2.42), one finds

$$
W^{res}_{2 \leftarrow 1} = \frac{2\pi^2}{\hbar(\Delta E_{12})^2 \Gamma_2} \sum_{s,q,s',q'} |A^2 \langle n_{s,q} - 1, n_{s',q'} + 1 | \epsilon^2 | n_{s,q}, n_{s',q'} \rangle |^2
$$

$$
\times \left\{ J^2 + \frac{2(\Delta E_{12})^2 J_2^2}{(\Delta E_{12})^2 + [2(\sqrt{2})\Gamma_2]^2} \right\}
$$

$$
\times [\delta(\Delta - \hbar\omega_{s,q}) \delta(\Delta + \Delta E_{12} - \hbar\omega_{s',q'})
$$

$$
+ \delta(\Delta - \hbar\omega_{s',q'}) \delta(\Delta - \Delta E_{12} - \hbar\omega_{s',q'})]. \tag{2.45}
$$

The phonon sums in (2.45) are decoupled. It is convenient to define a quantity $W_\uparrow$ which is the phonon-induced transition rate between $|j^*>$ and $|j^{**}>$

$$W_\uparrow = \frac{2\pi}{\hbar} \sum_{s,q} |A\langle n_{s,q} - 1 | \epsilon | n_{s,q}\rangle|^2 \, \delta(\Delta - \hbar\omega_{s,q}). \tag{2.46}$$

By construction, $W_\uparrow$ is related to Γ_2 by detailed balance

$$W_\uparrow = \frac{2}{\hbar} \Gamma_2 \exp(-\beta\Delta). \tag{2.47}$$

We neglect the variation of the phonon density of states between Δ and $\Delta + \Delta\dot{E}_{12}$. Using (2.35, 46), (2.45) then becomes

$$W_{2\leftarrow 1}^{res} = \left\{ J^2 + \frac{2(\Delta E_{12})^2 J_2^2}{(\Delta E_{12})^2 + [2(\sqrt{2})\Gamma_2]^2} \right\} \left[\frac{n(\Delta + \Delta E_{12}) + 1}{n(\Delta) + 1} + \frac{n(\Delta - \Delta E_{12})}{n(\Delta)} \right] \frac{W_\uparrow}{(\Delta E_{12})^2}. \tag{2.48}$$

This is our final expression for $W_{2\leftarrow 1}^{res}$. Application to systems such as ruby allows for a simplification at $k_B T \ll \Delta \pm \Delta E_{12}$

$$W_{2\leftarrow 1}^{res} \Big|_{k_B T \ll \Delta \pm \Delta E_{12}} = \left\{ J^2 + \frac{2(\Delta E_{12})^2 J_2^2}{(\Delta E_{12})^2 + [2(\sqrt{2})\Gamma_2]^2} \right\} \frac{1 + \exp(\beta\Delta E_{12})}{(\Delta E_{12})^2} W_\uparrow. \tag{2.49}$$

The first term in (2.49) is inversely proportional to the square of the site–site energy mismatch. This is a consequence of all the phonon interactions taking place at a single site, leading to an expression equivalent to a spectral overlap approach in the limit $\beta |\Delta E_{12}| \ll 1$. The second term has a more complicated behavior. Phonon absorption and emission take place at different sites (see Fig. 2.4, process C) by virtue of the exchange coupling between the ground and third level. It is conceivable that situations may arise where J [defined in (2.1)] vanishes, but J_2 is finite. Excitation transfer could then take place via phonon excitation to a level which does possess a finite exchange coupling (here J_2) with a similar level on a neighboring site, and then a subsequent phonon de-excitation at the transferred configuration.

The limit in which (2.49) is valid results in an excitation transfer rate proportional to $\exp(-\beta\Delta)$. This behavior was first reported by *Heber* and *Murman* [2.22] for Cr^{3+} in $LaAlO_3$. *Holstein* et al. [2.14] expected this transfer process to be the dominant mechanism for excitation transfer in dilute ruby. However, as discussed in detail in Chap. 5, and alluded to in Sect. 2.3.2, an additional contribution to the one-phonon-assisted excitation transfer arising from weakly coupled pairs appears to have a larger magnitude at low temperatures ($T < \Delta/k_B$) for ruby.

The relative importance of (2.48) clearly depends on the energy splitting to the third level, as the transfer rate falls off exponentially with temperature for $k_B T \ll \Delta$.

An explicit comparison of magnitude with the nonresonant one-site transfer process will be made at the end of Sect. 2.3.3d for the special case of dilute ruby.

d) One-Site Nonresonant Process

The previous section focused upon the near-divergence in the t matrix (2.39) at resonance ($\hbar\omega_{s,q}$, $\hbar\omega_{s',q'} = \Delta$). The purpose of this section is the calculation of the t matrix away from resonance, and the related phonon-assisted transfer rate. We remarked in the earlier section that three more processes, in addition to those considered in Sect. 2.3.3c and pictured in Fig. 2.4, need to be included for the nonresonant contribution. These involve reversing the time arrow of the phonons, so that phonon emission and absorption are interchanged from the processes in Sect. 2.3.3c. The additional terms in the t matrix are [compare (2.39)]

$$t_{f \leftarrow i} = |A|^2 \langle n_{s,q} - 1, n_{s',q'} + 1 \mid \epsilon^2 \mid n_{s,q}, n_{s',q'} \rangle \exp\left[i(q - q') \cdot r_1\right]$$

$$\times \left\{ \frac{J}{\Delta E_{12}(\Delta + \hbar\omega_{s',q'})} - \frac{J \exp\left[i(q - q') \cdot r\right]}{\Delta E_{12}(\Delta + \hbar\omega_{s,q})} + \frac{J_2 \exp\left(-iq \cdot r\right)}{(\Delta + \hbar\omega_{s,q})(\Delta + \hbar\omega_{s',q'})} \right\}. \tag{2.50}$$

The factors of $i\Gamma_2$ in each of the denominators are not necessary here because only regular behavior for phonon energies near Δ is exhibited in (2.50). We ignore any interference between the resonance (Sect. 2.3.3c) and nonresonant contributions to the transfer rate. These can be important (see, for example, [2.23, 24]) at temperatures comparable to Δ/k_B. However, the algebra is tedious, and the interested reader who wishes to calculate this effect will have no difficulty following procedures already established in the literature [2.23, 24].

Adding (2.39) and (2.50), we obtain

$$t_{f \leftarrow i} = |A|^2 \langle n_{s,q} - 1, n_{s',q'} + 1 \mid \epsilon^2 \mid n_{s,q}, n_{s',q'} \rangle \exp\left[i(q - q') \cdot r_1\right]$$

$$\times \left\{ \frac{J}{\Delta E_{12}} \left(\frac{1}{\Delta - \hbar\omega_{s,q}} + \frac{1}{\Delta + \hbar\omega_{s',q'}} \right) \right.$$

$$- \frac{J \exp\left[i(q - q') \cdot r\right]}{\Delta E_{12}} \left(\frac{1}{\Delta - \hbar\omega_{s',q'}} + \frac{1}{\Delta + \hbar\omega_{s,q}} \right) \tag{2.51}$$

$$\left. + J_2 \exp\left(-iq' \cdot r\right) \left[\frac{1}{(\Delta - \hbar\omega_{s,q})(\Delta - \hbar\omega_{s',q'})} + \frac{\exp\left[i(q + q') \cdot r\right]}{(\Delta + \hbar\omega_{s,q})(\Delta + \hbar\omega_{s',q'})} \right] \right\}.$$

This expression is complicated, but some reasonable limits can bring it into a simpler form. If we assume $J_2 \cong J$, and the limit

$$|\Delta E_{12}| \ll k_B T \ll \Delta \tag{2.52}$$

and

$$qr, qr' \gg 1 \tag{2.53}$$

then we can drop the term proportional to J_2 in (2.51), as well as the phonon energies in the denominators of (2.51). Inserting (2.51) in (2.7), we arrive at

$$W_{2\leftarrow1}^{\text{non-res }III} = \frac{16\pi J^2}{\Delta^2(\Delta E_{12})^2\hbar} \sum_{s,q,s',q'} |A^2\langle n_{s,q} - 1, n_{s',q'} + 1 \mid \epsilon^2 \mid n_{s,q}, n_{s',q'}\rangle|^2$$

$$\times \delta(\hbar\omega_{s',q'} - \hbar\omega_{s,q} - \Delta E_{12}). \tag{2.54}$$

The reader is warned that the temperature range in which (2.52, 53) are true may be rather narrow.

Using (2.6, 15), the Debye approximation, and neglecting ΔE_{12} in the delta function,

$$W_{2\leftarrow1}^{\text{non-res }III} = \frac{J^2|A|^4(k_BT)^7}{\pi^3\hbar^7\Delta^2(\Delta E_{12})^2\rho^2} \left(\sum_s \frac{\alpha_s}{v_s^5}\right)^2 I_6. \tag{2.55}$$

We can express Γ_2 in a similar manner. From (2.35), we write

$$\Gamma_2 = \frac{|A|^2\Delta^3}{4\pi\hbar^3\rho} \left(\sum_s \frac{\alpha_s}{v_s^5}\right) [n(\Delta) + 1]. \tag{2.56}$$

Inserting (2.56) into (2.55) and neglecting induced emission [the term $n(\Delta)$] because of the temperature limits, we finally obtain

$$W_{2\leftarrow1}^{\text{non-res }III} = \frac{16J^2(k_BT)^7\Gamma_2^2 I_6}{\pi\hbar\Delta^6(\Delta E_{12})^2}. \tag{2.57}$$

It is interesting to compare (2.57) with the resonant contribution (2.49). In the limit that $\theta_D/T \gg 1, I_6 \simeq 6!$. One has

$$W_{2\leftarrow1}^{\text{res}} / W_{2\leftarrow1}^{\text{non-res }III} = 3.5 \times 10^{-4} (\Delta/\Gamma_2) (\beta\Delta)^7 \exp(-\beta\Delta)$$

$$\times \left[1 + \frac{2(\Delta E_{12})^2 (J_2/J)^2}{(\Delta E_{12})^2 + (2\sqrt{2})^2 (\Gamma_2)^2}\right]. \tag{2.58}$$

The maximum value of this ratio is obtained for $T = \Delta/7k_B$, and equals

$$W_{2\leftarrow1}^{\text{res}} / W_{2\leftarrow1}^{\text{non-res }III} = 0.26 (\Delta/\Gamma_2) \left[1 + \frac{2(\Delta E_{12})^2 (J_2/J)^2}{(\Delta E_{12})^2 + (2\sqrt{2})^2 (\Gamma_2)^2}\right]. \tag{2.59}$$

For most cases, $\Delta/\Gamma_2 \gg 1$, so that in the region of $\Delta/7k_B$ the resonant process dominates. The nonresonant process is expected to dominate well below and above this temperature regime.

The transfer rates we have calculated in Sect. 2.3.3 exhaust the examples of two-phonon-assisted processes. We know of no situation where it is required to utilize higher-order processes to effect excitation transfer for small energy mismatches. However, when ΔE_{12} becomes comparable to or greater than the maximum phonon energy, multiple phonon interactions may be required to conserve energy. Such a situation would arise when excitation transfer occurs between different crystal field levels on the two sites. We have not worked out the details for this case, but the procedure is clear: the t matrix must be carried to higher order and the appropriate terms computed much as we have done here. Very large energy mismatch leads to the so-called exponential form for the excitation transfer rate, first developed by *Miyakawa* and *Dexter* [2.12] and exhibited experimentally by *Yamada* et al. [2.25]. Briefly, the form for the excitation transfer is

$$W_{2 \leftarrow 1}^{\text{multi-phonon}} \propto \exp(-\beta \Delta E_{12})$$

where β is related to a single-ion multiphonon relaxation rate exponent [2.12] α by $\beta = \alpha \cdot \gamma$, and γ is of order $(k_B \theta_D)^{-1}$.

The calculations presented above are only for two-site transfer rates. The measured time development of the emission in the case of fluorescent line narrowing depends on summation of the transfer rates across the spectral line. Very different forms of the emission profile will develop depending on whether the transfer rate is or is not dependent on the energy mismatch ΔE_{12}. These have been extensively discussed by *Holstein* at al. [2.26], and are developed in Chap. 3.

All of the processes considered above utilize a site—site coupling Hamiltonian H_{s-s} which arises from multipolar or exchange couplings. As the active ionic concentration is reduced, one might ask what residual excitation transfer mechanisms would remain after the short-range interactions become negligible. One such mechanism was originally suggested by Selzer (private communication). Photons carry the excitation from site to site, with phonons making up the site—site energy mismatch. This mechanism has been observed by *Selzer* and *Yen* [2.27] in dilute ruby. Its magnitude was calculated microscopically by *Holstein* et al. [2.28], and the details are summarized below. The principal example will be ruby, but the calculations are in fact quite general.

2.4 Phonon-Assisted Radiative Transfer

We treat excitation transfer by photons, assisted by phonons to conserve overall system energy. The problem can be formulated rather precisely. Introduce the electron—photon coupling Hamiltonian H_k for a photon of wave vector k (and frequency Ω_k). The coupling is linear in the photon amplitude, and is spatially referenced to the site for photon emission or absorption. Let site 1 be initially excited, site 2 initially in its ground electronic level. Then, site 1 can emit a photon of wave vector k, simultaneously changing its electronic state from excited to ground; the same photon can be absorbed at site 2, with a simultaneous change in electronic state from ground to excited. Phonons can be involved in the transfer process exactly as in Sect. 2.2.2, and make up the energy mismatch between sites 1 and 2. It should be stressed that it is a real photon which is exchanged. Virtual photons of course contribute to the Coulomb coupling between the sites, but the photon Green's function acts to force the photon involved in emission and absorption to be the same, and the t matrix forces them to have an energy equal to one of the individual site energies.

We shall show that the efficacy of this process depends on the degree of radiative trapping in the sample. Under conditions of extreme trapping, the excitation transfer process is 100% efficient. Every photon which is emitted is subsequently reabsorbed and reemitted until it couples with a phonon at a particular site, and excitation transfer takes place. The volume through which the photon propagates is roughly a hohlraum of radius equal to the photon trapping length. The problem can be complicated by energy mismatches of the order of the width of the optical absorption line because the trapping length is a strong function of photon frequency. Strong trapping may hold at the center of the line, but photons corresponding to the excitation energies of sites in the wings of the line may not be strongly trapped. This can weaken the efficacy of phonon-assisted photon transfer. In practice, one finds a shape dependent transfer rate for moderate trapping, and a dependence on the spatial position of the exciting laser beam relative to the sample surfaces in an FLN experiment [2.27].

We treat one-phonon-assisted photon transfer in the next section, and two-phonon-assisted photon transfer in Sect. 2.4.2.

2.4.1 One-Phonon-Assisted Process

The one-phonon-assisted photon transfer process is pictured in Fig. 2.5. Two different orderings are possible. The first, A, consists in order of a phonon emitted at site 1, followed by a photon emitted at the same site. A photon is then absorbed at the second site. The photon Green's function forces the two photons to be the same. The t matrix forces the photon energy to be equal to the second site transition energy. The energy conserving delta function in the golden rule requires that the phonon energy be equal to the energy mismatch between the two sites. The second, B, consists, in order, of a photon emitted at site 1, a photon absorbed at site 2, and a phonon emitted at site 2. The photon Green's function forces the two photons to be the same, while the t matrix forces the photon energy to be equal to the first site transition energy. Again, the energy conserving delta function in the golden rule forces the phonon energy to be equal to the energy mismatch between the two sites.

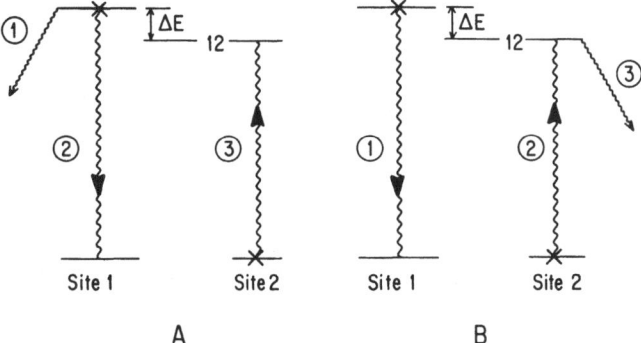

Fig. 2.5. One-phonon-assisted radiative transfer. The larger wiggly lines represent photons, the smaller lines phonons of wave vector q ($\hbar v_s |q| = \Delta E_{12}$, the energy mismatch between sites 1 and 2). The circled numbers represent the position of the step in the perturbation chain. The crosses denote initial electronic occupancies, the arrows the sense of the transition. (A) A phonon of wave vector q is emitted at site 1, a photon of wave vector $|k| = E_2/\hbar c$ is emitted at site 1, and the same photon is absorbed at site 2. (B) A photon of wave vector $|k| = E_1/\hbar c$ is emitted at site 1, the same photon is absorbed at site 2, and a phonon of wave vector q is emitted at site 2

The explicit expression for the t matrix for process A from the third order term in the expansion (2.8) is,

$$t_{f \leftarrow i}^A = \sum_k \frac{\langle 1, 2^*, n_{s,q} \pm 1 | H_k | 1, 2, n_{s,q} \pm 1 \rangle \langle 1, 2, n_{s,q} \pm 1 | H_k | 1^*, 2, n_{s,q} \pm 1 \rangle}{(\mp \hbar \omega_{s,q})(E_2 - \hbar\Omega_k - i\Gamma)}$$
$$\times \langle 1^*, 2, n_{s,q} \pm 1 | H_{\text{ph}}(1) | 1^*, 2, n_{s,q} \rangle. \tag{2.60}$$

Here, the photon energy is denoted by $\hbar\Omega_k$, and Γ is the linewidth of the excited level (we have ignored the width of the ground level for conciseness). The upper and lower signs correspond to phonon emission (for $\Delta E_{12} > 0$) and phonon absorption (for $\Delta E_{12} < 0$), respectively. We define

$$\langle 1, 2^* | H_k | 1, 2 \rangle \langle 1, 2 | H_k | 1^*, 2 \rangle = F(k) \exp(-ik \cdot r), \tag{2.61}$$

where $F(k)$ is the product of the electron–phonon coupling constants (each proportional to $k^{-1/2}$). Introducing the consequence of energy conservation ($\mp \hbar \omega_{s,q} = -\Delta E_{12}$), the t matrix (2.60) reduces to

$$t_{f \leftarrow i}^A = \frac{f_1 \langle n_{s,q} \pm 1 | \epsilon | n_{s,q} \rangle \exp(\mp iq \cdot r_1)}{-\Delta E_{12}} \sum_k \frac{F(k) \exp(-ik \cdot r)}{E_2 - \hbar\Omega_k - i\Gamma}. \tag{2.62}$$

A similar analysis for process B yields

$$t_{f \leftarrow i}^{B} = \frac{f_1 \langle n_{s,q} \pm 1 \mid \epsilon \mid n_{s,q} \rangle \exp{(\mp iq \cdot r_2)}}{\Delta E_{12}} \sum_{k} \frac{F(k) \exp{(-ik \cdot r)}}{E_1 - \hbar \Omega_k - i\Gamma}. \qquad (2.63)$$

These processes are appropriate to phonon emission and absorption in the excited level. Adding diagrams for phonon emission and absorption in the ground level simply means replacing f_1 by the difference $f_1 - g_1$, where g_1 is the ground-state ion–phonon coupling constant. The sums appearing in (2.62, 63) need evaluation. Writing

$$k_1 = E_1/\hbar c, \quad \kappa_1 = k_1 r;$$

$$k_2 = E_2/\hbar c, \quad \kappa_2 = k_2 r; \quad F(k) = F_0/k$$

with F_0 independent of k, the sums in question can be represented by an effective site–site exchange coupling

$$J_1^{eff} = (VF_0/2\pi\hbar cr) I(\kappa_1) \qquad (2.64)$$

with

$$I(\kappa_1) = \frac{1}{\pi} \int_0^\infty d\kappa \, \frac{\sin \kappa}{\kappa_1 - \kappa - i\Gamma r/\hbar c}, \qquad (2.65)$$

where V is the active volume of the crystal. In general, at low temperatures, Γ is sufficiently small that the last term in the denominator of (2.65) is much smaller than one. That is, for ruby at conventional temperatures, $\Gamma/\hbar c \ll 1$ cm^{-1}. For $\Gamma/\hbar c$ equal to 1 cm^{-1}, $\Gamma r/\hbar c = 7r$, where r is measured in cm. It will turn out that energy transfer will be important between sites separated by no more than a mean free path for the photon. For sensible concentrations of Cr in ruby, this will imply r's of the order of hundreds of microns, so that indeed $\Gamma r/\hbar c \ll 1$. At low temperatures, Γ is exponentially small [2.29] and substantially larger values of r result in the same inequality. Under this condition $I(\kappa_1)$ becomes [2.30]

$$I(\kappa_1) = (1/\pi) \sin \kappa_1 \, ci(\kappa_1) - \cos \kappa_1 [(1/\pi) \, si(\kappa_1) + 1] + i \sin \kappa_1, \qquad (2.66)$$

where

$$si(\kappa_1) = -\int_{\kappa_1}^\infty \frac{\sin t}{t} dt; \quad ci(\kappa_1) = -\int_{\kappa_1}^\infty \frac{\cos t}{t} dt. \qquad (2.67)$$

The asymptotic properties of these functions are [2.30]

$$\kappa_1 \ll 1, \quad \begin{aligned} &\mathrm{si}(\kappa_1) \simeq -\frac{1}{2}\pi + \kappa_1, \\ &\mathrm{ci}(\kappa_1) \simeq -\ln(1/\gamma\kappa_1) \quad (\gamma \text{ is Euler's constant});\end{aligned}$$

$$\kappa_1 \gg 1, \quad \begin{aligned} &\mathrm{si}(\kappa_1) \simeq -\cos\kappa_1/\kappa_1, \\ &\mathrm{ci}(\kappa_1) \simeq \sin\kappa_1/\kappa_1.\end{aligned}$$

These properties result in

$$\kappa_1 \ll 1, \quad I(\kappa_1) \simeq -1/2; \tag{2.68a}$$

$$\kappa_1 \gg 1, \quad I(\kappa_1) \simeq -\exp(-i\kappa_1). \tag{2.68b}$$

For ruby, where $E_1 = 14{,}400$ cm^{-1}, the inequality (2.68a) is equivalent to $r \ll$ 1,000 Å, while (2.68b) corresponds to $r \gg 1{,}000$ Å. When (2.62, 63) are summed, and inserted into (2.7), the usual cross or interference term arises. This term is large and effectively cancels the direct contributions for single-phonon emission or absorption in the case of nonradiative coupling. For phonon-assisted photon transfer, the interference term vanishes and no cancellation occurs. This behaviour is a consequence of the fact that the photon propagates for a trapping length (r) before it is reabsorbed. This length is typically tens or hundreds of microns. Energy conservation forces the phonon wave vector $q = \Delta E_{12}/\hbar v_s$. For an energy mismatch of 1 cm^{-1}, $q = 10^{-3}$ Å^{-1}. Thus, qr is very great, and the interference term averages to zero. The large magnitude of r for photon transfer is the reason why this process dominates at small active ion concentrations. Short-range nonradiative coupling Hamiltonians are no longer effective at low concentrations.

The remarkable efficacy of spatial photon transport in the trapped regime must be understood as a major component of any analysis of localization or delocalization of excitations in dilute optical systems. Experiments of the type of *Koo* et al. [2.4] are particularly prone to this difficulty. Radiative trapping must be absent, or very carefully considered, when searching for a mobility edge for excitation transfer within the inhomogeneously broadened line. *Koo* et al. [2.4] went to great lengths (e.g., grinding ruby into very fine particles and dispersing them one particle layer thick on a flat substrate) to avoid this problem.

The absence of an interference term means that (2.62, 63) enter the golden rule (2.7) as absolute squares. Noting that different photons play a role in (2.62, 63), this requires us to separate our consideration for each photon. The two photons may have different trapping lengths, and an accurate treatment must treat them individually.

It is shown in [Ref. 2.28, Appendix] that

$$|J_1^{eff}(E_1, r)| = \tau_R^{-1} \hbar\, I(\kappa_1)|/2\kappa_1$$

$$\sim \tau_R^{-1} \hbar/(r/\lambda_1), \tag{2.69}$$

where τ_R is the radiative lifetime (assumed the same for each of the two sites), and λ_1 the photon wavelength corresponding to a photon energy E_1. The photon coupling strength will sometimes vary across the optical line, especially in glasses [2.31]. In such cases, τ_R in (2.69) is replaced by

$$\tau_R = [\tau_R(1)\,\tau_R(2)]^{1/2}. \tag{2.70}$$

Inserting (2.69) into (2.64), and thence (2.62, 63) into (2.7), we obtain the excitation transfer rate for each of the photons.

Process A:

$$W_{2\leftarrow 1}(E_2) = \frac{2\pi}{\hbar} \frac{(\tau_R^{-1}\hbar)^2\,|I(\kappa_2)|^2}{2(\Delta E_{12})^2\,(2\kappa_2)^2} (f_1 - g_1)^2$$

$$\times \sum_{s,q} |\langle n_{s,q} \pm 1\,|\epsilon|\, n_{s,q}\rangle|^2\, \delta(\hbar\omega_{s,q} \mp \Delta E_{12}). \tag{2.71a}$$

Process B:

$$W_{2\leftarrow 1}(E_1) = \frac{2\pi}{\hbar} \frac{(\tau_R^{-1}\hbar)^2\,|I(\kappa_1)|^2}{2(\Delta E_{12})^2\,(2\kappa_1)^2} (f_2 - g_2)^2$$

$$\times \sum_{s,q} |\langle n_{s,q} \pm 1\,|\epsilon|\, n_{s,q}\rangle|^2\, \delta(\hbar\omega_{s,q} \mp \Delta E_{12}). \tag{2.71b}$$

It is possible to evaluate the phonon sums in (2.71) explicitly. It is rather more transparent if we relate them to properties of a specific physical system. Consider (again) the case of ruby. From (2.46),

$$W_\uparrow = \frac{2\pi}{\hbar} \sum_{s,q} A^2\,|\langle n_{s,q} - 1\,|\epsilon|\, n_{s,q}\rangle|^2\, \delta(\Delta - \hbar\omega_{s,q}). \tag{2.46}$$

The sums in (2.46, 71) span very different energy regions of the phonon density of states ($\Delta E_{12} \sim 1\ \text{cm}^{-1}$, $\Delta \sim 29\ \text{cm}^{-1}$). For the present purposes (a good approximation for ruby) we assume that the Debye approximation holds for both sums. Then, we can express (2.71) in terms of (2.46)

$$W_{2\leftarrow 1}(E_1)/W_\uparrow = \frac{(\tau_R^{-1}\hbar)^2}{2(\Delta E_{12})^2(2\kappa_1)^2}\frac{|I(\kappa_1)|^2}{}\left(\frac{f_2-g_2}{A}\right)^2\frac{|\Delta E_{12}|^3}{\Delta^3 n(\Delta)}\left\{\frac{n(|\Delta E_{12}|)+1}{n(|\Delta E_{12}|)}\right\}.$$

$$(2.72)$$

An analogous expression holds for $W_{2\leftarrow 1}(E_2)$ if one substitutes κ_2 for κ_1 in (2.72). This expression is of the same form as (2.16) for one-phonon-assisted nonradiative transfer if one substitutes (2.69) into (2.72). For the example of ruby, the rate $W_\uparrow$ is known from experiment [2.32, 33],

$$W_\uparrow = 0.9 \times 10^9 \, n(\Delta)\, \mathrm{s}^{-1}. \tag{2.73}$$

The ratio of phonon–ion coupling parameters, $(f_2-g_2)/A$, can be extracted from static strain experiments. Comparing slopes from the figure in [2.34]

$$(f_2-g_2)/A \simeq 4, \tag{2.74}$$

where f_2-g_2 is actually site independent. Every parameter in (2.72) is now known ($\tau_R = 3.6$ ms for ruby), so that $W_{2\leftarrow 1}(E_1)$ can be evaluated. However, another step must be taken to obtain a rate comparable to what is actually measured. The expression $W_{2\leftarrow 1}(E_1)$ gives the energy transfer rate between two sites for fixed initial and final energies E_1 and E_2, respectively. The photon energy is fixed at E_1. We must sum this rate over all sites for which this excitation transfer channel applies. This requires that we sum $W_{2\leftarrow 1}(E_1)$ over all the sites within a volume spanned by the mean free path for a photon with energy E_1, designated by $l(E_1)$. Defining the sum as $W_{E_1}(E_2-E_1)$, we write,

$$W_{E_1}(E_2-E_1) = \Sigma_2 \, W_{2\leftarrow 1}(E_1) = \int_0^{l(E_1)} W_{2\leftarrow 1}(E_1)4\pi r^2 n \, dr, \tag{2.75}$$

where n is the active impurity (Cr^{3+} in the case of ruby) concentration. The upper limit implicitly assumes that the sample dimensions exceed the mean free path for photons of energy E_1. If the reverse be true, the integration must be truncated by the sample boundaries. The rate of excitation transfer then becomes sample size and shape dependent.

Equation (2.75) expresses the excitation transfer rate for a photon of energy E_1. The overall transfer rate from sites with energy E_1 to sites with energy E_2 is simply

$$W(E_2-E_1) = \sum_{i=1,2} W_{E_i}(E_2-E_1). \tag{2.76}$$

We obtain an explicit expression for the photon mean free path from standard sources [2.35]

$$l^{-1}(E_1) = \frac{1}{2}\,\pi n \bar{\lambda}^2 g(E_1)/\tau_R, \tag{2.77}$$

where $g(E_1)$ is the inhomogeneously broadened line shape amplitude at energy E_1. For ruby, $g(E_1) \sim (0.1 \text{ cm}^{-1})^{-1}$ at the line center, leading to $l(E_1) = 3.8 \times 10^{17}/n$. For a concentration of 0.1 at.% $(n = 10^{19})$, this leads to $l(E_1) = 0.038$ cm. This large value for $l(E_1)$ justifies our previous remark that the photon mean free path, and hence the transfer range, is much longer than the wavelength of the phonon involved in making up the energy mismatch between sites. This condition enabled us to drop the interference term in our original expression for phonon-assisted radiative transfer [i.e., to ignore the cross term between t-matrix amplitudes (2.62, 63)]. The large magnitude of the upper limit of (2.75) also leads to the condition that $\kappa_1 \gg 1$. We shall therefore use (2.68b) in the explicit evaluation of (2.75) through (2.72). Carrying out the integration in (2.75), we find

$$W_{E_1}(E_2 - E_1) = \pi \times 10^9 \, (\tau_R^{-1}/\hbar)^2 [(f_2 - g_2)/A]^2 \, (k_B T/2\Delta^3) \, [nl(E_1)/k_1^2] \, \text{s}^{-1}$$

(2.78)

The remarkable feature of (2.78) is that the terms in the last square bracket equal $(2/\pi)\tau_R/g(E_1)$, independent of active ion concentration. We have therefore derived the rather remarkable result that the phonon-assisted photon transfer rate is independent of the concentration of the active species. Every photon has been assumed to have been reabsorbed within a length $l(E_1)$. As long as this length is small compared to the sample size, excitation transfer is 100% efficient. The concentration of active species cannot matter.

Finally, using known values for the parameters appearing in (2.78) quoted earlier, we obtain

$$W_{E_1}(E_2 - E_1) = 10^{-4} \, T \, \text{s}^{-1} \, (T \text{ in K})$$

(2.79)

for 0.1 at.% ruby. This rate is much too slow to be observed in ruby, where the radiative life time is 3.6 ms. The slowness of the transfer rate lies not with the interference terms, as for one-phonon-assisted nonradiative processes, but rather with the small phonon energy density of states for small energy mismatch. One needs to go to second order in H_{ph} in order to utilize phonons of energy $k_B T$ to effect excitation transfer (see Sect. 2.4.2).

Before going on to the case of two-phonon-assisted photon transfer, it is interesting to return to the calculation of the transfer rate (2.75). Clearly this is a rough approximation, and serves only to give an estimate for the total one-phonon-assisted energy-transfer rate [as exhibited in (2.79), for example]. The more correct procedure to follow in an excitation transfer calculation is to consider the energy transport equation utilizing both processes A and B. The difficulty lies with the upper limit for the integral appearing in the expression for the transfer rate [e.g., see (2.75)]. The transport equation will involve frequencies over a large spectral range, causing $l(E)$ to vary strongly. In general, the problem is a complicated one, but a strong simplification can be made in the case of a Gaussian form for $g(E)$ [2.36]. In such a case, it is possible to define a cutoff energy (energy of the line center, E_c)

E_0, such that photons with energy less than $E_c - E_0$, or energy greater than $E_c + E_0$, are not trapped, but are able to traverse the sample freely. The condition which fixes E_0 is simply $l(E_c + E_0) = l(E_c - E_0) = d$, the dimension of the sample. Even this condition is only approximate, as one should correctly take into account the spatial position of the emitting and absorbing sites realtive to the sample boundaries. We are assuming that the spatial excitation distribution is essentially uniform. This will give us a qualitative feel for the emission profile, but will not be numerically correct [2.37].

Let the emission probability function be denoted by $P(E, t)$, where t is the time. Define the quantity $W(E' - E) = W_{E'}(E' - E) + W_E(E' - E)$. Then

$$\frac{dP(E, t)}{dt} = - \int_{-E_0}^{E_0} W_{E'}(E' - E)\, g(E')\, P(E)\, dE'$$

$$- \int_{-\infty}^{\infty} W_E(E' - E)\, g(E')\, P(E)\, \theta(E_0 - |E|)\, dE' - \tau^{-1}(E)\, P(E)\, \mathscr{E}(E)$$

$$+ \int_{-\infty}^{\infty} W_E(E - E')\, g(E')\, P(E')\, \theta(E_0 - |E|)\, dE'$$

$$+ \int_{-E_0}^{E_0} W_{E'}(E - E')\, g(E')\, P(E')\, dE'. \tag{2.80}$$

The quantity $\mathscr{E}(E)$ is the "escape factor," approximately equal to unity for $|E| > E_0$, and of the order of the ratio of phonon sideband emission intensity to the zero-phonon emission intensity for $|E| < E_0$. The function $\theta(x)$ is zero if $x < 0$, and 1 for $x > 0$. The expression (2.80) for the time development of the emission probability makes a sharp distinction between photons with absolute energies less than E_0 (trapped), and those with energies greater than E_0 (escaping). This distinction is incorrect for a Lorentzian line shape, and the problem becomes much more complex.

Approximately, the trapped radiation is quenched by a factor $e^{-x/l(E)}$, where x is the distance from the sample surface. As a consequence, to an observer outside of the sample, the emitted radiation profile has the form

$$l(E) = \tau_R^{-1} [P(E, t)\, l(E)/d][1 - e^{-d/l(E)}]. \tag{2.81}$$

The factors multiplying $\tau_R^{-1} P(E)$ arise from the integration of $e^{-x/l(E)}$ over x. Because $l(E) \propto g(E)^{-1}$, the product $P(E, t)\, l(E)$ is actually independent of $g(E)$. In the limit of large thickness the emitted radiation will have a rectangular shape.

Unfortunately, (2.80) has not yet been solved analytically. The only approach appears to be numerical, iterating (2.80) in a manner similar to the procedure outlined in [2.26, 38].

Because the calculated one-phonon-assisted radiative transfer rate (2.79) is too small to account for experiments on dilute ruby, we generalize our treatment to include two-phonon processes. Ultimately, we shall require solution of an integral equation of the form (2.80), but again we shall not be able to provide an analytical solution. Nevertheless, we can calculate the individual excitation transfer rate between sites with specific energy difference $E_1 - E_2$. This will serve as a very useful first approximation for the solution of the integral equation representing the time development of the full emission profile.

2.4.2 Two-Phonon-Assisted Processes

There are four classes of two-phonon-assisted radiative transfer, in one-to-one correspondence to those treated for nonradiative coupling in Sect. 2.3.3. The energy level structure of dilute ruby appears to favor the one-site resonant process (analogous to Sect. 2.3.3c), and for clarity we shall treat this process in detail here. The remaining three processes can be calculated in a very similar manner, through we shall see that one must be careful for the special case of the two-site nonresonant process. We return to this point at the end of this section.

Even for the one-site resonant process, there are three ways for radiative excitation transfer to occur. These are pictured in Fig. 2.6. As for the single-phonon-assisted radiative transfer process calculated in Sect. 2.4.1, the photon energy is fixed by the site which is either not phonon coupled (site 2 in A; site 1 in B) or by the site at which the phonon resonance takes place (site 1 in C).

We first calculate the excitation transfer rate for the process A pictured in Fig. 2.6. The order of the steps is indicated in the figure. The calculation requires fourth-order perturbation theory, with the three intermediate states

$$|m_1> = |1^{**}, 2, n_{s,q} - 1, n_{s',q'}> \equiv |1^{**}, n_{s,q} - 1>$$
$$|m_2> = |1^*, 2, n_{s,q} - 1, n_{s',q'} + 1> \equiv |1^*, n_{s',q'} + 1>$$
$$|m_3> = |1, 2, n_{s,q} - 1, n_{s',q'} + 1, n_k + 1> \equiv |1>$$

where n_k denotes the photon occupation number, and we have introduced a shortened notation for convenience. The intermediate state energies are

$$E_{m_1} = E_1 + \Delta - \hbar\omega_{s,q} + i\Gamma_2$$
$$E_{m_2} = E_1 - \hbar\omega_{s,q} + \hbar\omega_{s',q'} + i\Gamma$$
$$E_{m_3} = -\hbar\omega_{s,q} + \hbar\omega_{s',q'} + h\Omega_k.$$

The t matrix is calculated from the fourth-order term in (2.8). Introducing the above notation, we find for process A of Fig. 2.6

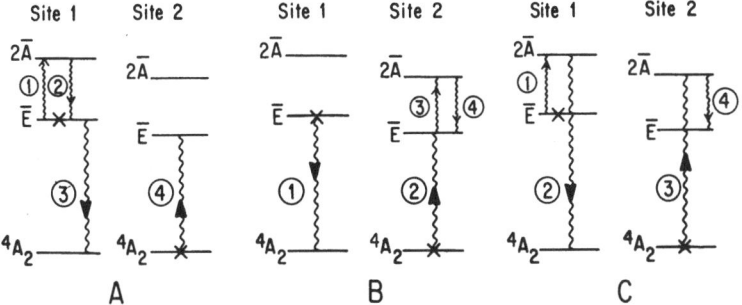

Fig. 2.6. Two-phonon-assisted radiative transfer – the resonant process, with application to ruby. The larger wiggly lines represent photons, the smaller lines phonons. The absorbed phonon has energy Δ, the R_1, R_2 line splitting (the energy difference between the $\bar{E}$ and $2\bar{A}$ excited levels) and the difference in phonon energies equals the energy mismatch between sites 1 and 2. The circled numbers represent the position of the step in the perturbation chain. The crosses denote initial electronic occupancies, the arrows the sense of the transition. (A) A resonant phonon of energy Δ is absorbed at site 1, another phonon is emitted at site 1, a photon of wave vector $|k| = E_2/\hbar c$ is emitted at site 1, and the same photon is absorbed at site 2. (B) A photon of wave vector $|k| = E_1/\hbar c$ is emitted at site 1, the same photon is absorbed at site 2, a resonant phonon of energy Δ is absorbed at site 2, and another phonon is emitted at site 2. (C) A resonant phonon of energy Δ absorbed at site 1, a photon of wave vector $|k| = (E_1 + \Delta)/\hbar c$ is emitted at site 1, the same photon is absorbed at site 2, and another phonon is emitted at site 2

$$t^A_{f \leftarrow i} = \sum_k \frac{\langle 2^* | H_k | 2 \rangle \langle 1 | H_k | 1^* \rangle \langle 1^*, n_{s'q'} + 1 | H_{\text{ph}}(1) | 1^{**}, n_{s',q'} \rangle}{[-\Delta + \hbar\omega_{s,q} - i\Gamma_2][-\Delta E_{12} - i\Gamma][E_1 - \Delta E_{12} - \hbar\Omega_k - i\Gamma]}$$
$$\times \langle 1^{**}, n_{s,q} - 1 | H_{\text{ph}}(1) | 1^*, n_{s,q} \rangle. \tag{2.82}$$

We ignore the width of the ground level in arriving at this result.

Using (2.64) together with (2.65), we may simplify (2.82) (ignoring $i\Gamma$ relative to ΔE_{12}, valid at low temperatures)

$$t^A_{f \leftarrow i} = \frac{J^{\text{eff}}_1(E_2, r)}{-\Delta E_{12}} \frac{\langle 1^*, n_{s'q'} + 1 | H_{\text{ph}}(1) | 1^{**}, n_{s',q'} \rangle \langle 1^{**}, n_{s,q} - 1 | H_{\text{ph}}(1) | 1^*, n_{s,q} \rangle}{-\Delta + \hbar\omega_{s,q} - i\Gamma_2}. \tag{2.83}$$

Similarly, for process B of Fig. 2.6, we find

$$t^B_{f \leftarrow i} = \frac{J^{\text{eff}}_1(E_1, r)}{-\Delta E_{12}}$$
$$\times \frac{\langle 2^*, n_{s',q'} + 1 | H_{\text{ph}}(2) | 2^{**}, n_{s',q'} \rangle \langle 2^{**}, n_{s,q} - 1 | H_{\text{ph}}(2) | 2^*, n_{s,q} \rangle}{\Delta E_{12} - \Delta + \hbar\omega_{s,q} - i\Gamma_2}. \tag{2.84}$$

For energy mismatch $\Delta E_{12} \ll E_1, E_2$ (the usual case), $J^{\text{eff}}_1(E_2, r) \simeq J^{\text{eff}}_1(E_1, r)$. Taking $\Delta E_{12} \ll k_B T$, we may combine (2.83, 84) to yield

$$t^A_{f \leftarrow i} + t^B_{f \leftarrow i} = - \frac{J^{\mathrm{eff}}_1(E_1, r)}{\Delta E_{12}} \frac{|A|^2 \langle n_{s',q'} + 1 | \epsilon | n_{s',q'} \rangle \langle n_{s,q} - 1 | \epsilon | n_{s,q} \rangle}{-\Delta + \hbar \omega_{s,q} - i \Gamma_2}$$

$$\times \exp\left[- i(q - q') \cdot r_1\right] \{1 - \exp\left[i(q - q') \cdot r\right]\}, \tag{2.85}$$

where we have introduced the off-diagonal ion—phonon coupling constant A [see (2.46)].

The t matrix for the resonant component of the process C pictured in Fig. 2.6 is given by

$$t^C_{f \leftarrow i} = J^{\mathrm{eff}}_2(E_1 + \Delta, r)$$

$$\times \frac{\langle n_{s',q'} + 1 | \epsilon | n_{s',q'} \rangle \langle n_{s,q} - 1 | \epsilon | n_{s,q} \rangle \exp\left[-i(q \cdot r_1 - q' \cdot r_2)\right]}{(-\Delta + \hbar \omega_{s,q} - i \Gamma_2)(\Delta E_{12} - \Delta + \hbar \omega_{s,q} - i \Gamma_2)}.$$
$$\tag{2.86}$$

The effective exchange coupling $J^{\mathrm{eff}}_2(E_1 + \Delta, r)$ derives from the photon matrix elements between the state $|1^{**}\rangle$ and the ground level $|1\rangle$. Following the same reasoning that led to (2.64), we find

$$J^{\mathrm{eff}}_2(E_1 + \Delta, r) = \tau^{-1}_{R^{**}} \hbar | I(\kappa_3) | / 2 \kappa_3. \tag{2.87}$$

Here, $\tau^{-1}_{R^{**}}$ is the radiative rate directly from $|1^{**}\rangle$ to the ground level. Generally, $E_1 \gg \Delta$, so that $\kappa_3 = \kappa_1 [1 + (\Delta/E_1)] \simeq \kappa_1$.

The denominators in (2.85, 86) insure that the phonon energy $\hbar \omega_{s,q}$ will lie close to Δ. For ruby, this leads to phonon wave vectors q and q' of the order of 10^7 cm^{-1}. When multiplied by the characteristic interaction distance (the photon mean free path) for 0.1 at.% ruby, 0.038 cm, the inequality $qr \simeq q' r \gg 1$ is well obeyed. The resulting rapid oscillations in the cross term in (2.85), and between (2.85, 86), allow us to drop them in the golden rule, (2.7). Carrying out the phonon sums, we obtain

$$W^{\mathrm{res}}_{2 \leftarrow 1} = 2 \left[|J^{\mathrm{eff}}_1(E_1, r)|^2 + \frac{(\Delta E_{12})^2 |J^{\mathrm{eff}}_2(E_1 + \Delta, r)|^2}{(\Delta E_{12})^2 + \Gamma^2_2} \right] \frac{W_\uparrow}{(\Delta E_{12})^2}$$

$$= \frac{\lambda^2}{8 \pi^2 r^2} \left\{ \left[(\tau^{-1}_R \hbar)^2 + \frac{(\Delta E_{12})^2 (\tau^{-1}_{R^{**}} \hbar)^2}{(\Delta E_{12})^2 + \Gamma^2_2} \right] \frac{W_\uparrow}{(\Delta E_{12})^2} \right\} \tag{2.88}$$

where λ is the photon wavelength.

We have not made separate the individual contributions to (2.88) arising from photons of energies $E_1, E_2, E_1 + \Delta$, and $E_2 + \Delta$, as was done in Sect. 2.4.1. The full transport equation [see (2.80)] requires such a separation. Nevertheless, we can estimate a magnitude for the two-phonon-assisted radiative transfer rate by making the rather strong assumption that the photon mean free path $l(E)$ is equal at all four photon energies. We can then carry out the integration over a volume whose radius is $l(E)$, exactly as for single-phonon-assisted radiative transfer [see the steps leading to (2.78)]. We find,

$$W_{res}(E_2 - E_1) = \frac{4}{\pi}\left[(\tau_R^{-1}\hbar\Delta\nu^*) + \frac{(\Delta E_{12})^2(\tau_{R**}^{-1}\hbar)(\hbar\Delta\nu^{**})}{(\Delta E_{12})^2 + \Gamma_2^2}\right]\frac{W_\uparrow}{(\Delta E_{12})^2}. \quad (2.89)$$

We have replaced $1/g(E)$ by $\Delta\nu^*$ and $\Delta\nu^{**}$ for the transitions between $|1^*>$ and $|1^{**}>$ and the ground state, respectively. Again, the radiative excitation transfer rate is explicitly independent of the active ion concentration.

It is interesting to evaluate (2.89) for dilute ruby. We take as before $\Delta E_{12} \sim \hbar\Delta\nu^* \sim 0.1$ cm^{-1} and $\tau_R^{-1}\hbar = 1.5 \times 10^{-9}$ cm^{-1} ($\tau_R = 3.6$ ms). Setting the second term in (2.89) equal to the first for simplicity, we obtain

$$W_{res}(E_2 - E_1) = 1.1 \times 10^2 \exp(-42/T)\,\mathrm{s}^{-1}. \quad (2.90)$$

This value is sufficiently large to be observable. The work of *Selzer* and *Yen* [2.27] exhibits an excitation transfer rate with the same temperature dependence as (2.90), with a coefficient within a factor of two of the calculated value. Remembering the various approximations which led to (2.90), this agreement must be regarded as excellent.

We have evaluated the one-site resonant radiative transfer process in detail because of its direct relevance to experiments in dilute ruby. In fact, all the nonradiative processes treated in Sect. 2.3.3 possess their radiative counterpart. One need only replace the exchange coupling J which appears in the relevant expressions of Sect. 2.3.3 with $J_1^{eff}(E, r)$ of (2.69) in this section to obtain the radiative excitation transfer counterpart. The relative importance of the various phonon-assisted radiative transfer processes would then be expected to scale the same as the nonradiative transfer processes. However, this will not be true for the two-site nonresonant process. This occurs because the efficacy of radiative transfer depends upon the ratio of the photon mean-free path to the same dimensions. Under conditions of photon trapping, one-site processes will be maximally efficient if the fluorescent radiation trapping length is short compared to the sample dimensions. However, the two-site nonresonant process involves photons whose energy is shifted by $\hbar\omega_{s,q}$ from the optical emission energy. As a consequence, these photons may possess trapping lengths considerably in excess of the fluorescent photons, and so may escape from the sample even in the so-called strong trapping regime. For this reason, systems in which the two-site nonresonant (nonradiative) process dominates at higher concentrations may not exhibit phonon-assisted radiative transfer in the dilute regime. This interesting difference has not yet been reported, but one must remember that to date only ruby has been investigated in the very dilute regime. And, for this system, the one-site resonant process is expected to dominate.

Some comments are in order at this stage regarding future experiments. Only ruby in the radiative transfer regime, and Cr^{3+} in LaAlO$_3$, have exhibited one-site (resonant) excitation transfer. All other studies appear to exhibit two-site nonresonant transfer. The latter is explicitly energy mismatch independent. The former varies as $(\Delta E_{12})^{-2}$. References [2.26, 28] exhibit very different time developments for the emission profile under these two different conditions. In particular, the inverse square dependence on the energy mismatch generates asymmetric emission profiles at intermediate times. The relative absence of the two-site nonresonant process for radiative transfer should allow one-site processes to become observable for a number of systems, allowing one to test for their characteristic energy dependence. Experiments along these lines would be most welcome.

2.5 Phonon Trapping

It is now generally recognized that strong ion–phonon interactions cause phonons resonant with ionic energy transitions to be strongly trapped. This effect was noted as early as 1941 by *Van Vleck* in his treatment of spin lattice relaxation [2.39], and was first reported for the excited states of ruby by *Geschwind* et al. [2.40]. They observed bottlenecking of the phonons which are resonant with the $|1^* >$ to $|1^{**} >$ (namely the $\bar{E}$ to $2\bar{A}$) transition. Resonant trapping of the same phonons (we shall henceforth call them Δ phonons) was present in the first monochromatic detection of terahertz phonons by *Renk* et al. [2.41] using short heat pulses in ruby. Experiments in the stationary case were first carried out by *Dijkhuis* et al. [2.42]. The optical experiments use the fast ($\sim 10^{-9}$ s) nonradiative decay of pumped Cr^{3+} impurities from the $|1^{**} >$ to the $|1^* >$ level to produce phonons of energy $\simeq \Delta$. When the population of the metastable $|1^* >$ level is low, the Δ phonon simply decays to the thermal reservoir. However, when the population of the $|1^* >$ level is sufficiently large, the Δ phonon produced by the $|1^{**} > \rightarrow |1^* >$ transition at one Cr^{3+} site is quickly trapped by another Cr^{3+} site in the $|1^* >$ state, the latter being excited to the $|1^{**} >$ state after absorption of the Δ phonon. The repopulation processes from the $|1^* >$ to the $|1^{**} >$ levels will cease when the Δ phonon escapes from the active volume, or becomes frequency shifted out of the absorption band. This phenomenon is usually detected by observing the otherwise absent fluorescence radiation from the $|1^{**} >$ level to the ground state. Knowing the population of the $|1^* >$ level and measuring the $|1^{**} >$ fluorescence intensity gives a measure of the occupation number for phonons of energy Δ.

2.5.1 Relationship to Radiative Trapping

The phonon trapping lenth for fixed Cr^{3+} ion concentration can be changed by adjusting the population of the $|1^* >$ state. This can be achieved in ruby either by pumping into the green band, allowing nonradiative decay to populate the $|1^* >$ level, or by pumping directly into the $|1^* >$ level from the ground state by using a tunable dye laser. Phonons of energy Δ are trapped because of the large phonon cross section for excitation of the $|1^* >$ to the $|1^{**} >$ state. This is quite analogous to the case of resonance radiation trapping in gases, first worked out in quantitative detail by *Holstein* [2.36]. Indeed, experiments in this field have relied heavily on the *Holstein* approach for their analysis.

As *Holstein* [2.36] showed, the absorption line shape is crucial to the trapping dynamics. In particular, escape from the active volume can only occur by virtue of spectral diffusion to the wings of the absorption line, where the trapping length becomes of the order of the sample size. The controlling factor for phonon escape is the spectral diffusion rate.

In order to obtain spectral diffusion, an energy shifting mechanism must be present. The mechanism for phonons of energy Δ in ruby is the subject of a great deal of discussion at the present time. Dipolar coupling between ions in the excited and ground Cr^{3+} states was suggested by *Dijkhuis* et al. [2.42]. The weakness of this interaction would allow only a very small energy shift for each Δ-phonon absorption and reemission step. A quite different energy shifting mechanism, allowing murch larger Δ-phonon energy shifts, has been proposed by *Egbert* et al. [2.43]. It is closely related to the processes for excitation transfer considered in this paper.

The actual experimental situation in ruby is complicated. The active volume over which the absorbing states $|1^*>$ are distributed is usually much smaller than the sample size. This is because a focused laser light excitation spot is necessary to obtain sufficient population of the $|1^*>$ level for Δ-phonon trapping. As a consequence, phonons can escape the active region either by frequency shifting or by simple spatial diffusion to the edge of the "spot." Each Δ-phonon energy packet would possess its own diffusion constant, with spatial diffusion to the edges of the spot depleting each packet separately.

Spatial diffusion dominates for small spot sizes and low population density of $|1^*>$ levels [2.44, 45]. Very recent experiments [2.43] at large spot size and excitation level exhibit cross-over to a regime where spectral diffusion appears to dominate over spatial diffusion. The most recent analysis [2.45] of *Dijkhuis* et al. reached a similar conclusion, though the frequency shifting mechanism they proposed differs from that of *Egbert* et al. [2.43].

We shall not discuss these differences of interpretation, which are themselves a subject for another review. Rather, because the *Egbert* et al. [2.43] mechanism is so close to the excitation transfer processes we have calculated, we include a treatment of it here.

2.5.2 Frequency Shifting Mechanism by Energy Transfer

An example of a frequency shifting mechanisms is pictured in Fig. 2.7A. In step 1, a Δ phonon can be absorbed by an ion excited to the $|1^*>$ level, taking it to the $|1^{**}>$ state. For a simple spin system, with a nondegenerate ground state, excitation transfer could then occur to another site, with the first site de-excited to the ground level, and the second site into its $|2^{**}>$. Subsequent emission of a phonon at site 2 returns the system to the first excited state, $|2^*>$. However, the transition $|1^*>$ to $|1>$ (similarly for site 2) is inhomogeneously broadened. For the special case of ruby, this broadening is much larger than the width of the trapping transition

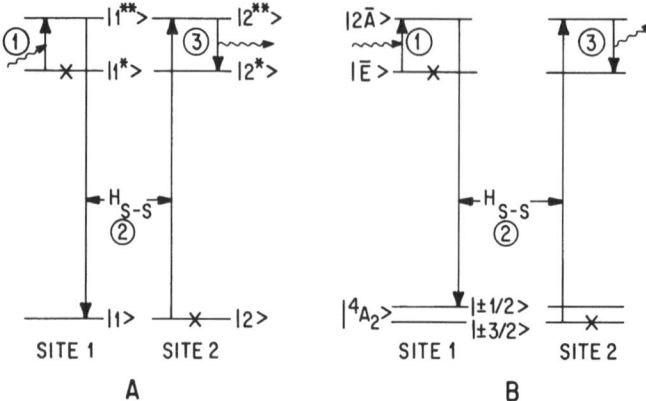

Fig. 2.7. Frequency shifting mechanism for resonantly trapped phonons of energy Δ, corresponding to Fig. 2.4C for resonant one site energy transfer. Additional processes, not shown, correspond to Figs. 2.4A, B. (*A*) The phonon frequency is shifted by the inhomogeneous width of the transitions $|1^{**}> \leftrightarrow |1>$ and $|2> \leftrightarrow |2^{**}>$. (*B*) The phonon frequency is shifted by a ground-state splitting. For ruby, this amounts to 0.38 cm^{-1}

$|1^*>$ to $|1^{**}>$, itself a measure of the Δ-phonon energy width. The Δ phonon emitted at step 3 can therefore have its energy changed from the incoming Δ phonon by an amount roughly equal to the inhomogeneous width of the $|1^*>$ to $|1>$ transition. This is likely to frequency shift the Δ phonon out of the trapping energy regime, set by the active volume size and the width of the $|1^{**}>$ to $|1^*>$ phonon emission transition. This mechanism allows for Δ-phonon escape. A concentration dependence of the Δ-phonon trapping lifetime at high intensities and spot sizes which agrees with this conclusion has been reported by *Egbert* et al. [2.43]. For the particular case of ruby, the ground state is a spin quartet, split by the trigonal crystalline field splitting into two doublets separated by 0.38 cm^{-1}. In such a case, not only can frequency shifting of the Δ phonon take place by virtue of the $|1^*>$ to $|1>$ state inhomogeneous width, it can also result from differing ground states playing a role in the excitation transfer process $|1^{**}>$ to $|2^{**}>$. This idea was first suggested by Geschwind (private communication), and allows large frequency shifting, 0.38 cm^{-1}, in the Δ-phonon energy everytime it occurs. It was this phonon frequency shifting process which was used by *Egbert* et al. [2.43] in the analysis of their data at large excitation densities. One such process is pictured in Fig. 2.7.

The scattering rate for a phonon resonant with the $|1^*>$ to $|1^{**}>$ transition can be calculated in a manner very similar to the one-site resonant excitation transfer process treated in Sect. 2.3.3c. One finds for the process analogous to that pictured in Fig. 2.4C

$$W^{phonon}_{\hbar\omega_{s,q}=\Delta} = \frac{(J^2+J_{\frac{1}{2}}^2)A^4}{4\pi(\Delta E_{12})^2} \frac{\omega_{s,q}^4}{\Gamma_2^2 \rho^2 v^7} N^* \tag{2.91}$$

were N^* is the number of excited states $\mid 1^* >$ per unit volume. We have assumed in deriving (2.91) that $J_2 \simeq J$, and that $\Delta E_{12} \gg \Gamma_2$. The coherence factors were treated in accordance with the discussion following (2.39). Use of (2.46, 47) in (2.91) results in the simple result,

$$W^{\text{phonon}}_{\hbar\omega_{s,q}=\Delta} = \frac{J_2^2 \pi}{(\Delta E_{12})^2} \frac{v^3}{\omega_{s,q}^2} N^*. \tag{2.92}$$

Remarkably, the frequency shifting rate is independent of the absorption linewidth, Γ_2. The result previously obtained by *Egbert* et al. [2.43] for the phonon frequency shifting rate caused by energy transfer is identical to (2.92), but begins from a different starting point (equating steady-state transition rates). Evaluation of this rate for ruby depends on a number of parameters. We take

$$J_2 = 10^{-3} \text{ cm}^{-1}$$

$$\Delta E_{12} = 10^{-1} \text{ cm}^{-1}$$

$$\omega_{s,q} = 6 \times 10^{12} \text{ s}^{-1}$$

At the extreme limit of experiment [2.43], $N^* = 10^{18} \text{ cm}^{-3}$. Using this number, and the values for the parameters listed above, we find

$$W^{\text{phonon}}_{\hbar\omega_{s,q}=\Delta} = 0.6 \times 10^6 \text{ s}^{-1} \tag{2.93}$$

This value is very close to that quoted by *Egbert* et al. [2.43], and appears sufficient to account for the needed frequency shifting rate at high power and large spot size. The expressions (2.91, 92) hold for both processes A and B of Fig. 2.7 as long as the difference in between sites 1 and 2 is small compared to the magnitude of the inhomogeneous width ΔE_{12} or 0.38 cm^{-1}, respectively. It represents a limiting value for the phonon "lifetime" as measured by experiments involving phonon trapping.

2.6 Conclusions

We have presented a detailed microscopic treatment of phonon-assisted excitation transfer for localized excitations in solids. We have shown how centers having different excitation energies can transfer excitations with the assistance of one- and two-phonon processes. One-phonon-assisted excitation transfers are inhibited between sites of similar ion–phonon coupling strengths because of an interference term. This effect is important when the spatial separation of the sites is less than the wavelength of the phonon which makes up the energy mismatch between the two sites. As a consequence, two-phonon-assisted transfer mechanisms are important. Their detailed properties were calculated. The participating phonons have energy

$\sim k_{\mathrm{B}}T$, eliminating the interference and thereby producing significant excitation transfer rates. It was shown that a marked difference occurs for two-phonon-assisted transfer processes with ion–phonon interactions taking place on a single site, as compared to both sites. The former map onto the *Förster–Dexter* formalism [2.6, 7], exhibiting an inverse square dependence on the energy mismatch between transferring sites. The latter is new and does not depend on the site–site energy mismatch. It is characterized by a temperature dependence which cannot be described in terms of the linewidth of any participating states.

These excitation transfer mechanisms all make use of the nonradiative site–site coupling Hamiltonian (multipolar or exchange couplings). These processes dominate at high ion concentration where short-range interactions are most important. At very low concentrations, another class of transfer mechanism takes over – phonon-assisted radiative transfer. This mechanism depends on the degree of entrapment of the fluorescent photons within the sample. Under severe trapping conditions, the excitation transfer rate uses every photon and is 100% efficient. As a consequence, the transfer rate is concentration independent. Therefore, assuming that one remains in the strong trapping regime, phonon-assisted radiative transfer always dominates over nonradiative transfer at sufficiently low concentrations. Both one- and two-phonon-assisted rates are calculated. The latter account for recent measurements of the excitation transfer rate of dilute ruby with no adjustable paramaters.

The fluorescent line narrowing experiments performed in recent years all exhibit evidence of the excitation transfer rates calculated in this chapter. This experimental method remains one of the most powerful techniques available for determination of specific excitation transfer mechanisms in optical systems.

There remains an area where fluorescent line narrowing measurements cannot determine the spatial transfer rate. Fluorescent line narrowing methods determine only the energy packet-to-packet transfer rate. They are incapable of determining the transfer rate within an energy packet of an inhomogeneously broadened line. This is an important limitation: resonant excitation transfer is one of the most interesting questions being explored today in condensed matter science. A tool for investigating excitation transfer within a packet is the transient grating technique [2.46, 47]. *Anderson* localization [2.1] can be detected using such techniques, for no spatial diffusion should take place (at $T = 0$ K) if the disorder is sufficiently large. The processes considered in this chapter are all finite temperature effects, and will "de-localize Anderson-localized states". As such, they serve as a guide to whether one is at sufficiently low temperatures to expect absence of diffusion from a strictly disorder point of view.

Finally, phonon entrapment is a topic which is also relevant to the calculations of this section. At large excitation intensities, one can frequency shift resonantly trapped phonons using excitation transfer mechanisms. One such mechanism was calculated explicitly in Sect. 2.5. These frequency shifting mechanisms can mask other phonon decay channels (e.g., anharmonic) being studied on either a pulse or cw basis. It is therefore important to take them into account in the analysis of high energy phonon lifetimes under trapped conditions.

References

2.1   P. W. Anderson: Phys. Rev. **109**, 1492 (1958)
2.2   P. D. Antoniou, E. N. Economou: Phys. Rev. B **16**, 3768 (1977), and references therein
2.3   A. Szabo: Phys. Rev. Lett. **25**, 924 (1970); **27**, 323 (1971)
2.4   J. Koo, L. R. Walker, S. Geschwind: Phys. Rev. Lett. **35**, 1669 (1975)
2.5   S. K. Lyo: Phys. Rev. B **3**, 3331 (1971)
2.6   D. L. Dexter: J. Chem. Phys. **21**, 836 (1953)
2.7   T. Förster: Ann. Phys. (Paris) **2**, 55 (1948)
2.8   Quoted in P. Kisliuk, N. C. Chang, P. L. Scott, M. H. L. Pryce: Phys. Rev. **184**, 367 (1969)
2.9   R. J. Birgeneau: J. Chem. Phys. **50**, 4282 (1969)
2.10 K. Sugihara. J. Phys. Soc. (Jpn.) **14**, 1231 (1959)
2.11 L. K. Aminov, B. I. Kochelaer: J. Exptl. Theor. Phys. (U.S.S.R.) **43**, 1303 (1962) [Engl. trans.: Sov. Phys.-JETP **15**, 903 (1962)];
      D. H. McMahon, R. H. Silsbee: Phys. Rev. **135**, A91 (1964);
      R. Orbach, M. Tachiki: Phys. Rev. **158**, 524 (1967)
2.12 T. Miyakawa, D. L. Dexter: Phys. Rev. B **1**, 2961 (1970)
2.13 T. F. Soules, C. B. Duke: Phys. Rev. B **3**, 262 (1971);
      see also M. Kohli, M. L. Huang Liu: Phys. Rev. B **9**, 1008 (1974)
2.14 T. Holstein, S. K. Lyo, R. Orbach: Phys. Rev. Lett. **36**, 891 (1976)
      A full exposition of this method is planned for publication in Rev. Mod. Phys.
2.15 R. Orbach: In *Optical Properties of Ions in Crystals*, ed. by H. M. Crosswhite, H. W. Moos (Interscience, New York 1967) p. 445
2.16 P. M. Selzer, D. S. Hamilton, W. M. Yen: Phys. Rev. Lett **38**, 858 (1977)
2.17 A. A. Kaplyanskii, A. K. Przkeviskii: Sov. Phys. – Solid State **9**, 190 (1967);
      L. F. Mollenauer, A. L. Schawlow: Phys. Rev. **168**, 309 (1968)
2.18 D. S. Hamilton, P. M. Selzer, W. M. Yen: Phys. Rev. B **16**, 1858 (1977);
      D. S. Hamilton: Ph. D. Thesis, University of Wisconsin, Madison (1977) unpublished
2.19 R. Orbach, H. J. Stapleton: In *Electron Paramagnetic Resonance*, ed. by S. Geschwind (Plenum Press, New York 1972) p. 121
2.20 R. Orbach: Proc. Roy. Soc. A **264**, 458 (1961)
2.21 W. Heitler: *The Quantum Theory of Radiation* (Clarendon Press, Oxford 1957) p. 196
2.22 J. Heber, H. Murman: Z. Phys. B **26**, 145 (1977)
2.23 A. M. Stoneham: Phys. Status Solidi **19**, 787 (1967)
2.24 S. K. Lyo: Phys. Rev. B **5**, 795 (1972)
2.25 N. S. Yamada, S. Shionoya, T. Kushida: J. Phys. Soc. Jpn. **32**, 1577 (1972)
2.26 T. Holstein, S. K. Lyo, R. Orbach: Phys. Rev. B **15**, 1693 (1977); Colloq. Int. C.N.R.S. **255**, 185 (1977)
2.27 P. M. Selzer, W. M. Yen: Opt. Lett. **1**, 90 (1977)
2.28 T. Holstein, S. K. Lyo, R. Orbach: Phys. Rev. B **16**, 934 (1977)
2.29 M. Blume, R. Orbach, A. Kiel, S. Geschwind: Phys. Rev. **139**, A314 (1965)
2.30 I. S. Gradshteyn, I. M. Ryzhik: *Table of Integrals, Series, and Products* (Academic Press, New York 1965)
2.31 C. Brecher, L. A. Riseberg, M. J. Weber: Appl. Phys. Lett. **30**, 475 (1977)
2.32 S. Geschwind, G. E. Devlin, R. L. Cohen, S. R. Chinn: Phys. Rev. **137**, A1087 (1965)
2.33 J. E. Rives, R. S. Meltzer: Phys. Rev. B **16**, 1808 (1977)
2.34 A. L. Schawlow: In *Advances in Electronics*, ed. by J. R. Singer (Columbia Univ. Press, New York 1961) p. 50
2.35 Allan C. G. Mitchell, Mark W. Zemansky: *Resonance and Excited Atoms* (Cambridge Univ. Press, Cambridge 1961) p. 99
2.36 T. Holstein: Phys. Rev. **72**, 1212 (1947)

2.37 This approximation is discussed at some length by J.-P. Barrat: J. Phys. Radium (1959). An abbreviated version has been published in J. Phys. (Paris) **20**, 541 (1959); **20**, 633 (1959); **20**, 657 (1959). The approximation amounts to taking each atom at the center of a sphere of radius d, and is more correct the larger the ratio $d/l(E)$

2.38 T. Holstein, S. K. Lyo, R. Orbach: Phys. Rev. B **15**, 4693 (1977)

2.39 J. H. Van Vleck: Phys. Rev. **59**, 724, 730 (1941)

2.40 S. Geschwind, G. E. Devlin, R. L. Cohen, S. R. Chinn: Phys. Rev. **137**, A1087 (1965)

2.41 K. F. Renk, J. Deisenhofer: Phys. Rev. Lett. **26**, 764 (1971);
      K. F. Renk, J. Peckenzell: J. Phys. (Paris) **33**, C-4-103 (1972)

2.42 J. I. Dijkhuis, A. van der Pol, H. W. de Wijn: Phys. Rev. Lett. **37**, 1554 (1976)

2.43 W. C. Egbert, R. S. Meltzer, J. E. Rives: In Phonon Scattering in Condensed Matter, ed. by H. J. Maris (Plenum, New York 1980) p. 365
      R. S. Meltzer, J. E. Rives, W. C. Egbert: Phys. Rev. B **25**, 3026 (1982)

2.44 R. S. Meltzer, J. E. Rives: Phys. Rev. Lett. **38**, 421 (1977)

2.45 J. I. Dijkhuis, K. Huibregtse, H. W. de Wijn: Phys. Rev. B **20**, 1835, 1844 (1979)

2.46 R. W. Hellwarth: J. Opt. Soc. Am. **67**, 1 (1977);
      D. M. Bloom, G. C. Bjorklund: Appl. Phys. Lett. **31**, 592 (1977);
      A. Yariv, D. M. Pepper: Opt. Lett. **1**, 16 (1977)
      P. F. Liao, D. M. Bloom: Opt. Lett. **3**, 4 (1978)

2.47 J. R. Salcedo, A. E. Siegman, D. D. Dlott, M. D. Fayer: Phys. Rev. Lett. **41**, 131 (1978);
      D. S. Hamilton, D. Heiman, J. Feinberg, R. W. Hellwarth. Opt. Lett. **4**, 124 (1979);
      H. J. Eichler, J. Eichler, J. Knof, C. Noach: Phys. Status Solidi A **52**, 481 (1979);
      P. F. Liao, L. M. Humphrey, D. M. Bloom, S. Geschwind: Phys. Rev. B **20**, 4145 (1979)

2.48 S. Chu, H. M. Gibbs, S. L. McCall, A. Passner: Phys. Rev. Lett. **45**, 1715 (1980)

2.49 P. E. Jessop, A. Szabo: Phys. Rev. Lett. **45**, 1712 (1980)

2.50 H. M. Gibbs, S. Chu, S. L. McCall, A. Passner: In Coherence and Energy Transfer in Glasses, ed. by P. Fleury and B. Golding (Plenum Press, New York 1984)

3. Dynamics of Incoherent Transfer

D.L. Huber

With 12 Figures

In the material presented in Chap. 2 emphasis was placed on the microscopic aspects of inter-atomic energy transfer. Expressions were given for the rates characterizing the transfer of excitation between individual atoms or molecules. In this chapter we discuss the connection between experimental studies of the time evolution of the fluorescence and the underlying microscopic transfer processes. We consider two classes of experiments: time resolved fluorescence line narrowing and the time evolution of the integrated fluorescence in the presence of traps. In each case we first discuss the general theory. Various approximations are then outlined, followed by a brief discussion of the use of the theory in the interpretation and analysis of experiments. Special features of the analysis which pertain to transport in one-dimensional systems are discussed in Appendix 3.A. Back-transfer from the traps is analyzed in Appendix 3.B.

3.1 Rate Equations

We consider an array of identical "donor" atoms (or molecules) and focus our attention on a pair of levels. One of these is the ground state; the other is an excited state having a lifetime which is long in comparison with a time characteristic of the inter-atom transfer process. The latter, typically, is identified with the reciprocal of the transfer rate between atoms separated by the average nearest-neighbor distance. Since we are interested in effects arising from incoherent transfer, where phase information is not preserved, we can characterize the system by a set of probability functions $[P_n(t)]$. Here $P_n(t)$ is the probability that atom (or molecule) n is in the excited state at time t. We further assume that the number of excited atoms in the crystal at any given time is much less than the total number of atoms. In such a situation we can neglect "excluded volume" effects associated with having two neighboring atoms in the excited state at the same time. This being the case we can consider the limiting case of a single excited atom so that $P_n(t)$ becomes the probability that atom n is excited and all other atoms are in their ground state.

The evolution of the $P_n(t)$ is governed by a set of coupled rate equations which take the form

$$dP_n(t)/dt = -(\gamma_R + X_n + \sum_{n' \neq n} W_{nn'}) P_n(t) + \sum_{n' \neq n} W_{n'n} P_{n'}(t). \qquad (3.1)$$

Here the terms multiplying $P_n(t)$ characterize processes which remove atom n from its excited state. The parameter γ_R is the reciprocal of the lifetime of the excited state in the absence of transfer processes. We will assume that it is the same for all atoms. The parameter X_n denotes the total transfer rate to "acceptor" atoms or traps in the vicinity of the nth atom. By transfer to traps we mean any inter-atomic transfer process which removes excitation from the level of interest. Such processes can arise, for example, in transfer to impurities, or perturbed donors (e.g. exchange-coupled pairs). In addition, when cross-relaxation effects are present the donor atoms themselves act as traps. Since the distribution of traps will vary from atom to atom we append the subscript n to the total transfer rate. We will also assume that backtransfer from the traps to the donors is sufficiently slow that it can be neglected on the time scale of interest. The behavior in situations where this is not the case is discussed briefly in Appendix B. The sum $\Sigma_{n'} W_{nn'}$ denotes the total transfer rate from atom n to all other donor atoms in the array, $W_{nn'}$ being the transfer rate from atom n to atom n'. The term $\Sigma_{n'} W_{n'n} P_{n'}(t)$ describes the inverse process in which excitation is transferred from atom n' to atom n.

Because of perturbations arising from random strains, impurities, etc., the difference in energy between the excited and ground states, $E_{n'}$, will vary from atom to atom. Provided the distribution in these energy differences has a width which greatly exceeds the homogeneous linewidth, its shape will be mirrored in the excited state $\rightarrow$ ground state fluorescence spectrum observed after broadband excitation. An important question not answered by such a measurement is the extent to which the E_n of neighboring atoms are correlated as would be the case were the inhomogeneous linewidth to arise from random *macroscopic* strains. In our analysis we will assume that this does not happen. Instead, we hypothesize the opposite limit, *microscopic* strain broadening, where there is no correlation in the E_n of neighboring atoms.

The existence of a distribution in E_n can have an important effect on the transfer process. As discussed in Chap. 2 the transfer rates themselves can be functions of the difference $E_n - E_{n'}$. In addition, they must satisfy the detailed balance condition associated with thermal equilibrium at temperature T

$$W_{nn'} \exp(-E_n/k_B T) = W_{n'n} \exp(-E_{n'}/k_B T). \qquad (3.2)$$

From (3.2) we conclude that the transfer rates will be approximately symmetric $(W_{nn'} \approx W_{n'n})$ only when the temperature is high enough so that $k_B T \gg |E_n - E_{n'}|$. We analyze situations where there is a lattice of sites which are occupied at random by donor atoms and traps. The probability of a site being occupied by a donor is c_D; the corresponding probability for occupation by a trap or acceptor atom is c_A. In all cases we have $c_D + c_A \leq 1$. In addition to depending on the energy mismatch, $E_n - E_{n'}$, and the relative separation $r_{nn'} = r_n - r_{n'}$, the transfer rate can also depend on the orientation of $r_n - r_{n'}$ relative to the crystal or molecular axes. Although orientational effects can be included in the theory, in the interest of simplicity we will leave them out.

3.2 Fluorescence Line Narrowing

As discussed elsewhere in this book time resolved fluorescence line narrowing
(TRFLN) experiments can provide detailed information about incoherent energy
transfer in inhomogeneously broadened systems. In this section we will outline
a microscopic theory, based on the rate equations given in Sect. 3.1, which relates
the time dependence of the fluorescence to the inter-atomic transfer rates. The
essential feature of the TRFLN experiment is the use of a pulsed, narrowband
source, typically a laser, to excite atoms whose resonance frequencies span a
small segment of the inhomogeneous line [3.1]. After the source is turned off
the fluorescence evolves in time a manner shown schematically in Fig. 3.1. Initially there is only the sharp line spectrum coming from the atoms which were
excited by the incident light. However as time passes atoms whose resonant frequencies are outside the bandwidth of the source begin to fluoresce. The broadband fluorescence arises from the transfer of excitation to ions which were not
excited by the light. By measuring the time decay of the narrow component one
can obtain detailed information about the microscopic transfer processes.

3.2.1 General Theory and Exact Results

In our analysis of the TRFLN experiments we will begin with a somewhat simplified model which is applicable to $Pr_xLa_{1-x}F_3$, a system which has been studied
in great detail using this technique [3.1, 2]. After presenting exact results and
various approximations we will indicate how the approximations can be extended
so as to be applicable to more general situations. In the model we 1) neglect the

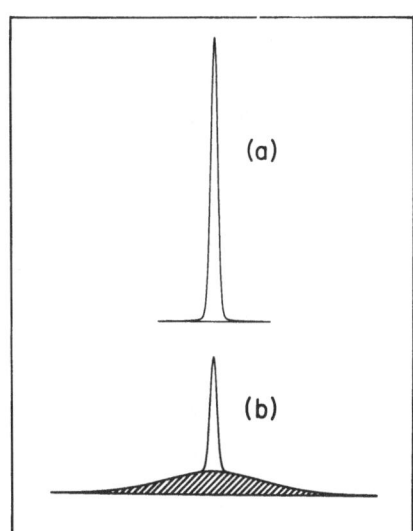

Fig. 3.1a, b. Schematic diagram of the time
development of the fluorescence in a fluorescence line narrowing experiment. (a) $t = 0$; (b)
$t > 0$. The shaded area indicates the background
fluorescence coming from ions which were not
initially excited

presence of traps ($c_A = 0$), 2) assume that $k_B T \gg$ inhomogeneous linewidth so that the transfer rates are symmetric, and 3) consider only transfer mechanisms which lead to rates which are independent of the energy mismatch between atoms. The latter assumption, together with the assumption of microscopic strain broadening, leads to a background fluorescence which has the same profile as the inhomogeneous line observed in fluorescence following broadband excitation.

Under these conditions we define a dimensionless, time-dependent function, $R(t)$, to be the ratio of the intensity of the narrowband fluorescence (with the interpolated background subtracted) to the total intensity over the entire inhomogeneous line

$$R(t) = \frac{\text{narrowband intensity at time } t}{\text{total intensity at time } t}. \tag{3.3}$$

The function $R(t)$ has a simple physical interpretation. It is the conditional probability that an atom which is excited at $t = 0$ is still excited at a later time t. In [3.2] it was shown that $R(t)$ could be written in the form

$$R(t) = \int_0^\infty \exp(-\lambda t) \, \rho(\lambda) \, d\lambda. \tag{3.4}$$

Here $\rho(\lambda)$ is the normalized density of eigenvalues of the relaxation matrix Λ characterizing the rate equations for a system where $\gamma_R = X_n = 0$

$$dP_n(t)/dt = -\sum_{n'} \Lambda_{nn'} \, P_{n'}(t), \tag{3.5}$$

with $$\Lambda_{nn'} = \delta_{nn'} \sum_{n''} W_{nn''} - (1 - \delta_{nn'}) \, W_{nn'}. \tag{3.6}$$

The calculation of $\rho(\lambda)$ and hence $R(t)$ poses no problem when the donors occupy every site on the lattice ($c_D = 1$). Because of the translational symmetry of the system we have [3.2]

$$\rho(\lambda) = 1/N \sum_k \delta[\lambda - W(0) + W(k)], \tag{3.7}$$

where the sum on k is over the N points in the Brillouin zone associated with the donor lattice [3.3]. The function $W(k)$ is given by the equation

$$W(k) = \sum_{n'} W_{nn'} \exp(ik \cdot r_{nn'}), \tag{3.8}$$

where $r_{nn'} = r_n - r_{n'}$ is a vector connecting atoms n and n'. Inserting (3.7) into (3.4) we obtain the result

$$R(t) = 1/N \sum_k \exp\{-[W(0) - W(k)]t\}. \tag{3.9}$$

Although the function $R(t)$ can only be expressed in closed form in a few simple cases (e.g., simple cubic lattice with nearest-neighbor transfer) one can identify three regimes in the evolution of $R(t)$: an early-time exponential regime, an asymptotic diffusion regime, and a transition regime connecting the two. In the exponential regime we have

$$R(t) \approx \exp\left[-W(0)t\right], \qquad (3.10)$$

while in the diffusion regime $R(t)$ is given by

$$R(t) = [8\pi^{3/2}\Omega^{-1}(D_xD_yD_z)^{1/2}t^{3/2}]^{-1}. \qquad (3.11)$$

Here Ω is the volume per donor and D_x, D_y, D_z are the components of the diffusion tensor which are defined by the sums

$$D_x = 1/2 \sum_{n'} W_{nn'} x^2_{nn'}, \text{etc.}$$

3.2.2 Approximations

The calculation of the function $R(t)$ in situations where the donor array is disordered by the presence of optically inactive atoms is a problem of tremendous complexity. When $c_D < 1$ there is no longer any translational symmetry. As a result the analysis outlined in (3.7-9) is not applicable. The most general methods of obtaining information about $\rho(\lambda)$ and $R(t)$ in disordered systems involve numerical techniques applied to finite arrays [3.4-7]. These calculations all begin with a finite lattice of N sites. A number $c_D N$ of donors are distributed at random and the corresponding relaxation matrix is calculated assuming periodic boundary conditions. At this point either of two approaches is followed: equation-of-motion techniques [3.4-7] or matrix diagonalization [3.6]. The former is equivalent to integrating the rate equations with initial conditions $P_{n'}(0) = \delta_{n'n_0}$, where the initially excited donor is picked at random. Either $R(t)$ is calculated directly [3.7] or the eigenvalue distribution is obtained and $R(t)$ is obtained from (3.4) [3.6]. The matrix methods involve the calculation of the eigenvalues of $\overleftrightarrow{\Lambda}$ using standard numerical routines for matrix diagonalization.

The equation-of-motion techniques have the advantage that comparatively large (10^3-10^4) numbers of donors can be easily handled. They suffer from the limitation that the long time or equivalently small λ behavior is relatively inaccurate because of the accumulation of errors in the integration of the differential equations. The accuracy of the matrix approach is much higher being limited only by the accuracy of the diagonalization routine. However this is offset by the fact that ensembles with only relatively small numbers of donors $(\lesssim 500)$ can be studied. As a consequence it is often necessary to repeat the calculations with different configurations of donors in order to obtain meaningful results.

Although the numerical methods give the most accurate estimates of $R(t)$ they are complicated to use and provide relatively little physical insight. Because of this it has been worthwhile to develop simple short-time approximations to $R(t)$. Such approximations are particularly valuable in the analyses of experimental data, which have been limited to times such that $R(t) \lesssim 0.1$ because of noise and lifetime effects.

The most useful of these approximations takes the form of a product over sites [3.2, 6]

$$R(t) = \prod_l [1 - c_D + c_D \exp(-W_{0l}t) f(W_{0l}t)], \tag{3.12}$$

where W_{0l} is the transfer rate from a donor at site 0 to a donor at site l. The probability that site l is occupied is c_D while the probability that it is unoccupied is $1-c_D$. The form of the function $f(W_{0l}t)$ depends on the donor concentration. For $c_D \gtrsim 0.5$ a reasonable fit to the data out to times such that $R(t) \approx 0.05$ is obtained with

$$f(W_{0l}t) = 1 + (1/2)(W_{0l}t)^2. \tag{3.13}$$

For $c_D \lesssim 0.2$ an equally good fit over the same range is obtained with

$$f(W_{0l}t) = \cosh(W_{0l}t). \tag{3.14}$$

In Fig. 3.2 we show the approximate and exact values for $R(t)$ for the case of a fully occupied ($c_D = 1$), simple cubic lattice with nearest-neighbor transfer. Figure 3.3 displays the corresponding results for a face centered cubic lattice with

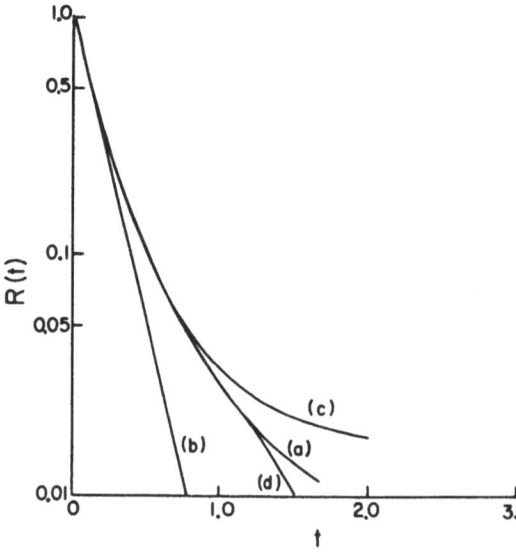

Fig. 3.2. $R(t)$ vs t in a simple cubic lattice with all sites occupied. Transfer only between nearest neighbors. Time is measured in units of the reciprocal of the nearest-neighbor transfer rate. (a) Exact result; (b), (c) and (d) refer to $R(t)$ calculated from (3.12) with $f(W_{0l}t) = 1$, $\cosh(W_{0l}t)$, and $1 + 1/2 (W_{0l}t)^2$, respectively. [3.6]

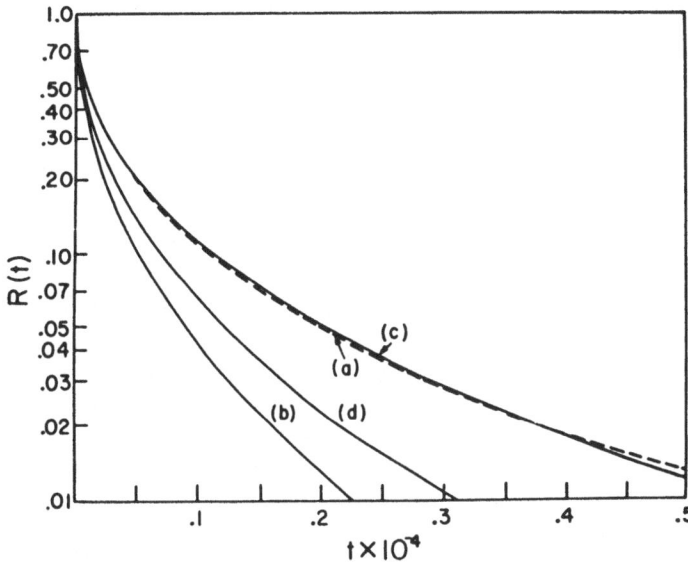

Fig. 3.3. $R(t)$ vs t in a fcc lattice with $c_D = 0.01$. $W_{nn'} \propto r_{nn'}^{-6}$. Time is measured in units of the reciprocal of the nearest-neighbor transfer rate. (a) broken line) exact result; (b), (c) and (d) refer to $R(t)$ calculated from (3.12) with $f(W_{0l}t) = 1$, $\cosh(W_{0l}t)$, and $1 + 1/2(W_{0l}t)^2$, respectively. [3.6]

$c_D = 0.01$, where $W_{nn'} \propto r_{nn'}^{-6}$, (dipole-dipole transfer). In the case of Fig. 3.2 the exact result is $R(t) = \exp(-6Wt) I_0 (2Wt)^3$, where I_0 is a modified Bessel function. The exact results for the dilute system were obtained by matrix diagonalization. In each case we also display the curves obtained when $f(W_{0l}t)$ is set to equal 1. In both figures the appropriate approximations are in excellent agreement with the exact values out to times such that $R(t) \approx 0.02$. In the range $0.2 < c_D < 0.5$ neither approximation works particularly well for $R(t) < 0.1$ since there is a crossover from behavior best approximated by (3.13) to that where (3.14) is the appropriate approximation.

The approximations embodied in (3.13, 14) have a simple physical interpretation. Recalling our comment that $R(t)$ is the conditional probability that an atom excited at $t=0$ is still excited at time t we see that with $f(W_{0l}t) = 1$ we have the exact result for $R(t)$ in the absence of any backtransfer of excitation to the site initially excited (site 0). It is equivalent to a result first obtained by *Golubov* and *Konobeev* [3.8]. The choice $f(W_{0l}t) = \cosh(W_{0l}t)$ is appropriate to a situation where either there is a relatively isolated donor, $W_{0l}t \ll 1$, all l, in which case $\cosh(W_{0l}t) \approx 1$ or there is a single donor nearby at site, say l. Multiple exchange of excitation between 0 and $\bar{l}$ is taken into account as well as one-way transfer from the pair to more distant sites, the latter at the rate $\Sigma_{\bar{l} \neq l} W_{0\bar{l}}$ for both members of the pair. Such an approximation is particularly well suited to dilute systems where the most numerous configurations involve either single atoms or near-neighbor pairs.

Equation (3.13) is obtained by keeping only the first two terms in the expansion of the cosh. Broadly speaking it is appropriate to situations where there is only one exchange of excitation between 0 and l. At high concentrations multiple exchange of excitation between a pair of atoms is less important than processes involving three or more atoms. However the latter contribute to $R(t)$ only in order $(Wt)^3$ and higher and hence do not have an appreciable effect on the short-time behavior of the function. Put somewhat differently (3.13) is useful in a situation where the excitation (on the average) hops back to the initially excited state at most once before leaving the area.

In concluding this section we have two additional comments. First, it must be emphasized that our results for $R(t)$ have been obtained in the low excitation limit where the fraction of donors initially excited, p, is much less than unity. *Alexander* [3.9] has shown that for arbitrary values of p (3.4) is replaced by

$$R(t) = p + (1-p) \int_0^\infty \exp(-\lambda t) \rho(\lambda) d\lambda. \tag{3.15}$$

Secondly, in the low concentration regime, $c_D \ll 1$, we can obtain a simple approximation to $R(t)$ by writing the right-hand side of (3.12) as the exponential of a sum of logarithms and expanding the logarithms to first order in c_D. Treating the transfer rate as a continuous function of the separation, the sum on l is then converted to an integral over r_{0l}. For $W_{0l} = \beta/r_{0l}^\nu$ we obtain the result

$$R(t) = \exp -(4\pi/3) n_D (\beta t)^{3/\nu} \Gamma(1 - 3/\nu) 2^{(3/\nu - 1)}], \tag{3.16}$$

where n_D is the number of donors per unit volume and $\Gamma(x)$ denotes the gamma function. Equation (3.16) which is obtained with $f(W_{0l}t) = \cosh(W_{0l}t)$ differs from the corresponding result obtained with no backtransfer $[f(W_{0l}t) = 1]$ by the factor $2^{(3/\nu - 1)}$ in the exponential. For the case $\nu = 6$ (dipole-dipole transfer) it is essentially equivalent to the pair approximation of *Godzik* and *Jortner* [3.10] and the three-body approximation of *Gochanour* et al. [3.11].

3.2.3 Further Approximations

The approximations discussed in the preceding section apply only to systems with a negligible concentration of traps. In addition, it is assumed that the transfer rates are symmetric and independent of the energy mismatch between donors. The generalization of (3.12-14) to systems with traps and asymmetric, energy-dependent transfer rates is straightforward [3.12]. In place of (3.12) we have

$$R(t) = \prod_l \int dE_l P(E_l) \Big\{ 1 - c_A - c_D + c_A \exp[-X_{0l}(E_0)t]$$
$$+ c_D \exp[-W_{0l}(E_0, E_1)t] f(E_0, E_1; t) \Big\}, \tag{3.17}$$

assuming no correlation in the energy levels of different donors (microscopic strain broadening). Here $P(E_l)$ is the normalized inhomogeneous line-shape function, E_0 is the (excited state) energy of the atom initially excited, which is located at site 0, and E_l is the corresponding energy for a donor at site l. The symbol X_{0l} denotes the transfer rate from the donor at 0 to the acceptor at l, and W_{0l} (E_0, E_l) is the transfer rate from a donor with energy E_0 located at 0 to a donor with energy E_l located at l. In analogy with (3.14) when $c_D \lesssim 0.2$ $f(E_0, E_l;t)$ is given by [3.12]

$$f(E_0, E_l;t) = \frac{W_{l0}(E_l, E_0) \exp\left[W_{0l}(E_0, E_l)t\right] + W_{0l}(E_0, E_l) \exp\left[-W_{l0}(E_l, E_0)t\right]}{W_{0l}(E_0, E_l) + W_{l0}(E_l, E_0)}$$

$$(3.18)$$

where $W_{l0}(E_l, E_0)$ is the transfer rate from a donor at l having energy E_l to a donor at 0 which has energy E_0. Corresponding to (3.13), when $c_D \gtrsim 0.5$ f takes the form

$$f(E_0, E_l;t) = 1 + 1/2\, W_{0l}(E_0, E_l)\, W_{l0}(E_l, E_0)t^2,$$

$$(3.19)$$

which is obtained by expanding (3.18) in powers of t. In general we expect (3.17-19) to have the same range of validity as (3.12-14), i.e., out to times such that $R(t) \approx 0.05$.

As an application of these equations we consider the example of a system with a Gaussian inhomogeneous line shape with symmetric donor-donor transfer rates which vary inversely with the square of the energy mismatch [3.13]. When the initially excited ions are at the center of the line we have

$$R(t) = \prod_l \int dE_l\, (2\pi\Delta)^{-1/2} \exp\left[-(E_0-E_l)^2/(2\Delta)\right]$$

$$\{1-c_D + c_D \times \exp\left[-A_{0l}t/(E_0-E_l)^2\right] \cdot \cosh\left[A_{0l}t/(E_0-E_l)^2\right]\},$$

$$(3.20)$$

for $c_A = 0$, $c_D \ll 1$, $A_{0l}/(E_0-E_l)^2$ being the donor-donor transfer rate. Evaluating the integral we obtain the result

$$R(t) = \prod_l \{1-c_D + c_D \exp\left[-(2A_{0l}t/\Delta)^{1/2}\right] \cosh(2A_{0l}t/\Delta)^{1/2}\}.$$

$$(3.21)$$

Note that in contrast to (3.12, 14) $R(t)$ is a function of $t^{1/2}$.

3.2.4 Development of the Background Fluorescence

When the transfer rates are symmetric and independent of the energy mismatch and there is microscopic strain broadening the background fluorescence coming from atoms outside of the bandwidth of the source will have a profile which is similar to the inhomogeneous line shape. However when these conditions are not satisfied the fluorescence can have a rather complicated time and frequency dependence.

In the past, theoretical studies of this problem have been based on an integral equation introduced by *Montegi* and *Shionoya* [3.14]. However in [3.7] it was pointed out that their approach has at best only qualitative validity since the integral equation takes into account only the variation of the donor-donor transfer rates with energy mismatch and overlooks the dependence on the separation between the atoms.

A short-time approximation which takes into account both of these effects is discussed in [3.12]. In this approach the background fluorescence at frequency E/h and time t which is generated by a source at frequency E_0/h, $I_B(E,t;E_0)$, is given by

$$I_B(E,t;E_0) = C(1-\exp[-c_D U(t)])U(t)^{-1} P(E)$$

$$\times \sum_l [1-\exp[-W_{0l}(E_0,E)t] \cdot f(E_0,E;t)], \tag{3.22}$$

for $c_D \lesssim 0.2$. Here C is a constant and f is given by (3.18). The function $U(t)$, which is related to $R(t)$ through the equation

$$R(t) = \exp[-c_D U(t)], \tag{3.23}$$

is expressed as

$$U(t) = \prod_l \int dE_l P(E_l) \{1-\exp[-W_{0l}(E_0,E_l)t] \cdot f(E_0,E_l;t)\}. \tag{3.24}$$

Equations (3.22, 23) are expected to be good approximations for $c_D U(t) \lesssim 0.5$ and to be reasonably accurate out to times such that $c_D U(t) = 3$, which corresponds to $R(t) = 0.05$.

In a system with symmetric transfer ($W_{nn'} = W_{n'n}$) (3.22, 24) simplify considerably. We have

$$I_B(E,t;E_0) = C (1-\exp[-c_D U(t)]) U(t)^{-1} P(E) \sum_l \exp(-W_{0l}t)$$

$$\times \sinh(W_{0l}t), \tag{3.25}$$

with

$$U(t) = \sum_l \int dE_l P(E_l) \exp(-W_{0l}t) \sinh(W_{0l}t). \tag{3.26}$$

In the case where $W_{0l} = \beta(|E_0-E_l|)/r_{0l}^\nu$, where β is an arbitrary function of $|E_0-E_l|$, we convert the sums on l to integrals thus obtaining the results

$$I_B(E,t;E_0) = C\{(1 - \exp[-c_D U(t)])U(t)^{-1}P(E)C_\nu(\beta(|E_0 - E|)t)\}, \tag{3.27}$$

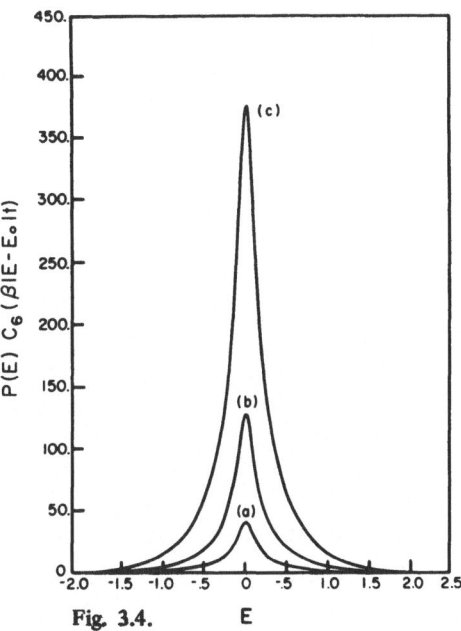

Fig. 3.4.   E

Fig. 3.4. $P(E)C_6(\beta(|E_0-E|)t)$ vs E for $E_0 = 0$. $P(E) = \pi^{-1/2}\exp(-E^2)$, $\beta = R_{min}^6/[(E_0 - E)^2 + 0.01]$, fcc lattice. Time is measured in units of the reciprocal of the nearest-neighbor transfer rate. (a) $t = 1$, $U(t) = 22$; (b) $t = 1$, $U(t) = 70$; (c) $t = 100$, $U(t) = 216$. [3.12]

Fig. 3.5. Same as Fig. 3.4 except $E_0 = 1$. (a) $t = 5$, $U(t) = 28$; (b) $t = 50$, $U(t) = 87$; (c) $t = 500$, $U(t) = 261$. [3.12]

Fig. 3.6. Same as Fig. 3.4 except $E_0 = 2$. (a) $t = 10$, $U(t) = 14$; (b) $t = 100$, $U(t) = 45$; (c) $t = 1000$, $U(t) = 140$. [3.12]

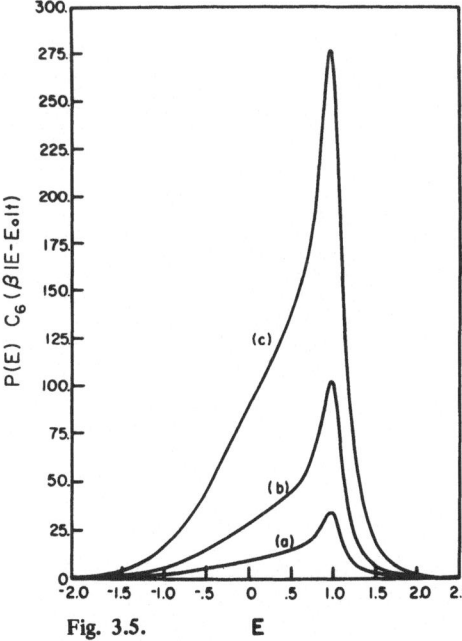

Fig. 3.5.   E

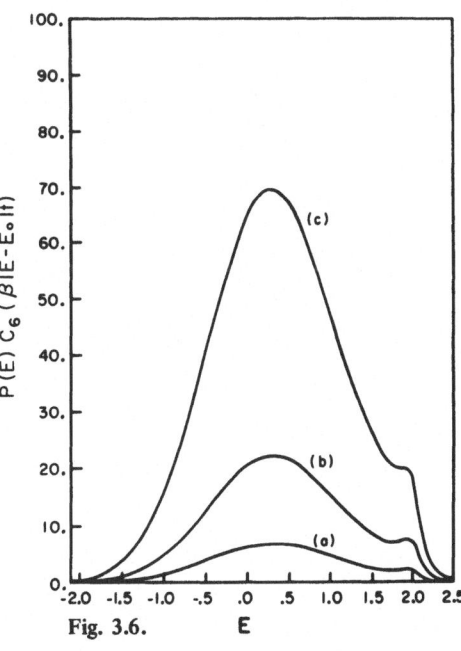

Fig. 3.6.   E

and

$$U(t) = \int dE P(E) C_\nu(\beta(|E-E_0|)t), \qquad (3.28)$$

with

$$C_\nu(x) = (4\pi n_L/3)\, x^{3/\nu}\, \Gamma(1-3/\nu) 2^{(3/\nu-1)}. \qquad (3.29)$$

in which n_L is the number of lattice sites per unit volume.

In Figs. 3.4-6 we show $P(E)C_\nu(\beta(|E_0-E|)t)$ for a system where $P(E) = \pi^{-1/2}\exp(-E^2)$, $\nu = 6$, and $\beta = R_{\min}^6/[(E_0-E)^2 + 0.01]$ where $R_{\min}$ is the nearest-neighbor distance on an fcc lattice. Particularly noticeable is the difference in the fluorescence patterns depending on whether the incident light excites atoms at the center or the wings of the line. Such effects arise from the interplay of $P(E)$ and the energy factor in the transfer rates.

3.2.5 Experimental Studies

In this section we discuss experimental studies of time resolved fluorescence line narrowing which have been carried out in $Pr_{0.2}La_{0.8}F_3$ [3.2] and PrF_3 [3.15].

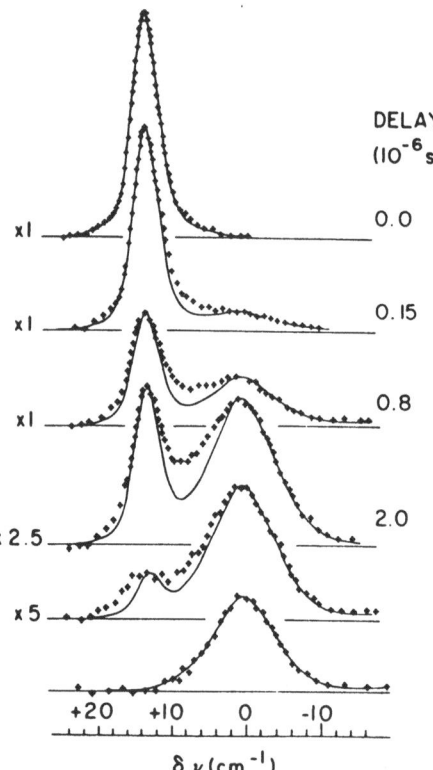

Fig. 3.7. Time resolved emission spectra for the $^3P_0 \rightarrow (^3H_6)_1$ fluorescence in $Pr_{0.2}La_{0.8}F_3$. The incident light is 12 cm^{-1} on the high-energy side of line center and the sample temperature is 14 K. The solid curve is a theoretical fit. [3.2]

The emphasis here is on the analysis of the decay curves using the approximate equations given in Sect. 3.2.2. We begin with the dilute system. In Fig. 3.7 we show the time evolution of the emission spectrum associated with the $^3P_0 \rightarrow$ $(^3H_6)_1$ transition of the Pr ion. The incident light is 12 cm^{-1} on the high side of the line center and the sample temperature is 14 K. The temperature dependence of the sharp line fluorescence and the presence of a background fluorescence which has the shape of the inhomogeneous line support the interpretation that the inter-atom transfer takes place via a process which is symmetric and independent of the energy mismatch [3.2, 13].

By fitting the fluorescence profile (the theoretical curves are shown as solid lines) one obtains $R(t)$. In order to extract the microscopic transfer rates from $R(t)$ one must know how the transfer rates vary with inter-atomic separation. One way of finding this out is to construct log-log plots of $R^{-1}(t)$ vs t. Assuming the donor-donor transfer rate is given by β/r^ν one obtains the appropriate value of ν from the limiting slope at long times. As shown in Fig. 3.8 ν is equal to 6.5 indicating dipole-dipole transfer is dominant in this system.

The second step in the analysis is to obtain a value of β by fitting the decay curves using (3.12-14). The results of such fits are shown in Fig. 3.9 where we compare the experimental data with theoretical curves which were calculated by evaluating the product in (3.12) over $16 \cdot 10^3$ lattice sites. The three theoretical curves correspond to $f(W_{0l}t) = 1$, $f(W_{0l}t) = \cosh(W_{0l}t)$, and $f(W_{0l}t) = 1 + 1/2$ $(W_{0l}t)^2$. It is evident that the best overall fit is obtained with the cosh function. The value of β corresponding to the solid curve is equivalent to a nearest-neighbor transfer rate of $0.37 \cdot 10^6$ s^{-1} (14 K). The fit to the data, while quite reasonable, is not exact at short times, as discussed in [3.2]. This may be caused by the neglect of short-range transfer processes involving quadrupolar and exchange interactions which can be relatively important in the early-time regime.

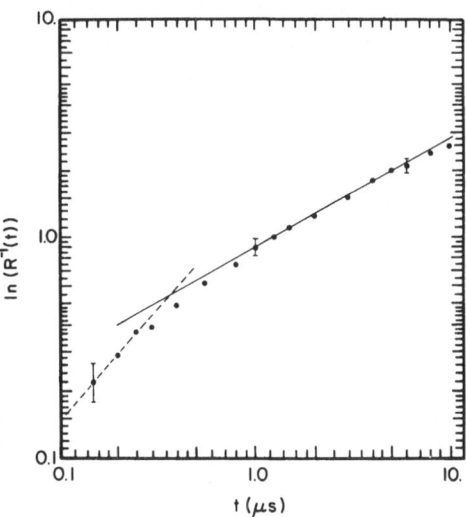

Fig. 3.8. Values of $R(t)$ extracted from the data shown in Fig. 3.7. $\ln[R(t)^{-1}]$ is plotted against t. The solid and broken lines indicate regions to $t^{3/6.5}$ and t behavior, respectively. [3.2]

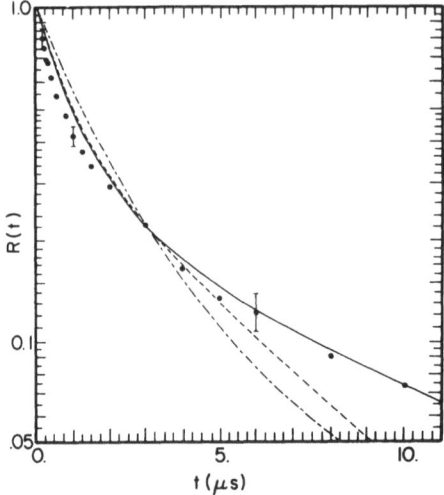

R(t)

t (μs)

Fig. 3.9. Comparison of the measured and calculated values of $R(t)$ in $Pr_{0.2}La_{0.8}F_3$ assuming an r^{-6} donor-donor transfer rate. $(-\cdot-\cdot-)\ f(W_{0l}t) = 1;\ (\text{——})\ f(W_{0l}t) = \cosh (W_{0l}t);\ (---)\ f(W_{0l}t) = 1 + 1/2(W_{0l}t)^2$. The theoretical curves have been normalized to agree with the data at $t = 3 \cdot 10^6$ s. [3.2]

A similar analysis has been carried out for the $^3P_0$ fluorescence in PrF_3 [3.15]. Since $c_D = 1$ in this system $R(t)$ is best approximated by taking $f(W_{0l}t) = 1 + 1/2\ (W_{0l}t)^2$. Over the time interval of observation $R(t)$ is approximately an exponential function in contrast to the nonexponential behavior observed in the dilute system (Fig. 3.9). This difference is a consequence of the fact that all Pr sites are equivalent and have twelve "nearest" neighbors. Each Pr ion transfers "out" at the same rate. Because of the large coordination number there is little backtransfer within the lifetime of the excited state. From (3.12) it is evident that in the region where $R(t)$ is exponential the decay rate is equal to $c_D \Sigma_l\ W_{0l}$. Using this expression with $c_D = 1$ and $W_{0l} = \beta/r_{0l}^6$ one obtains a nearest-neighbor transfer rate equal to $(0.4 \pm 0.05) \cdot 10^6\ s^{-1}$ at $T = 14$ K in agreement with the estimate obtained from the dilute system.

In addition to the Pr studies a similar analysis of the incoherent transfer in ruby has been carried out [3.16]. However the interpretation of the data in this system is complicated by a lack of detailed knowledge about the excited states involved in the transfer process. According to the interpretation given in [3.16] the incoherent transfer takes place between extended states which have finite amplitude on a number of chromium sites.

3.3 Fluorescence in the Presence of Traps, Exact Results

One of the oldest problems in the field of fluorescence involves the study of the influence of traps on the time development of the donor fluorescence following broadband excitation. In their simplest form the experiments involve the excitation of a small fraction of the donors by a source whose bandwidth is much greater than the inhomogeneous linewidth. After the source is turned off the intensity of the donor fluorescence, integrated over the line, is measured as a function of time. Provided the lifetime of the excited state is known information about donor-donor and donor-acceptor transfer can be obtained by fitting the decay curve to appropriate theoretical expressions.

In this section we analyze the decay of the donor fluorescence using the rate equations given in Sect. 3.1. The integrated intensity at time t is proportional to the number of excited donors $N_D(t)$. We write this number as

$$N_D(t) = N_D^0 \, \exp(-\gamma_R t) \, f(t), \tag{3.30}$$

where N_D^0 is the number of donors excited at the time the source is turned off ($t=0$) and $f(t)$ denotes the fraction of excited donors that would be present were the excited state to have an infinite radiative lifetime. It is assumed that $f(t)$ evolves in time according to (3.1) with $\gamma_R = 0$.

As discussed in a recent paper [3.17] exact calculations of $f(t)$ based on the rate equations are possible in two limiting cases: no donor-donor transfer and (infinitely) rapid donor-donor transfer. In the former limit we have

$$f(t) = \prod_l \, [1 - c_A + c_A \, \exp(-X_{0l}t)]. \tag{3.31}$$

In (3.31), which is a generalization to all values of c_A of a result first obtained by *Förster* [3.18], X_{0l} denotes the transfer rate from a donor at site 0 to an acceptor at site l. This equation has a simple physical interpretation. It is the average over all configurations of acceptors of $\exp(-X_n t)$, where X_n is the total donor-acceptor transfer rate associated with the nth donor. The quantity $1 - c_A$ is the probability that site l is unoccupied by an acceptor; if site l is occupied by an acceptor it contributes a factor $\exp(-X_{0l}t)$ to $\exp(-X_n t)$.

In the rapid transfer limit it is assumed that all donors are in contact with one another and that the donor-donor transfer takes place so quickly that for $t > 0$ all donors have an equal probability of being excited. In this case $f(t)$ has a simple form [3.17, 19]

$$f(t) = \exp(-c_A \sum_l X_{0l}t). \tag{3.32}$$

3.3.1 Early Models

The behavior of $f(t)$ in the regime intermediate between the diffusion and rapid transfer limits is extremely complicated. Typically, $f(t)$ is initially nonexponential but evolves into an exponential form as $t \to \infty$ (see Fig. 3.10). Traditional theories of $f(t)$ fall into two categories, a diffusion model originally developed by *De Gennes* [3.20] to explain the quenching of the nuclear magnetization by paramagnetic impurities, and a hopping model introduced by *Burshtein* [3.21].

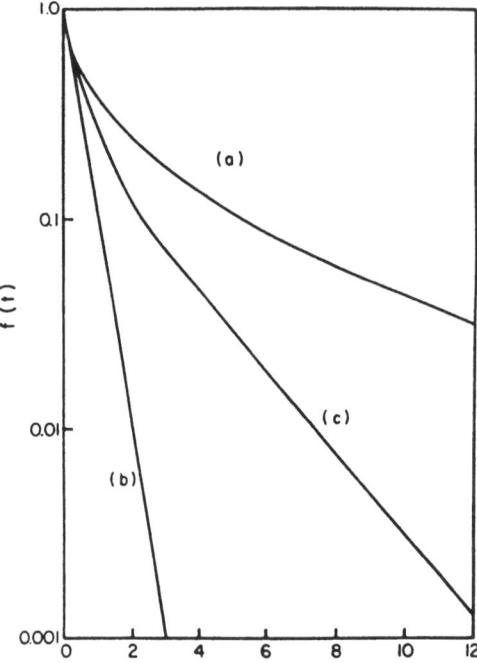

Fig. 3.10. Schematic semi-log plot of $f(t)$ vs t. Curve (a) is $f(t)$ in the absence of donor-donor transfer, (3.310); (b) is $f(t)$ in the rapid transfer limit, (3.32); (c) is appropriate to a situation intermediate between (a) and (b) where $f(t)$ approaches an exponential in the limit $t \rightarrow \infty$

In the diffusion model, which was first applied to fluorescent decay by *Yokota* and *Tanimoto* [3.22], the asymptotic behavior of the density of excited donors, n_D (r,t), obeys the diffusion equation

$$\partial n_D (r,t)/\partial t = D \nabla^2 n_D (r,t) - \sum_j X(r-r_j)n_D(r, t), \qquad (3.33)$$

where D is the diffusion constant associated with the donor-donor transfer and $X(r-r_j)$ is the transfer rate from a donor at r to an acceptor at r_j.

In the limit as $t \rightarrow \infty$ the diffusion model predicts that $f(t)$ varies as $\exp(-\Lambda_D t)$ where Λ_D is given by

$$\Lambda_D = 4\pi D n_A a_S, \qquad (3.34)$$

to lowest order in n_A, the number of acceptors per unit volume. The quantity a_S is the scattering length [3.23] obtained from a one-particle Schrödinger equation for a particle of "mass" $(2D)^{-1}$ in a repulsive potential $X(r)$ in units where $\hbar = 1$. For $X(r) = \alpha/r^\mu$ we have

$$a_S = (\alpha/D)^{1/(u-2)}(u - 2)^{-2/(u-2)}\Gamma[1 - 1/(u - 2)]\Gamma[1 + 1/(u - 2)]^{-1}, \qquad (3.35)$$

where $\Gamma(x)$ denotes the gamma function. The behavior of $f(t)$ at intermediate values of the time is also discussed in [3.22], where the authors developed an approximation for $f(t)$ which interpolates between the short and long time limits (cf. (3.65)).

In the hopping model $f(t)$ obeys the equation

$$f(t) = f_0(t) \exp(-t/\tau_0) + (1/\tau_0) \int_0^t \exp[-(t-t')/\tau_0]$$

$$\cdot f_0(t-t')f(t')dt',$$

(3.36)

in which $f_0(t)$ denotes $f(t)$ in the absence of donor-donor transfer (3.31) and τ_0 is phenomenological "hopping" time. In the limit $t \to \infty$ (3.36) has the solution $\exp(-\Lambda_H t)$ where the decay rate Λ_H is given by

$$\Lambda_H = c_A \sum_l X_{0l} (1 + X_{0l}\tau_0)^{-1},$$

(3.37)

to lowest order in c_A.

The summand $X_{0l}(1 + X_{0l}\tau_0)^{-1}$ in (3.37) has a simple physical interpretation as the renormalized donor-acceptor transfer rate. The renormalization factor $(1 + X_{0l}\tau_0)^{-1}$ arises from the competition between donor-donor and donor-acceptor transfer. When $X_{0l}\tau_0 \gg 1$ it takes a relatively long time for the excitation to hop to the vicinity of a trap so that the effective donor-acceptor transfer rate is very small. On the other hand when $X_{0l}\tau_0 \ll 1$ the excitation quickly reaches the trap. As a consequence the transfer takes place at the unrenormalized rate, X_{0l}. When $X_{0l}\tau_0 \ll 1$ for all l $f(t)$ approaches the exact result appropriate to the rapid transfer limit, (3.32).

3.3.2 Average T-Matrix Approximation

It is apparent that both the diffusion and the hopping models have serious limitations. The recognition of these limitations has stimulated the development of a more fundamental theory of the transfer which is based on the microscopic rate equations [3.17]. The analysis in [3.17] pertains to the Laplace transform of $f(t)$ which we denote by $\hat{f}(s)$, viz.,

$$\hat{f}(s) = \int_0^\infty dt \ \exp(-st) f(t).$$

(3.38)

In the limit of low trap concentration, $c_A \ll 1$, $\hat{f}(s)$ is given by

$$\hat{f}(s) = [s + c_A \sum_{ll'} t_{ll'}(s)]^{-1},$$

(3.39a)

where the t matrix obeys the equation

$$t_{ll'}(s) = X_{0l}\delta_{ll'} - \sum_{l''} X_{0l}g_{ll''}(s) \ t_{l''l'}(s).$$

(3.40a)

When the donor array has translational symmetry ($c_D = 1$) the function $g_{ll'}(s)$ has the form

$$g_{ll'}(s) = 1/N \sum_k \exp[ik \cdot (r_l - r_{l'})]/(s + \sum_{n'} W_{nn'} \times \{1 - \cos[k \cdot (r_n - r_{n'})]\}).$$

(3.41)

In the event that the donor array is dilute or otherwise disordered $g_{ll'}(s)$ is replaced by its configurational average. Note that $t_{ll'}(s)$ is nonzero only when X_{0l} and $X_{0l'}$ both differ from zero. Also, if we define $T_l(s)$ as being equal to $\sum_{l'} t_{ll'}(s)$ then we have

$$\hat{f}(s) = [s + \sum_l T_l(s)]^{-1},$$

(3.39b)

where T_l satisfies the equation

$$T_l(s) = X_{0l} - \sum_{l'} X_{0l} g_{ll'}(s) T_{l'}(s).$$

(3.40b)

As pointed out in [3.17], (3.39a) reproduces the results of the two phenomenological models discussed in Sect. 3.3. In order to obtain an approximation equivalent to the hopping model we neglect the off-diagonal elements of the t matrix thus obtaining the equation

$$\hat{f}(s) = \{s + c_A \sum_l X_{0l}/[1 + X_{0l} g_0(s)]\}^{-1},$$

(3.42)

where $g_0(s) \equiv g_{ll}(s)$. We obtain agreement with the equations of the hopping model by approximating $g_0(s)$ as

$$g_0(s) = (s + 1/\tau_0)^{-1},$$

(3.43)

and identifying τ_0 with $g_0(0)$.

The function $g_0(s)$ which appears in (3.42, 43) is closely connected with the intensity of the narrow line fluorescence observed in time resolved fluorescence line narrowing experiments. In particular, $g_0(s)$ is the Laplace transform of $R(t)$ in the limit $c_A \rightarrow 0$

$$g_0(s) = \int_0^\infty dt\, e^{-st} R(t).$$

(3.44)

When the donor array is dilute we can obtain a simple approximation for τ_0 by approximating $R(t)$ as indicated in (3.16). With $W_{nn'} = \beta/r_{nn'}^\nu$, we have

$$g_0(0) = \tau_0 = \Gamma(\nu/3 + 1)\{\beta n_D^{\nu/3}[(4\pi/3)\Gamma(1 - 3/\nu)2^{3/\nu - 1}]^{\nu/3}\}^{-1}.$$

(3.45)

As discussed in [3.17] results equivalent to the diffusion model are obtained when the summations in (3.39a, 40a) are replaced by integrations. In such a transcription, $g_{ll'}(s)$ becomes a continuous function of position, $g(r,r';s)$, defined by

$$g(r,r';s) = (2\pi)^{-3} \int dk \, \exp[ik \cdot (r-r')] \, (s + Dk^2)^{-1}. \tag{3.46}$$

When the donor array has translational symmetry the diffusion constant is inferred from the denominator of (3.41)

$$D = 1/6 \sum_{n'} W_{nn'} (r_n - r_{n'})^2, \tag{3.47}$$

assuming cubic symmetry.

The calculation of the diffusion constant of a disordered array of donors is a formidable problem. However in situations where $c_D \ll 1$ we have $D \propto \beta n_D^{(v-2)/3}$ since D has the units $(\text{length})^2/\text{time}$ and the length scale is set by the average separation between donors, the lattice parameter being of no importance. Under these circumstances the result obtained by generalizing the equation of *Trlifaj* [3.24] is a reasonable first approximation

$$D = [2(v-5)]^{-1} (4\pi/3)^{(v-2)/3} \beta n_D^{(v-2)/3}. \tag{3.48}$$

Recently *Godzik* and *Jortner* [3.10] and *Gochanour* et al. [3.11] have calculated D for the case $v = 6$. Using diagrammatic techniques they obtained results which are smaller than that given in (3.48) by a factor ≈ 0.9.

Since neglecting the off-diagonal elements of the t matrix is equivalent to the assumption that there is no communication between different donors in the neighborhood of a trap it is apparent that the hopping model is appropriate when the number of donors in the sphere of influence of an acceptor is small. In contrast the diffusion model requires that there be a large number of donors in the sphere of influence. The determination of the number of donors in the sphere of influence is straightforward when $W_{nn'}$ has a finite range. However it is somewhat more complicated in the case where both $W_{nn'}$ and $X_{nn'}$ vary as the inverse power of the separation, i.e., $W_{nn'} = \beta/r_{nn'}^v$, $X_{nn'} = \alpha/r_{nn'}^\mu$. In this situation we find that the hopping model is appropriate when [3.17]

$$n_D^{(1-v/u)} (\alpha/\beta)^{3/u} \lesssim 1, \tag{3.49}$$

while the diffusion model is to be used when

$$n_D^{(1-v/u)} (\alpha/\beta)^{3/u} \gg 1. \tag{3.50}$$

Note that in the event $u=v$ we have $\alpha/\beta \gg 1$ for diffusion and $\alpha/\beta \lesssim 1$ for hopping [3.21].

It must be emphasized that the equations for $\hat{f}(s)$ given in this section apply only to systems where the number of traps is small in comparison with the number of donors. High concentration effects are much more difficult to analyze. However progress has been made in the case $c_D, c_A \ll 1$ where both donor-donor and donor-acceptor transfer rates vary as r^{-6} with $\alpha \lesssim \beta$ (so that the off-diagonal elements of the t matrix can be neglected) [3.25]. The calculations of [3.25] exploit the similarity between the rate equations and the equations of motion for the electron operators in the tight binding model of an alloy with diagonal disorder. Use is made of the coherent potential approximation or CPA [3.26] to calculate $\hat{f}(s)$. The equations have the form

$$\hat{f}(s) = [s + X_{CPA}(s)]^{-1}, \tag{3.51}$$

where $X_{CPA}(s)$ is the solution to a nonlinear self-consistent equation.

From (3.51) we infer the asymptotic behavior

$$f(t) \sim \exp[-X_{CPA}(0)\,t]. \tag{3.52}$$

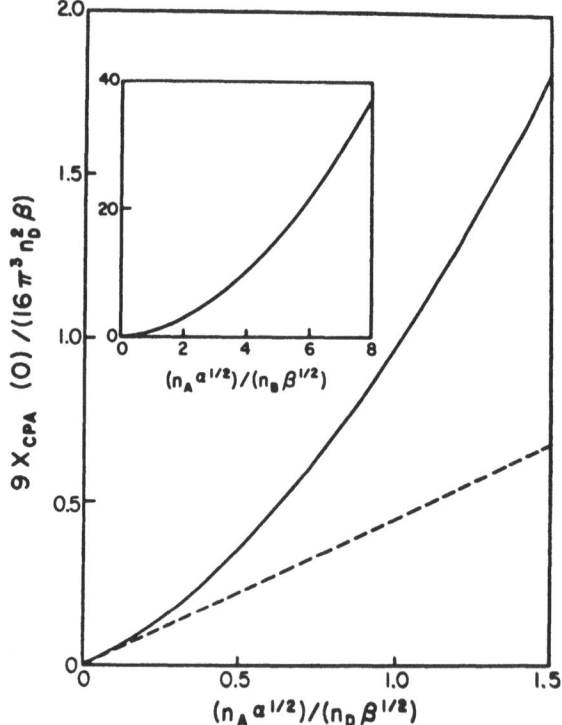

Fig. 3.11. $X_{CPA}(0)/(16\pi^3 n_D^2 \beta/9)$ vs $(n_A \alpha^{1/2})/(n_D \beta^{1/2})$. The broken line is the linear approximation, (3.53). The inset shows the behavior for larger values of $(n_A \alpha^{1/2})/(n_D \beta^{1/2})$. [3.25]

In the limit of low acceptor concentration $X_{CPA}(0)$ is given by

$$X_{CPA}(0) = (4/9)\,\pi^{7/2} n_D n_A\,(\alpha\beta)^{1/2}, \tag{3.53a}$$

in agreement with the average t-matrix approximation with $g_0(0)$ given by (3.45).

The behavior of $X_{CPA}(0)$ as a function of the ratio $(n_A \alpha^{1/2})/(n_D \beta^{1/2})$ is shown in Fig. 3.11. From the figure it is evident that the linear approximation fails for $(n_A \alpha^{1/2})/(n_D \beta^{1/2}) \gtrsim 0.5$.

It should be mentioned that the result (3.53a) is obtained when the sum in (3.42) is approximated by an integral over the range $0 < r < \infty$. If we cut the integral off at the lower limit r_c, where r_c is chosen such that $X_{CPA}(0)$ reproduces the rapid transfer result, (3.32), we obtain

$$X_{CPA}(0) = (4/9)\pi^{7/2} n_D n_A (\alpha\beta)^{1/2} [1-(2/\pi)\tan^{-1} (r_c^6/\alpha\tau_0)^{1/2}], \tag{3.53b}$$

where τ_0 is given by (3.45) with $\nu = 6$.

3.3.3 Experimental Studies

It is beyond the scope of this chapter to survey the vast literature pertaining to experimental studies of fluorescence in the presence of traps. However we will discuss three papers in which a careful analysis of the data was carried out using equations similar to those given in Sects. 3.3.1 and 2. The first of these is by Voronko et al. [3.27] who studied the time dependence of the fluorescence from the $^4F_{3/2}$ state in $Nd_x La_{1-x}F_3$ over the range $0.003 \leqslant x \leqslant 1$. In this system the energy transfer takes place via the dipole-dipole interactions. The Nd ions also act as traps through the mechanism of cross relaxation. By carrying out measurements at different concentrations and temperatures the authors were able to identify various features of the decay curves.

At high concentrations, $x > 0.1$, the authors saw ultrarapid energy transfer, which they referred to as "supermigration." In this situation, which corresponds to (3.32), the decay was exponential for all values of the time. For $x < 0.1$ the decay curves had features which were associated with three distinct time regimes. For $0 < t < t_1$ the decay was exponential in time

$$f(t) = \exp(-At), \tag{3.54}$$

for $t_1 < t < t_2$ $f(t)$ varied as

$$f(t) = \exp(-Bt^{1/2}), \tag{3.55}$$

and for $t > t_2$ $f(t)$ again varied exponentially

$$f(t) = \exp(-Ct). \tag{3.56}$$

These results have a simple physical interpretation. Equations (3.54, 55) are the short and long time limits of the *Golubov-Konobeer* expression [3.8]

$$f(t) = \prod_l [1 - c_A + c_A \exp(-X_{0l} t)] \tag{3.57}$$

which describes transfer to traps in the absence of donor-donor transfer. Equation (3.56) corresponds to the asymptotic behavior discussed in Sects. 3.3.1 and 2.

An estimate of the "critical concentration" at which there is a crossover into supermigration is obtained by equating the average donor-donor transfer rate $\beta/\bar{r}^6$ to the maximum donor-acceptor transfer rate α/r_{min}^6, $\bar{r}$ and r_{min} being the average donor-donor separation and minimum donor-acceptor separation, respectively. We obtain

$$\alpha/r_{min}^6 = \beta/\bar{r}^6. \tag{3.58}$$

Since the Nd ions act as both donors and traps we have $\bar{r} \simeq (c_{Nd})^{-1/3} r_{min}$ where c_{Nd} is the fraction of lanthanum sites occupied by neodymium ions. Using this relation in (3.58) we obtain for the critical concentration

$$c_{Nd}^* = (\alpha/\beta)^{1/2}. \tag{3.59}$$

Experimentally it was found that $c_{Nd}^* \approx 0.1$ so that from (3.59) we have $\alpha/\beta \approx 0.01$, a result which justifies the use of the hopping model in the analysis of the data.

The boundaries between the three regimes characterized by (3.54-56) are diffuse. However we can obtain a rough estimate of t_1 from the equation

$$c_A \sum_l X_{0l} t_1 \approx 1, \tag{3.60}$$

which is equivalent to identifying t_1^{-1} with the average donor-acceptor transfer rate. An equally rough estimate of t_2 is obtained by equating the arguments of the exponentials in (3.55, 56) which leads to

$$t_2 = (B/C)^2. \tag{3.61}$$

Using (3.53a) for C and the Förster value, $(4\pi^{3/2}/3)n_{Nd}\alpha^{1/2}$, for B we have

$$t_2 = (9^{-1}\pi^4 n_{Nd}^2 \beta)^{-1},$$ (3.62)

where n_{Nd} is the neodymium concentration.

There are two additional comments to be made in conjunction with this analysis. As noted, the neodymium ions served both as acceptors and donors so that the acceptor concentration was equal to the donor concentration. Although it would appear that an analysis based on (3.37, 53), which are linear in the acceptor concentration, would be invalid, this is not the case. In the discussion at the end of Sect. 3.3.3, it was pointed out that the CPA calculation of [3.25] shows that the linear theory is valid as long as $(n_A\alpha^{1/2})/(n_D\beta^{1/2}) < 0.5$. Since in the Nd system $\alpha/\beta \approx 0.01$ the linear theory is appropriate even when $n_A = n_D$.

The second point concerns the interpretation given to the donor-donor transfer. By fitting the decay curves the authors of [3.27] were able to obtain estimates of α and β in terms of the Förster-Dexter theory of energy transfer. On the basis of the discussion given in Chap. 2 it would appear that such an analysis is questionable and that a reinterpretation of the data in terms of phonon-assisted transfer rates may be required.

The analysis of *Voronko* et al. was carried out for a system where the applicability of the hopping model (or the neglect of the off-diagonal elements of the t matrix) was justified. *Weber* [3.28] has reported an analysis of the fluorescence in Cr^{3+} doped europium phosphate glass which illustrates the analogous behavior in the opposite limit where the diffusion model is appropriate. In this system the Eu^{3+} ions are the donors and the Cr^{3+} ions act as traps. As in [3.27] the distinction was made between the three limiting cases shown in Fig. 3.10. An extensive study of the asymptotic behavior in the regime between the zero and rapid donor-donor transfer limits was carried out. The data were interpreted using the diffusion model and assuming dipole-dipole transfer.

In the diffusion model the rate governing the exponential decay of $f(t)$ is given by (3.34) with

$$a_S = 0.68\,(\alpha/D)^{1/4}.$$ (3.63)

As a consequence the asymptotic slope of $\ln f(t)^{-1}$ has the value

$$\Lambda_D = 8.55\,n_A\alpha^{1/4}D^{3/4}.$$ (3.64)

An estimate of α was obtained by fitting the Eu^{3+} decay curve to the approximate formula of *Yokota* and *Tanimoto* [3.22]

$$f(t) = \exp\{-(4\pi^{3/2}/3)\, n_A\, (\alpha t)^{1/2}[(1 + 10.87x + 15.50x^2)$$

$$\div (1 + 8.743x)]^{3/4}\},\qquad (3.65)$$

with $x = D\alpha^{-1/3}t^{2/3}$. The values of $\alpha = 8 \cdot 10^{-38}$ cm^{6}s^{-1} and $D = 2.5 \cdot 10^{-10}$ cm^{2}s^{-1} were obtained at a temperature of 77 K. A brief analysis of the temperature dependence of the diffusion constant was also carried out. The variation of D with temperature was calculated assuming a simple model for the transport process, and it was shown that the data could be fit by taking $\Lambda_D \propto D(T)^{3/4}$ as predicted by (3.64).

It should be noted that in the diffusion limit, just as in the hopping limit, there are times t_1 and t_2 marking the boundaries between the early exponential and nonexponential regimes and between the nonexponential and asymptotic exponential regimes. The parameter t_1 is given by (3.60). In the case of t_2 we can obtain an estimate by equating $f(t)$ as given by the Förster theory $\{f(t) = \exp[-(4\pi^{3/2}/3)n_A\,(\alpha t)^{1/2}]\}$ to $\exp(-\Lambda_D t)$. We find

$$t_2 = 0.75\,\alpha^{1/2}D^{-3/2}. \qquad (3.66)$$

Finally, we note that according to (3.59) when $\alpha > \beta$ the critical concentration is greater than one. This result we interpret as indicating that the rapid transfer limit cannot be reached when $\alpha \gg \beta$.

Recently *Hegarty* et al. [3.29] have studied the decay of the integrated intensity in $Pr_{0.95}Nd_{0.05}$. At different temperatures they found behavior resembling the three curves in Fig. 3.10. In this sytem the Nd ions act as traps for the $^3P_0$ excitation in the Pr array. At low temperatures the donor-donor transfer is so slow that over the interval of the measurement $f(t)$ has the form $\exp[-Bt^{1/2}]$ appropriate to a situation where there is dipole transfer between donors and acceptors and negligible donor-donor transfer [3.18]. As the temperature increases the donor-acceptor rate remains approximately constant ($\alpha = 5.46 \cdot 10^{-38}$ cm^{6}s^{-1}) while the donor-donor rate increases rapidly. When this happens the asymptotic intensity begins to vary as $\exp[-Ct]$. The variation of the limiting slope C with $K = \Sigma_l W_{0l}/\Sigma_l X_{0l}$, the ratio of the strength of the donor-donor transfer to the donor-acceptor transfer, is shown in Fig. 3.12. For $K < 0.1$ C is proportional to $K^{3/4}$ while for $K > 1$ C approaches a constant value. These results are consistent with the analysis in Sects. 3.3 and 3.3.1. When K is small the diffusion model is appropriate and C is given by (3.34, 35). We then have $C \propto D^{3/4} \propto K^{3/4}$, the last step following from (3.47). Beyond $K = 1$ C approaches the rapid transfer limit (3.32).

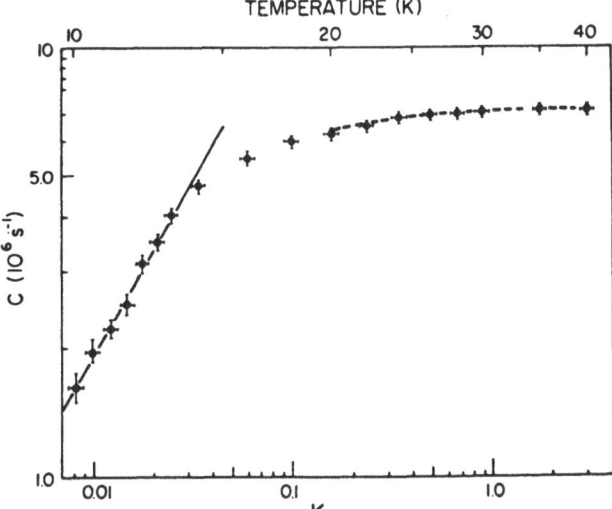

Fig. 3.12. Limiting slope characterizing the asymptotic decay of the $^3P_0$ fluorescence in $Pr_{0.95}Nd_{0.05}F_3$. The data are plotted against $K = \Sigma_l W_{0l}/\Sigma_l X_{0l}$, the ratio of the donor-donor to donor-acceptor transfer rates, and against temperature. The solid line corresponds to $C \propto K^{3/4}$. The broken line indicates the approach to the rapid transfer limit and is obtained by fitting to (3.42). [3.29]

3.4 Summary and Outlook

In this chapter we have summarized the results of recent theoretical studies of incoherent transfer in optically active materials. The theory provides a bridge between experimental studies of the time development of the fluorescence and the microscopic rate equations. By fitting the data to theoretical decay curves one can obtain estimates of the interatomic transfer rates.

We have given brief analyses of time-resolved fluorescence line narrowing and the time evolution of the integrated fluorescence in the presence of traps. From the point of view of experimental difficulty measurements of the time evolution of the integrated intensity are the simpler of the two measurements. However the discussion of Sect. 3.2 shows that the analysis of the decay curves in the asymptotic regime is made complicated by the presence of both donor-donor and donor-acceptor transfer rates. In order to obtain unique values for both of these rates it is necessary also to have data from the nonexponential regime where $f(t)$ can be approximated by (3.57).

Alternatively one can measure the intensity in a fluorescence line narrowing experiment. By fitting the decay curves to (3.12), or its generalization (3.17), one obtains estimates of the donor-donor transfer rates. These can then be used in the analysis of the time dependence of the integrated intensity.

It is apparent that we are on the threshold of a new era in the field of luminescence in condensed matter. With the availability of tunable, narrow-band sources and sensitive detection equipment on one hand and the development of microscopic theories of energy transfer and luminescence on the other, we can expect rapid progress in this area in the near future.

Appendix 3.A

The purpose of this appendix is to summarize some of the results that have been obtained recently in the theory of incoherent transfer in one-dimensional systems with nearest-neighbor transfer. We begin with fluorescence line narrowing. The numerical evaluation of the function $R(t)$ using (3.4) poses no problem in one dimension even when the transfer rates are random functions of position along the chain. The distribution of the eigenvalues of the relaxation matrix can be calculated for chains of 100,000 atoms with little difficulty using Sturm-sequence techniques [3.30].

The analytic behavior of $R(t)$ in disordered one-dimensional systems has also been investigated [3.31-33]. *Bernasconi* et al. [3.31] studied how the long-time variation of $R(t)$ depended on the behavior of the distribution of (nearest-neighbor) transfer rates $j(W)$. They found that when $j(W)$ is such that $\bar{W}^{-1} = \int dW\, W^{-1} j(W)$ exists then

$$R(t) \approx t^{-1/2}, \quad t \to \infty, \tag{3.A1}$$

i.e., the behavior is diffusive. If $j(0)$ is finite then we have

$$R(t) \sim (\ln t/t)^{1/2}, \quad t \to \infty, \tag{3.A2}$$

and if $j(W)$ diverges as $W^{-\alpha}$, $0 < \alpha < 1$, then $R(t)$ varies as

$$R(t) \sim t^{-(1-\alpha)/(2-\alpha)}, \quad t \to \infty. \tag{3.A3}$$

The behavior of $R(t)$ in systems where a fraction $1-p$ of the transfer rates are zero has also been studied [3.32, 33]. Since the absence of transfer between pairs of atoms breaks the chain up into finite segments $R(t)$ approaches the finite value $1-p$ as $t \to \infty$ [3.33]. As discussed in [3.33] the approach to the asymptotic limit is a complicated mathematical problem with analytic results being available only in the weak ($p \ll 1$) [3.32] and strong ($1-p \ll 1$) [3.33] disorder limits.

Appendix 3.B

In this appendix we discuss the behavior of $f(t)$ when allowance is made for back-transfer from the traps. We consider the situation $c_A \ll 1$ where the average t-matrix approximation is applicable and $\hat{f}(s)$ is given by (3.39). In this case we can treat the interaction of an array of donors located at sites l with a single acceptor at site 0. The relevant rate equations are

$$dP_l(t)/dt = -(\sum_{l'} W_{ll'} + X_{0l})P_l(t) + \sum_{l'} W_{ll'}P_{l'}(t) + X^B_{0l}P_A(t), \tag{3.B1}$$

and

$$dP_A(t)/dt = -(\overline{\gamma}_A + \sum_l X^B_{0l})\, P_A(t) + \sum_l X_{0l}P_l(t). \tag{3.B2}$$

Here $P_A(t)$ denotes the probability that the trap is filled at time t, where $\overline{\gamma}_A = \gamma_A - \gamma_R$ is the "radiative lifetime" of the trap and X^B_{0l} is the rate of transfer from the trap to the donor at site l. The latter is related to the donor-acceptor transfer rate, X_{0l}, by the detailed balance equation (3.2).

Combining the Laplace transforms of (3.B1) and (3.B2) we obtain the result

$$\begin{aligned}
s\,\hat{P}_l(s) - P_l(0) = -(\sum_{l'} W_{ll'} + X_{0l})\hat{P}_l(s) + \sum_l W_{ll'}\hat{P}_{l'}(s) \\
+ X^B_{0l} \sum_{l'} X_{0l'}(s + \overline{\gamma}_A + X^B_T)^{-1}\hat{P}_{l'}(s),
\end{aligned} \tag{3.B3}$$

where $X^B_T = \sum_l X^B_{0l}$ is the total backtransfer rate. From (3.B3) we infer that the T matrix for the problem obeys the equation

$$t_{ll'}(s) = V_{ll'}(s) - \sum_{l'',l'''} V_{ll''}(s)\, g_{l'',l'''}(s)\, t_{l'''l'}(s), \tag{3.B4}$$

in which the nonlocal interaction $V_{ll'}(s)$ is given by

$$V_{ll'}(s) = X_{0l}\delta_{0l'} - X^B_{0l}\, X_{0l'}\, (s + \overline{\gamma}_A + X^B_T)^{-1}. \tag{3.B5}$$

We consider two limits. If the acceptor interacts with a single donor located at l we have

$$t^{(s)}_{ll} = \frac{X^{\mathrm{eff}}_{0l}(s)}{1 + X^{\mathrm{eff}}_{0l}(s)\, g_0(s)}, \tag{3.B6}$$

where the effective donor-acceptor transfer rate is given by

$$X^{\mathrm{eff}}_{0l}(s) = X_{0l}\frac{(s + \overline{\gamma}_A)}{(s + \overline{\gamma}_A + X^B_{0l})}. \tag{3.B7}$$

In the rapid transfer transfer limit $t_{ll'}(s) \approx V_{ll'}(s)$. As a consequence we find

$$\hat{f}(s) = [s + c_A\, X_T(s + \bar{\gamma}_A)/(s + \bar{\gamma}_A + X_T^B)]^{-1}, \tag{3.B8}$$

where $X_T = \Sigma_l\, X_{0l}$. From (3.B8) it is evident that $f(t)$ is the sum of two exponentials with decay rates

$$\tfrac{1}{2}(\bar{\gamma}_A + X_T^B + c_A\, X_T) \pm \tfrac{1}{2}[(\bar{\gamma}_A + X_T^B + c_A\, X_T)^2 - 4c_A\, X_T\bar{\gamma}_A]^{1/2},$$

in contrast to a situation without backtransfer where $f(t) = \exp(-c_A X_T t)$.

References

3.1 W.M. Yen: J. Lumin. 18/19, 639 (1979) (and references therein)
3.2 D.L. Huber, D.S. Hamilton, B. Barnett: Phys. Rev. B16, 4642 (1977)
3.3 C. Kittel: *Introduction to Solid State Physics*, 4th ed., (John Wiley, New York 1971) Chap. 2
3.4 R. Alben, M.F. Thorpe: J. Phys. C8, L275 (1975); M.F. Thorpe, R. Alben: J. Phys. C9, 2555 (1976). The connection between the spin wave problem considered here and in [3.5] and the rate equations is discussed in the appendix to [3.2]
3.5 R. Alben, S. Kirkpatrick, D. Beeman: Phys. Rev. B15, 346 (1977)
3.6 W.Y. Ching, D.L. Huber, B. Barnett: Phys. Rev. B17, 5025 (1978) also M. Fibich, D.L. Huber: Phys. Rev. B20, 5369 (1979)
3.7 S.K. Lyo, T. Holstein, R. Orbach: Phys. Rev. B18, 1637 (1978)
3.8 S. I. Golubov, Y. V. Konobeer: Fiz. Tverd. Tela (Leningrad) 13, 3185 (1971) [Sov. Phys.-Solid State 13, 2679 (1972)]
3.9 S. Alexander: Unpublished (cited in [3.7])
3.10 K. Godzik, J. Jortner: Chem. Phys. 38, 227 (1979), Chem. Phys. Lett. 63, 429 (1979). See also S.K. Lyo: Phys. Rev. B20, 1297 (1979); A. Blumen, J. Klafter, R. Silbey: J. Chem. Phys. 72, 5320 (1980)
3.11 C.R. Gochanour, H.C. Anderson, M.D. Fayer: J. Chem. Phys. 70, 4254 (1979). The connection between the diagrammatic approach followed in this reference and other methods of calculating R(t) involving continuous random walk arguments has been discussed by J. Klafter and R. Silbey: Phys. Rev. Lett. 44, 55 (1980)
3.12 W.Y. Ching, D.L. Huber: Phys. Rev. B18, 5320 (1978)
3.13 T. Holstein, S.K. Lyo, R. Orbach: Phys. Rev. Lett. 36, 891 (1976)
3.14 N. Montegi, S. Shionoya: J. Lumin. 8, 1 (1973)
3.15 D.S. Hamilton, P.M. Selzer, W.M. Yen: Phys. Rev. B16, 1858 (1977)
3.16 P.M. Selzer, D.L. Huber, B.B. Barnett, W.M. Yen: Phys. Rev. B17, 4979 (1978)
3.17 D.L. Huber: Phys. Rev. B20, 2307 (1979)
3.18 M. Th. Förster: Z. Naturforsch. 4a, 321 (1949). See also M. Inokuti, F. Hirayama: J. Chem. Phys. 43, 1978 (1965)
3.19 D. Fay, G. Huber, W. Lenth: Opt. Commun. 28, 117 (1979)
3.20 P.G. de Gennes: J. Phys. Chem. Sol. 7, 345 (1958)
3.21 A.I. Burshtein: Zh. Eksp. Teor. Fiz. 62, 1695 (1972). [Sov. Phys. - JETP 35, 882 (1972)]. See also L.D. Zusman: Zh. Eksp. Teor. Fiz. 73, 662 (1977). [Sov. Phys. - JETP 46, 347 (1977)]
3.22 M. Yokota, O, Tanimoto: J. Phys. Soc. (Jpn.) 22, 779 (1967)

3.23 L.D. Landau, E.M. Lifshitz: *Quantum Mechanics: Non-Relativistic Theory* (Addison-Wesley, Reading 1958) pp. 403-407

3.24 M. Trlifaj: Czech. J. Phys. 8, 510 (1958)

3.25 D.L. Huber: Phys. Rev. **B20**, 5333 (1979)

3.26 R.J. Elliott, J.A. Krumhansl, P.L. Leath: Rev. Mod. Phys. **46**, 465 (1974)

3.27 Y.K. Voronko, T.G. Mamedov, V.V. Osiko, A.M. Prokhorov, V.P. Sakun, I.A. Shcherbakov: Zh. Eksp. Teor. Fiz. **71**, 478 (1976). [Sov. Phys.-JETP **44**, 251 (1976)]

3.28 M.J. Weber: Phys. Rev. **B4**, 2932 (1971)

3.29 J. Hegarty, D. L. Huber, W. M. Yen: Phys. Rev. B **23**, 6271 (1981)

3.30 D.L. Huber: Phys. Rev. **B8**, 2124 (1973) and references therein. The connection between the spin wave calculation in the reference and the rate equations is established in the appendix to [3.2]

3.31 J. Bernasconi, S. Alexander, R. Orbach: Phys. Rev. Lett. **41**, 185 (1978)

3.32 S. Alexander, J. Bernasconi, R. Orbach: Phys. Rev. **B17**, 4311 (1978)

3.33 J. Heinrichs, N. Kumar: Phys. Rev. **B20**, 1377 (1979)

4. General Techniques and Experimental Methods in Laser Spectroscopy of Solids

P. M. Selzer

With 7 Figures

In the previous chapters the reader has been exposed to the motivation for studying the dynamics of impurity ions and molecules in solids and to the theoretical problems related to these studies. In summary, it may be stated that two types of information are generally sought – the relaxation mechanisms of an individual optical center due to its interactions with the host, and the propagation of the optical excitation within the solid due to the mutual interaction between different centers. Typically, the former process gives rise to homogeneous line broadening and the latter to temporal evolution of the individual fluorescence lines.

In this chapter, different, recently developed experimental methods which allow observation of the various dynamical processes are discussed. Most emphasis is devoted to the technique of fluorescence line narrowing since this is a very versatile method for observing both relaxation and energy migration; it is also the methodology with which the author has greatest familiarity. However, much of what is said in this discussion is equally applicable to other techniques. Later sections of the chapter are devoted to several additional spectroscopic methods in both the frequency and time domain. Most of these techniques have only become generally viable with the advent of high-resolution, tunable dye lasers, i.e., since the early 1970's.

Because the homogeneous linewidth and the relaxation (or dephasing) time of an optical center are related by the expression $\Delta\nu = (2\pi\tau)^{-1}$,[1] one may presumably extract complementary information by working in either the frequency or the time domain. While direct measurement of dephasing times has been recently demonstrated as practicable in solids [4.1], when observing sub-nanosecond relaxation phenomena it is often more convenient to work in the frequency domain, where the limitations on instrumental spectral resolution start to become favorable compared to their temporal response. In the other extreme, working in the frequency domain is the only way to observe dynamics associated with spectral diffusion which may be much slower than the dephasing process. Unfortunately, in many cases the homogeneous linewidth is many orders of magnitude smaller than the strain (inhomogeneous) broadening associated with the system

---

[1] Note that τ as used here is a generalized lifetime describing the exponential decay of the probability for being in a given state. If the relaxation of the system is due to dephasing, the dephasing time, T_2, is related to τ by: $\tau = T_2/2$. This arises because T_2, by definition, refers to an amplitude, not a probability, decay.

under investigation. This inhomogeneous broadening can range from a small fraction of a wave number for the best crystals to many hundreds of wave numbers for amorphous hosts. The problem, then, is often one of probing deep within a large inhomogeneous envelope.

Extracting relaxation rates and observing spectral diffusion are not the only reasons for overcoming inhomogeneous broadening in solids. The hyperfine structure of various transitions is also frequently hidden within the strain-broadened linewidth. Thus, if we could eliminate strain broadening, we could determine both the hyperfine structure and the dynamical processes connecting these levels.

Faced with inhomogeneous broadening in gases (Doppler broadening), atomic and molecular spectroscopists have developed some clever methods of extracting the homogeneous linewidth [4.2]. Strain broadening in solids is somewhat analogous to Doppler broadening in gases, and several of these techniques have subsequently been applied to solid-state systems. For years atomic beams have been used to reduce the Doppler linewidth, however, there is clearly no analogous technique for solids. Since the early 70's, a number of other schemes have been devised for gaseous spectroscopy, mostly involving nonlinear effects. Saturation or "hole burning" in optical transitions, first applied inside a laser cavity [4.3] and later outside [4.4-6], is now a common method of eliminating the Doppler component. More recently, various refinements and extensions of this technique have been developed, such as laser polarization spectroscopy [4.7] and double modulation [4.8]. Most of these methods require a strong, high-resolution beam which slightly depopulates the ground state of a particular velocity group, v_g, of atoms or molecules. A weak, counter-propagating probe beam of the same optical frequency then samples the ground-state population of the velocity group $-v_g$. (The sign is opposite due to the reversed direction of the probe beam.) In the highest resolution experiments, pump and probe have been provided by a cw tunable dye laser [4.9]. As the laser is tuned across the line, the two beams interact with the same subset of atoms with $v_g = 0$ near line center, and at this point, the probe beam experiences a decrease in absorption due to the bleaching action of the pump. In this way, the width of the hole can be scanned using only one tunable laser. As will be seen in Sec. 4.2.1, the analogous technique in solids requires two tunable sources in general — one to burn a hole at the chosen frequency, the other to interrogate the hole width.

Two-photon spectroscopy [4.10] is yet another method of extracting the homogeneous width of a gaseous transition. Once again, counter-propagating beams are used, only with this technique, atoms in *any* velocity group can simultaneously absorb two *counter-propagating* photons, each at half the resonant frequency. In this way the first-order Doppler effect is completely eliminated as the laser is scanned across the absorption line. It should be apparent that inhomogeneous broadening in solids will not be overcome with this method: counter-propagation accomplishes nothing since the broadening depends on static strains and as the laser is scanned, it interacts with only one subset of optical centers at a time.

The least common method of homogeneous linewidth extraction in gases has been fluorescence line narrowing [4.11]. Here as with hole burning, only a portion of the ground-state population is excited — that particular velocity component which interacts with a high-resolution laser pump beam. Instead of scanning the width of the hole, however, the fluorescent emission from the excited level is detected. Because detection of fluorescence is far more sensitive than absorption, a much weaker pump source may be used, making this essentially a linear technique.

Each of the above-mentioned methods has relative merits, depending on the particular system being studied and the experimental tools available. Fluorescence line narrowing differs from the former absorption techniques in that it requires a high-resolution spectrometer to analyze the fluorescence in addition to a high-resolution laser. The stringent collimation requirement for a Fabry-Perot interferometer frequently used for this type of measurement is generally incompatible with the typical geometry of gas-phase spectroscopy, and hence, this technique has had only limited applicability. In condensed matter, on the other hand, where the geometry is much more compact, fluorescence line narrowing has been the most widely applied method for probing the inhomogeneous profile.

4.1 Fluorescence Line Narrowing in Solids

The technique of fluorescence line narrowing (FLN) was first applied to gases by *Feld* and *Javan* in 1969 [4.11] and subsequently to ionic solids by *Szabo* [4.12] and to molecular solids by *Personov* et al. [4.13]. It is in principle a very simple method for overcoming inhomogeneous broadening and has provided numerous interesting details about ions and molecules in solids. In this section a phenomenological view of FLN is presented, followed by a discussion of the advantages and disadvantages of this method and some specific experimental details.

4.1.1 Phenomenological Model

The simplest model for strain broadening in solids is one in which the inhomogeneous linewidth represents merely a probability distribution of homogeneous packets, spread in energy by the variations in the crystal field at different optically active impurity sites. In some instances, the validity of this model has been challenged [4.14, 15], but it is likely to be a reasonable approach for quite low concentrations of impurities (less than 1%) and relatively weak impurity-impurity coupling.

There is then a simple pictorial way to represent this model as shown in Fig. 4.1 [4.16-19]. Rather than drawing all the energy levels with conventional horizontal lines, we draw only the ground state horizontal and all the other lines with a slope to represent the degree of strain broadening. The abscissa is labeled "impurity sites" to indicate that a continuum of different sites is being represented, each in a slightly different crystal environment which alters its energy level structure; the projection of the slope onto the energy axis gives the inhomogeneous linewidth. The absolute slope of the lines is arbitrary; it is only the relative slope which is important and this is experimentally determinable [4.19, 20]. It should be noted that these slopes are in principle equivalent to the slopes of the Tanabe-Sugano (T-S) diagrams described in Chap. 1, only in this case the picture is more general (yet more qualitative) in that it applies to noncubic symmetries as well. It also pertains to very small changes in crystal field parameters compared to the scale of the T-S diagrams. Figure 4.1 does not include the vibronic levels associated with electronic transitions in molecular solids, but this omission does not alter the suitability of the model for molecular systems [4.16].

Using these diagrams, FLN is very easy to visualize. If the system is pumped with a narrow-band laser with a linewidth much smaller than the inhomogeneous broadened width, $\Delta\nu_3$, then the laser can only interact with the subset of impurities labeled A. When these centers re-radiate to some intermediate level or back to the ground state, they do so with a linewidth which under many circumstances will be much narrower than the strain-broadened width and under certain conditions may indeed reflect the true homogeneous linewidth of a particular transition.

This picture may be refined somewhat to include additional features observed in FLN of solids. First of all, a finite width is given to each line as in Fig. 4.1b to represent the homogeneous linewidth associated with each site. In the most general case, this width should be allowed to vary from site to site [4.20, 21].

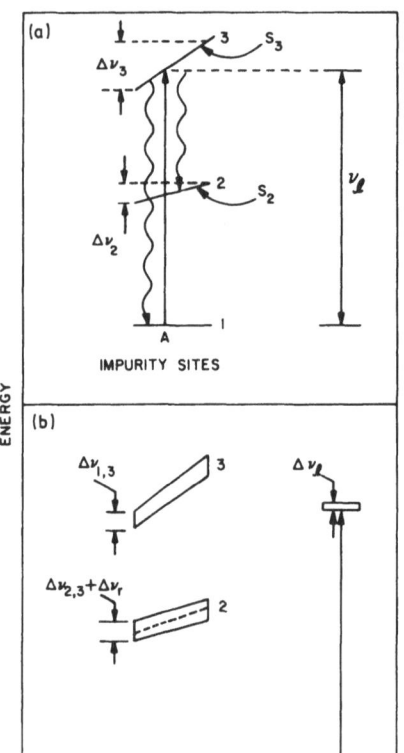

Fig. 4.1a, b. Pictorial model of FLN. (a) Projection slopes of levels 2 and 3 onto the vertical (energy) axis depict inhomogeneous linewidths $\Delta\nu_2$ and $\Delta\nu_3$, respectively; (b) Finite width given to each level, as explained in text

The intermediate level in Fig. 4.1b is given yet an additional width to indicate the concept of accidental coincidence: Most of the solids normally examined have relatively low symmetry with several parameters necessary to characterize the crystal field. It is therefore possible that impurities in very different crystalline environments may accidentally have two of their energy levels in exact or near resonance; however, the probability that they have more than two levels in coincidence becomes considerably smaller. Hence, when we detect resonant fluorescence we have preselected sites which *must* have two levels in coincidence, however the possible spread in intermediate levels leads to some residual strain broadening when detecting nonresonant emission. It is this residual inhomogeneous broadening which is represented by the additional width of the intermediate level in Fig. 4.1b, and which has always been observed to varying degrees in nonresonant FLN studies of ionic systems [4.20,22,23]. It is also a proposed explanation for the relatively broad 0-0 fluorescence linewidths in some molecular systems after excitation into one of the vibronic levels [4.24]. In ionic crystals the width of the residual component can be generally correlated to the width of the strain-broadened absorption line (or the relevant emission line under broadband excitation), but often the presence of weak satellite lines which become folded into the main lines as the inhomogeneous linewidth increases [4.20] makes a quantitative study of this phenomenon difficult.

Using this simple pictorial model it is then possible to predict the linewidths which one should observe under different experimental conditions. In a typical case where the laser linewidth is significantly less than the inhomogeneous absorption width, the laser excites one energy distribution given by the convolution of the laser width, $\Delta\nu_l$, and homogeneous width of the particular transition being pumped (in our example, $\Delta\nu_{1,3}$). Each excited impurity radiates back to the ground state with this same width, $\Delta\nu_{1,3}$, and therefore, the total resonant fluorescence linewidth should be the convolution of $\Delta\nu_l$ and twice $\Delta\nu_{1,3}$.

The situation becomes more complicated for nonresonant fluorescence because both the differences in slope of the levels and the residual inhomogeneous width, $\Delta\nu_r$, must be accounted for. It is straightforward to show that in this case, as long as $\Delta\nu_l$ is less than $\Delta\nu_3$, the total experimental linewidth is a convolution of $(1-\alpha)(\Delta\nu_l + \Delta\nu_{1,3})$, $\Delta\nu_{2,3}$, and $\Delta\nu_r$, where α is the ratio of terminal level to initial level slopes, S_2/S_3. This ratio can be experimentally determined by plotting the fluorescence spectrum as a function of laser excitation energy [4.19,20].

4.1.2 Sources of Residual Broadening

It should be apparent, then, that only the detection of resonant fluorescence will allow extraction of the true homogeneous linewidth.[2] This method has been

---

[2]Because of accidental coincidence, the best we are able to do is measure some weighted average homogeneous width of all the (potentially different) sites in resonance with the laser.

successfully used in a number of different systems [4.20, 25, 26], but the results must be interpreted with considerable caution. There exist several reasons why the width of a resonant transition may still have an inhomogeneous component. Perhaps the most obvious one relates to spectral diffusion. In the treatment so far, there has been no consideration of any interaction between different impurities; yet, if the concentration of these impurity ions or molecules is sufficiently high, it is well known that a mutual interaction will cause the optical excitation to propagate from impurity to impurity. In one extreme, this propagation can be pictured as a coherent, wavelike motion, known as a Frenkel exciton. In the other extreme the process may occur via an incoherent, random hopping, leading to diffusive-type behavior. Various models for these different modes of propagation, particularly the latter, have been discussed at length in the preceding chapters. Suffice it to say that the incoherent process will typically lead to alteration of the fluorescence profile in time if pulsed excitation is used, as the optical energy flows from the distribution of initially excited donors to the acceptors which have slightly shifted energy levels. If on the other hand cw excitation is used, a spectrum will be recorded which represents some equilibrium distribution of both donor and acceptor ion sites. Thus if any spectral dynamics are occurring, the narrowed component will exhibit the homogeneous width only if pulsed excitation is used and if the fluorescence is sampled at a time delay sufficiently short such that no energy transfer has taken place. In several cases of strongly coupled systems at high impurity concentrations no line narrowing has been observed even when sampling at nanosecond delays [4.27, 28], although the linewidths measured were probably too large to be homogeneous.

There are yet more subtle reasons why the narrowed component of a resonant transition may still not reflect the true homogeneous width arising from relaxation processes. One possible mechanism involves hyperfine or superhyperfine interactions. If the optical transition is accompanied by a spin flip in the impurity or a neighboring nucleus, then there will be a distribution of terminal states with differing energies. Particularly in the case of superhyperfine interactions, this distribution is not likely to be resolved by the laser [3] and will appear as an additional width. Note that this width does not have to arise from any dynamical processes and therefore gives us no information about relaxation or dephasing.

Another possible mechanism involves random fluctuations of the host matrix which may be very slow compared to those producing dephasing. These may be flip-flops of neighboring nuclear spins or perhaps anharmonic vibrations as found in amorphous materials. If these fluctuations produce detectable shifts of the optical energy levels via small charges in crystal field parameters, this will appear as an additional component of the linewidth when averaged over the measurement period. The frequency of fluctuation has no direct influence on the linewidth as

---

[3] These will generally be a near continuum of superhyperfine levels because of the interaction with different ligands at different distances.

long as it is faster than the observation time of the experiment. This is to be contrasted with a homogeneous broadening mechanism arising from dephasing (defining the dephasing time as the time it takes for two dipolar oscillators to get out of phase with each other by one radian under the influence of some disturbance): The weaker the disturbance, the higher the frequency of interaction necessary to produce the same dephasing rate. Typically, in pulsed experiments gate widths and delays of tens or hundreds of microseconds are used for low concentration samples in order to obtain adequate signal to noise and to reduce the influence of scattered laser light. Hence, fluctuations on this time scale (far too slow to produce observable homogeneous broadening in most cases) could contribute to a larger, time-averaged linewidth than that which would be measured with a shorter delay and narrower gate. Quite obviously, in a cw experiment this problem would be worse, where now any fluctuations occurring during the radiative lifetime could contribute to the observed width.

In principle, one way to determine the significance of such a process, which may be considered as *dynamic* inhomogeneous broadening, is through the analysis of line shape. It is well known that a homogeneous line resulting from lifetime or coherence limiting processes will have a Lorentzian line shape [4.29]. Conversely, many processes which result in a statistically random distribution of homogeneous packets on a microscopic scale (either static or as in this case, dynamic) will give rise to a Gaussian shape [4.30]. With adequate signal to noise such an analysis should be straightforward.

In some recent experiments where time-domain and frequency-domain results have been compared [4.31, 32], the latter have generally implied faster relaxation rates than the former (see Chap. 5). It is conceivable that dynamic inhomogeneous broadening may indeed be responsible for the submegahertz residual linewidths at very low temperatures. Unfortunately, the extreme narrowness of the lines has prevented accurate line-shape determination [4.25]. On the other hand, the comparatively large narrowed linewidths found for rare earths in amorphous hosts are Lorentzian [4.33] and are therefore likely to be truly homogeneous.

This type of inhomogeneous broadening may also provide a possible explanation for the lack of any detectable line narrowing in the liquid systems. It has been found [4.34] that dilute rare earth chelate dyes in various solvents maintain their relatively large linewidths (> 100 cm^{-1}) all the way down to the freezing temperature. However at slightly lower temperatures in the glassy phase, significant narrowing is observed. It is doubtful that the large linewidths in the liquid phase would be homogeneous, implying subpicosecond dephasing in these rather weakly interacting systems. A more plausible explanation would be that the width derives from an environment which is changing at least as fast as the time scale of the measurement.

4.1.3 Spectral Dynamics

Not only does FLN allow the extraction of the homogeneous linewidth and/or the hyperfine structure [4.35] under some conditions but it also permits the direct observation of energy migration and spectral dynamics. In this latter role, the fluorescence wavelength within the inhomogeneous profile acts as a way of labeling impurities which are excited after some time delay by *nonresonant* inter-actions with the initially pumped sites. The focus of these experiments has been the study of ionic solids because the short radiative lifetimes of typical singlet transitions in molecular solids will generally not permit the same type of high resolution, time resolved measurements in these systems. In principle, this type of energy transfer experiment is somewhat similar to the detection of any other donor-acceptor transfer using different donor and acceptor impurity species, of which numerous examples abound in the literature [4.36-42]. However, several differences do exist other than the obvious fact that in FLN experiments, the donor and acceptor are the same impurity species.

One difference lies in the scale of the energy mismatch, ΔE. For energy trans-fer within the inhomogeneous background, one observes *near-resonant* transfer in contrast to the more conventional experiments where donor-acceptor energy mis-matches may be hundreds of wave numbers. Thus, in the near-resonant case, it is possible to span the range from $kT < \Delta E$ to $kT >> \Delta E$ with only moderate changes in temperature and the concomitant line broadening. These experiments also allow the study of the ΔE dependence of the transfer process on a much finer scale than is possible with the large, discrete energy mismatches associated with dif-ferent impurity speces. Another difference is in the relative concentration of donors and acceptors. In the FLN studies, the acceptors (i.e., all the ions which fluoresce at slightly shifted energies within the inhomogeneous line profile) greatly outnumber the donors, whereas the reverse is generally true in the more conven-tional experiments using different species. A final and perhaps most significant difference is the motivation behind these experiments. The studies using different donor and acceptor species generally assume that the energy transfer mechanism between donor and acceptor is known (typically one-phonon emission), and the acceptors are really used as a probe to indirectly measure the donor-donor energy migration [4.38]. No attempt is made to distinguish resonant (coherent) and near-resonant (incoherent) processes. In the FLN experiments, true resonant transfer is not directly observable,[4] but a study of the near-resonant interaction can reveal the importance of this latter mechanism in the propagation of energy over large spatial distances. Ideally, both types of experiments should be per-formed in the same system to more fully understand spectral and spatial energy transport [4.43].

---

[4]This is due to the fact that in true resonant energy transfer, there is no energy shifting mech-anism to produce spectral changes.

A great deal of recent experimental and theoretical effort has been devoted to understanding the various mechanisms responsible for the near-resonant energy transfer observed by FLN, much of which is reviewed in other chapters of this book. This discussion will be restricted to a brief review of the precautions required in the performance and interpretation of these experiments.

The detection of near-resonant energy transfer is predicated upon observing changes in the fluorescence profile of a particular transition as a function of time after pulsed excitation. Yet under certain circumstances such changes may occur even without any ion-ion interactions. For example, the laser used for FLN excitation may contain a weak broadband component which can pump the entire inhomogeneous line, or else the full line may be excited via phonon-assisted absorption.[5] Normally these processes will produce a weak background which is present even at zero delay. If the ions comprising the narrowed component decay faster than the broadband background (which is possible due to variations of lifetime across the absorption line) [4.20, 21], it will erroneously appear that the latter is being fed by the former when fluorescence ratios are taken.

Another closely related phenomenon is that of "pseudodiffusion" [4.20]. The narrowband laser can, because of accidental coincidence, interact with ions in very different crystal field environments. It has also been observed that ions in different sites may have different lifetimes and different homogeneous or residual linewidths [4.20]. Hence, if two distinct ion sites possessing different linewidths and lifetimes are simultaneously excited by the laser, the resulting fluorescence linewidth may increase, decrease, or distort in time due to differential changes in the populations of the two sites. Once again no ion-ion interactions need be present. Note that these problems are accentuated in amorphous materials, where large variations in linewidths and lifetimes across the inhomogeneous profile are seen (refer to Chap. 6).

One way to determine with certainty the presence of energy transfer is to measure an absolute increase in the acceptor population, as has been observed in several different systems [4.17, 27, 44]. This may require raising the temperature to a high enough value that the energy transfer rate is significantly faster than the radiative decay rate. However, even if an absolute rise of the background cannot be observed because the radiative decay rate is too fast, there are other ways of distinguishing the various effects. For example, pseudodiffusion would not be expected to exhibit much if any temperature dependence whereas near-resonant energy transfer often shows a strong temperature dependence and can be nearly frozen out at sufficiently low temperatures. Pseudodiffusion is also much more likely to occur when pumping in the wings of the inhomogeneous absorption because it is here that the anomalous ion sites are usually found [4.15, 20]. Thus with sufficient care, it is in most cases possible to determine the source of the ob-

---

[5]Phonon-assisted absorption would consist of laser excitation of the weak phonon sidebands of those ions not directly resonant with the laser.

served spectral dynamics. If energy transfer is indeed occurring, it may then be studied as a function of excitation wavelength, laser intensity, polarization, magnetic field, temperature, dopant concentration, host material, and any other relevant parameters which will help clarify the nature of the process.

Two additional sources of potential complications in energy transfer experiments are worth noting before proceeding to a discussion of specific experimental details. The first is the presence of nonlinear effects. Although the oscillator strengths of the typical transitions in rare earth and transition metal ions are low (on the order of $10^{-6} - 10^{-8}$), a tightly focused, high intensity laser may still be able to excite a significant fraction of the ions in a given set of sites. (This phenomenon is considered more quantiatively in the section on saturation spectroscopy). The excited ions may then interact in such a way that will alter the energy transfer properties. For example, the interaction between two nearby excited ions will be very different than that between a neighboring excited and ground-state ion. An enhanced donor decay may result which would disguise the true donor-to-acceptor transfer process. A related phenomenon, bi-exciton decay, has recently been observed in MnF_2 [4.45] and will be discussed in the next chapter. Quite obviously studying the laser intensity dependence of the energy transfer process can assess the contribution of nonlinear phenomena.

Another possible influence on the energy transfer process may be the presence of phonons produced during the pumping cycle or during the decay of some intermediate level. Since nonresonant energy transfer is phonon assisted and can be strongly temperature dependent [4.46], any localized heating produced by the laser will be likely to enhance the transfer process. It should be noted that laser generated nonthermal phonons can also be effective in enhancing the transfer if they happen to be of the required energy. It has been suggested [4.47] that the resonant phonon-assisted process may be particularly susceptible to this influence. To date no direct evidence exists for this type of laser generated interference, and it presumably can be minimized by keeping the intensity as low as possible and the samples well thermally anchored. Also, pumping the fluorescing level directly rather than via a higher state should greatly reduce the effectiveness of this process.

4.1.4 Experimental Details

The experimental setup for FLN measurements requires a varying level of sophistication, depending on the system examined and the information sought. The basic requirements in any system are, of course, a high-resolution laser, a sample and cryostat, a spectrometer, and detection electronics. For measuring homogeneous linewidths in low concentration systems where no dynamics or other complications mentioned earlier are present, cw lasers are more desirable than pulsed because of the inherently higher resolution (1 MHz linewidth now routinely obtained with commercial units). However, it has been shown above that only resonance fluorescence transitions will provide unambiguous information about the homogeneous linewidth, and one is therefore confronted with a problem of extracting weak fluorescence in the presence of strong scattered laser light. For systems which have lifetimes on the order of milliseconds, this problem may be solved by using the double chopper method employed by *Erickson* [4.75]. Two chopper blades with teeth 180° out of phase, one before the sample and one between the sample and the detector, alternately shield the detector while the laser excites the sample and then block the laser while the fluorescence is measured. For faster lifetimes, this function could presumably be performed with two properly synchronized electro-optic shutters.

As usual the higher the resolution required, the more difficult the experiment. Room temperature linewidths in glasses can be easily resolved with simple grating spectrometers whereas at low temperatures the fluorescence linewidth can be on the order of a few megahertz (or less), requiring very high finesse Fabry-Perot interferometers. If several closely spaced lines, such as hyperfine levels, are present in the fluorescence spectrum, the numerous orders of a high-resolution Fabry-Perot may simultaneously overlap more than one line and thereby cause ambiguous results. It may then be necessary to use tandem interferometers to isolate one line. Two tandem interferometers have been routinely used for FLN [4.15, 17] and scattering measurements [4.48], but a very powerful and versatile instrument is obtained with the use of three pressure-scanned interferometers, as in the Pepsios instrument [4.49]. This device employs three independent pressure chambers housing Fabry-Perots with appropriately chosen spacers. The chambers can be separately pressure tuned in order to obtain overlap of the three passbands, and then slaved together to scan the instrument across the fluorescence line of interest.

After being dispersed by the interferometer, the fluorescence is generally detected with a photomultiplier tube (PMT) and analyzed with appropriate electronics. For strong signals a boxcar integrator or lock-in detector is often employed. Weaker signals may require photon counting instruments; recent advances in their speed and versatility have made such devices as digital boxcars easy and convenient to use.

Many of the interesting relaxation phenomena occur at relatively low temperatures, often in the liquid helium range, and they exhibit interesting temperature-dependent effects. It is therefore frequently desirable to have a variable temperature cryostat capable of spanning the range between 2 K and room temperature. Three general types of cryostats exist: In the simplest design, the sample sits in the inner liquid reservoir and is cooled either by contact with the liquid or by cold gas blown into the dewar. A more sophisticated design employs a sample chamber separate from the main liquid reservoir with a valve connecting the two through which liquid or vapor can pass. A third design uses a cold finger, one end of which is in contact with an appropriate cryogen and the other end of which protrudes into the vacuum space and to which the sample is affixed. All these designs generally require some type of controller in order to regulate the temperature drift to less than 1 K for the duration of a measurement.

Each type of dewar has relative merits depending on the needs of the experimentor. While the cold finger dewars are often the most convenient to use and require only one set of windows between the sample and the outside, they are more susceptible to inaccurate temperature readings because thermal contact is made only through those surfaces attached to the cold finger. Normally, the sample is bonded with a relatively high thermal conductivity substance such as GE 7031 varnish, but an irregularly shaped piece may still have poor thermal contact. The temperature gradient problem becomes more severe with higher doped or amorphous samples, as the thermal conductivity is lower in these materials than in relatively pure crystals. For example, pure Al_2O_3 crystals may have a thermal conductivity as high as 5 W/cm · K at 4 K, whereas most glasses have a value three to four orders of magnitude lower at the same temperature. With low thermal conductivity and poor thermal contact to the cold finger, exposure to typical amounts of 300 K radiation or even laser generated heating can cause gradients of several degrees across the sample. The other types of dewars, where the sample is cooled on all sides by gas or liquid, provide better thermal contact, but these heating effects still should not be ignored.

Different thermometers are desirable depending on the temperature range and the necessary accuracy of measurement. Carbon resistors are inexpensive and very accurate in the liquid He range, but other types of thermometers have more desirable properties. Diode thermometers, although less accurate than carbon, carbon/glass, or germanium, are usable throughout the entire range from 2 K to above room temperature. They are also quite small, which permits attachment directly to the sample, and the grounded cathode type of diode can have the temperature sensing element in direct contact with the surface to be measured. This latter property is to be contrasted with the encased germanium or carbon/glass thermometers where thermal contact is made through the leads and through He exchange gas which fills the case. Capacitor thermometers are preferred if high magnetic fields are present. Regardless of the thermometer used, appropriate cryogenic techniques [4.50] such as minimizing the drive power, anchoring the leads, and shielding the thermometer from room temperature radiation should be employed for accurate measurements at low temperatures.

In situations where spectral dynamics are present, pulsed measurements become necessary. Pulsed laser systems, while of poorer resolution than their cw counterparts, have an advantage of being obtainable at considerably less expense. Numerous designs also exist for home-made models [4.51]. For purposes of FLN, the N_2-laser-pumped dye laser based on the *Hänsch* design [4.52] has been widely used and is a very versatile instrument capable of spanning the entire visible region. The remaining equipment necessary for pulsed experiments is very similar to that for cw work with only a few exceptions, and a typical pulsed experimental setup is shown in Fig. 4.2.

Energy transfer measurements do not require knowledge of the homogeneous linewidth and can therefore be performed on nonresonant transitions. In this case, scattered laser light is no longer of much concern. For resonance fluorescence

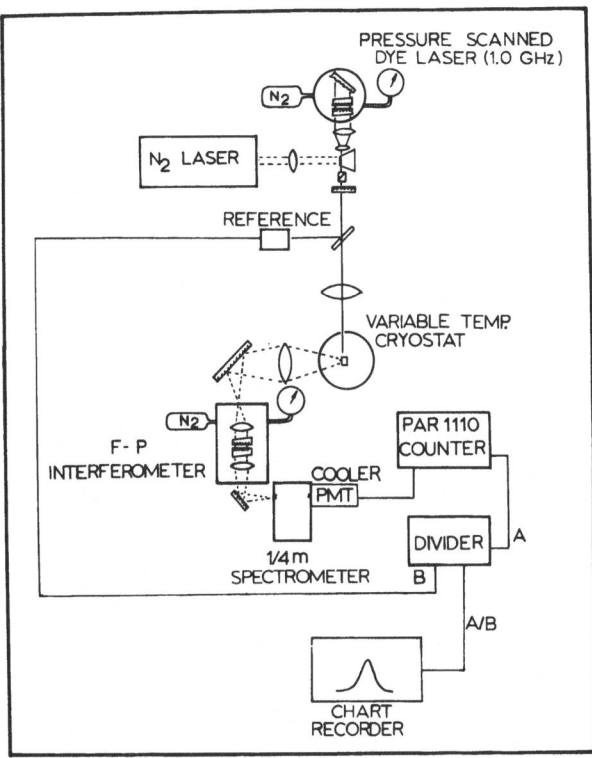

Fig. 4.2. Pulsed FLN experimental apparatus, using pressure scanned dye laser and pressure scanned Fabry-Perot etalon as analyzer. The 1/4 meter spectrometer acts as a broadband filter to isolate the one fluorescence line of interest

measurements, time resolved detection allows the temporal separation of scattered light and fluorescence. With well-polished, uniform samples and careful attention to the reduction of scattering, the photomultiplier will not be saturated by the laser light, and the signal can be electronically gated (with a boxcar or gated photon counter) almost immediately after termination of the laser pulse. If scattering is too severe, then the detector itself must be protected.

Gating the detector may be accomplished in two ways. The first involves covering the tube during the excitation of the sample. As in the cw case, this can be accomplished through mechanical or electrooptical shuttering, depending on the necessary temporal resolution. A second way involves shunting one or several stages of the photomultiplier dynode chain in order to momentarily reduce the tube gain during the laser pulse. This latter method is somewhat easier and requires less equipment than electronic shuttering, but is not as fast. A circuit developed for this purpose [4.53] is shown in Fig. 4.3, and has been successfully used in several experiments. The recovery time for the tube using this circuit is approximately 300 ns, and the on/off ratio is 10^4.

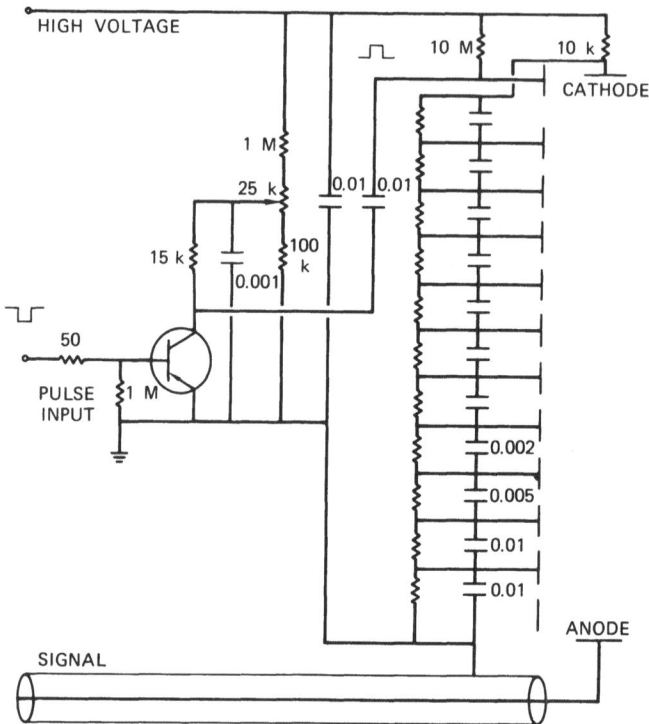

Fig. 4.3. Gated photomultiplier circuit for blinding the tube during laser excitation. With this circuit, the tube is normally off, but can be gated on for as long as 10 ms, using a ~ -0.5 V gate pulse. Transistor is a Motorola MPS-V60, and 100 Ω resistors and 0.001 μF capacitors are used unless otherwise indicated. This circuit is suitable for venetian blind (e.g., EMI 9558) or box and grid (e.g., RCA 7265) dynode structures but must be modified for squirrel cage tubes (e.g., RCA 1P28) [4.53]

For accurate linewidth or energy transfer analysis, it is desirable to have the data stored digitally for further computer averaging and processing. In fact, with presently available microcomputer technology, control of the laser, spectrometer, data collection, and display can all be automated for ease of operation and reproducibility.

4.2 Saturation Spectroscopy in Solids

This section briefly covers two types of saturation experiments in solids which are primarily useful for homogeneous linewidth studies. They both involve frequency domain measurements, and much of the equipment required for these experiments is very similar to that used in fluorescence line narrowing.

4.2.1 Hole Burning

The success of hole burning in solids requires that the pump laser create a large enough depletion in the ground-state population that a probe beam can detect the difference in absorption between pumped and unpumped sites. Adequate signal to noise dictates that this difference, δ, be on the order of 10^{-2}. With a laser of very narrow spectral width, a hole having the homogeneous linewidth, $\Delta\nu_h$, will be created. Since the probe beam also interacts with sites having a spread of $\Delta\nu_h$, the interrogated hole will reflect a width of $2\Delta\nu_h$.

In a simple, nondegenerate two-level system, δ is given by the following formula under cw pumping conditions [4.54]:

$$\delta \equiv (n - n_0)/n_0 = B_{1,2}\, \rho_\nu / A_{1,2} \qquad (4.1)$$

where n is the ground-state population of the sites in resonance with the laser in the presence of excitation, n_0 is the ground-state population without excitation, $B_{1,2}$ and $A_{1,2}$ are the Einstein coefficients, which are proportional to the oscillator strength of the transition, and ρ_ν is the energy density of the laser per unit frequency interval. The ratio of the Einstein coefficients depends only on the frequency of the transition and is given by

$$A_{1,2}/B_{1,2} = (\hbar\omega^3 \bar{n}^3)/\pi^2 c^3 \qquad (4.2)$$

where ω is the angular frequency of the transition and $\bar{n}$ is the index of refraction. Thus, we find that for a transition in the visible range and a laser linewidth of 10 MHz, an intensity on the order of 10-100 mW/cm^2 should produce the required degree of saturation.

For pulsed rather than cw excitation with a pulsewidth very short compared to the lifetime of the excited state, (4.1) must be replaced by

$$\delta = B_{1,2}\rho_\nu \Delta t \qquad (4.3)$$

where Δt is the pulsewidth of the laser. We then find that with a typical radiative lifetime for a two-level ionic system (such as ruby) on the order of 1 ms, a 10 ns laser pulse having a 1 GHz linewidth will require 10^7 times the intensity to produce the same degree of saturation. While both levels of intensity are readily available from cw and pulsed dye lasers respectively, the pulsed experiments generally require fairly tight focusing, and beam overlap between pump and probe then becomes more difficult. Therefore cw lasers, with their high spectral resolution and longer effective pumping time, are generally more suitable for saturation experiments in ionic systems.

Because of the higher oscillator strength of the allowed transitions in molecular systems, pulsed hole burning is more easily accomplished in these materials. Furthermore, many of the molecular systems will undergo photochemical hole burning [4.55, 56]. Due to photochemistry of selected molecules which interact with the laser beam, the holes may persist indefinitely, allowing the pump laser to be subsequently used as a probe of the hole width.

In more complex, multilevel systems with possible nonradiative coupling between levels, the required intensity level for saturation may be increased or decreased. For example, if a number of rapid de-excitation channels exist whereby an excited ion may return to the ground state, the saturation requirements will be higher. On the other hand, the existence of some long-lived intermediate metastable levels may result in saturation at lower intensity. These arguments are not pertinent to pulsed excitation if the laser pulse is shorter than any interlevel dynamics and the pulse separation long enough to allow complete relaxation to the ground state.

Whereas in gases a single laser beam split into two counter-propagating ones can be used to both burn and interrogate a hole in the $v = 0$ velocity group, there is no analogous way to employ a single laser for the study of solids. In general, these latter experiments must employ two different tunable lasers — one to burn a hole in the absorption line and a second to scan across it in order to measure its width. (In the special case of persistent holes, the same laser can be used in both capacities.) However, for hole widths less than 20 MHz, the need for two lasers has been circumvented through use of a modulation scheme [4.57]. In a cw experiment, amplitude modulation of the laser creates a sideband at the modulation frequency, and therefore scanning the modulation rate effectively creates a probe beam which scans across one side of the hole. Through use of heterodyne detection, it is possible to obtain the amplitude of the sideband as a function of modulation frequency. This technique eliminates the requirement for two tunable lasers but instead necessitates relatively complicated, high-speed modulation and electronics. The limited bandwidth can also be a drawback when confronting broad lines, multiple holes, and background effects. A Stark-field sweeping technique, which rapidly (relative to the radiative lifetime) scans the burned holes past the fixed frequency pump, has also been developed [4.58] and is suitable in systems with long lifetimes and reasonably large Stark shifts.

Hole burning and FLN are to some extent complementary techniques, each with its own advantages. For resolving narrow linewidths in systems with no dynamics, hole burning is more desirable because it eliminates the need for a high-resolution spectrometer and is therefore limited only by the laser linewidth itself. Thus, resolving linewidths which are a small fraction of a megahertz becomes possible [4.32]. Quite obviously, this technique is also more desirable in studying weakly fluorescing transitions. In addition hole burning eliminates one source of spurious broadening in FLN mentioned in Sect. 4.1.2 — the situation where fluorescence can occur to any one of several unresolved superhyperfine levels. (However, if it is the hyperfine [4.18] or superhyperfine structure which is to be studied, then FLN would be preferred.) It should be noted that hole burning is still susceptible to the problem of "dynamic inhomogeneous broadening" which can occur during the creation of the hole, causing it to spread. Furthermore, the greater laser power requirements in a hole-burning experiment may produce power broadening of the line, which occurs when the induced transition rate becomes of the same order as the dephasing rate of the system. Finally, in systems where slow spectral dynamics are occurring, pulsed hole burning experiments [4.59] using a delayed probe are more difficult and less effective than time-resolved FLN.

The necessary equipment for hole-burning experiments is very similar to that covered in the previous section on FLN with only a few exceptions. As mentioned above, the most versatile setup would include two tunable lasers of equally high resolution. (While it is not imperative that the two lasers be of the same frequency, use of a pump at one transition frequency and a probe at another will result in additional complications due to accidental coincidence and multiple homogeneous widths.) Because in these absorption experiments one is detecting small changes in relatively intense signals, photodiodes rather than photomultiplier tubes (PMT's) make more appropriate detectors. A typical experimental arrangement using two cw lasers, similar to that employed in [4.60], is shown in Figure 4.4: Chopping the pump beam permits ac detection and reduces sample heating by the laser. Differential detection is made possible by using a reference beam which passes through an unpumped portion of the sample.

If necessary, a lens of sufficient focal length can be used to match beam waist to the sample size, with a thin sample being more desirable. The limit in detectable absorption with this type of system may be on the order of 10^{-3}.

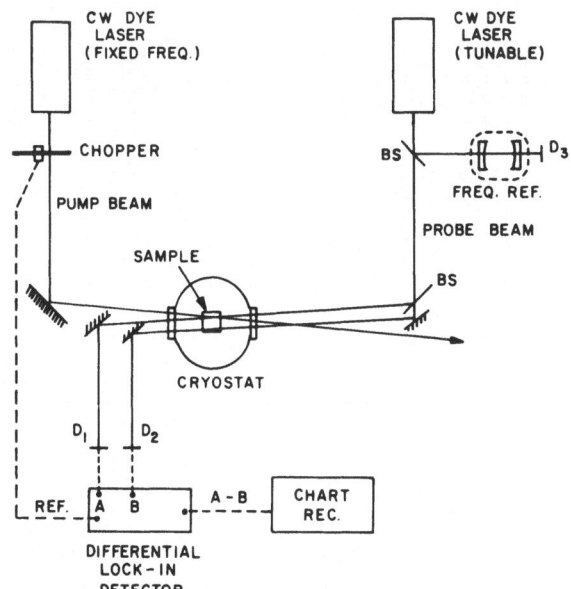

Fig. 4.4. A cw hole-burning apparatus using two dye lasers. A portion of the probe beam is passed through a Fabry-Perot interferometer as a frequency reference. The remaining part is split in order to facilitate differential detection

4.2.2 Polarization Spectroscopy

A technique closely related to that of hole burning is polarization spectroscopy. First used for homogeneous linewidth measurements in gases by *Wieman* and *Hänsch* [4.7], it has been more recently applied to liquids and solids by *Song* et al. [4.61]. Rather than looking for a change in ground-state population induced by the pump laser, this method uses the probe laser to detect a nonlinear bire-fringence in the sample produced by the pump.

An experimental setup of the type employed in [4.61] is shown in Fig. 4.5. In this arrangement an intense pump beam at frequency ν_1 which is linear or elip-tically polarized overlaps a counter-propagating probe beam within the sample. The probe beam at frequency ν_2 passes through crossed polarizers, p_1 and p_2, and then is detected with a monochrometer and PMT. In general, the linear or major eliptical axis of the pump beam is at $45°$ with respect to the probe polarization, and any frequency-dependent birefringence induced in the sample by the pump will then result in a frequency-dependent polarization component orthogonal to the original probe polarization. This component will be transmitted through the second (crossed) polarizer, p_2, and detected by the PMT as ν_2 is tuned through the

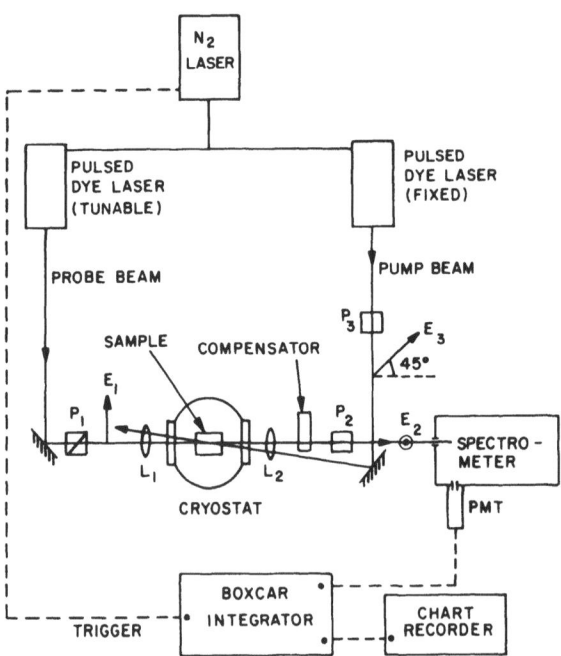

Fig. 4.5. Experimental apparatus for polarization spectroscopy in solids, employing two pulsed dye lasers (excited by the same pump laser). Polarization of the beams is determined by orien-tation of polarizers p_1, p_2, and p_3, as shown. Any additional birefringence is nulled using a Soliel-Babinet Compensator

$\nu_2 = \nu_1$ resonance. As opposed to the analogous technique used in gas-phase spectroscopy, the counter-propagating geometry is not necessary but is merely a convenient way to prevent stray pump light from striking the detector. The spectrometer serves to reduce any stray light and background fluorescence. As with hole burning, this type of experiment may be either pulsed or cw, using a boxcar integrator or lock-in detector, respectively, to analyze the PMT output.

The theory of the nonlinear processes which give rise to the induced birefrigence is derived in [4.61]. As ν_2 is tuned through resonance with ν_1 one should obtain a Lorentzian curve which has a width (FWHM) of twice the homogeneous linewidth of the particular transition under examination. Because this technique ideally involves a deviation-from-null measurement, it is more sensitive than hole burning and therefore is better suited to weaker transitions and/or lower pump beam powers. However, adequate results may be more difficult to obtain in cases where linear birefringence due to imperfect optics or the sample itself is much larger than the nonlinear birefringence induced by the pump laser (although this birefringence may be nulled within reason by a Soleil-Babinet compensator placed before p_2). Excessive resonance fluorescence or stimulated emission may also interfere with the signal in the study of systems with strong ground-state transitions. Once again, this type of absorption technique is potentially of higher resolution than FLN (since the spectrometer serves merely as a "broadband" filter) and is particularly suited to the study of strongly quenched transitions.

4.3 Coherent Transient Spectroscopy

The observation of coherent transients in solids differs from the above-mentioned techniques in a number of respects. While all of the others involve linewidth measurements, this class of experiments detects the relaxation or dephasing time (commonly referred to as T_2, borrowing from NMR terminology) directly. All of the different variations of coherent transient spectroscopy, such as optical free induction decay, optical nutation, and photon echoes, are based on the same experimental principles: At some point in time a coherent state of the system under investigation is prepared with the laser, and through various methods the decay of coherence due to dephasing is then monitored as a function of time. Note that this is the only technique mentioned so far which relies on the coherence of the laser light itself to produce the desired effect. All of the previous methods employ the laser merely as an intense, high spectral resolution light source. Furthermore, while the previous techniques had been developed for gas-phase spectroscopy and later applied to solids, use of coherent transients was first demonstrated in solids [4.62] and only later was it used to study relaxation in gases. The details of this class of experiments have been exhaustively covered in a number of recent publications [4.63-66], and consequently this section will be restricted to some general descriptive comments.

It is easiest to picture the coherent transient phenomena classically: The coherent, monochromatic laser beam resonantly excites one subset of atomic oscillators which are all initially in phase with the exciting light. The radiation emitted by these dipole oscillators will constructively interfere in the forward direction, and therefore if the laser were very suddenly switched off, these dipoles would continue to radiate forward until the amplitude of oscillation damps to zero or until the oscillators get out of phase with each other. Generally the dephasing rate dominates and gives rise to the homogeneous linewidth, and consequently it is this rate which we seek to determine. Note that in this model, each oscillating dipole produces an oscillating electric field, but the intensity of light radiated in the forward direction is proportional to the total field squared and consequently to the square of the number of oscillators excited (and thus, to the square of the excitation intensity). This quadratic intensity dependence arises from the fact that all of the oscillators were driven coherently by the laser and not from any mutual interaction between oscillators [4.67].

Photon echoes, the earliest of the optical coherent transients to be observed, were first detected in ruby by *Kurnit* et al. [4.62] in 1964, and an experimental setup of the type used in this experiment is shown in Fig. 4.6. Analogous to the spin echo technique developed by *Hahn* [4.68], this method requires two optical pulses of the proper intensity and duration, separated by time, t, to produce an "echo" from the sample — a third weak pulse in the forward direction — at a time $2t$. The decay of the echo pulse amplitude as a function of pulse spacing is presumably a measure of the dephasing rate of the system. However, it has been found later [4.69] that producing pulses by electronically gating a cw laser gave

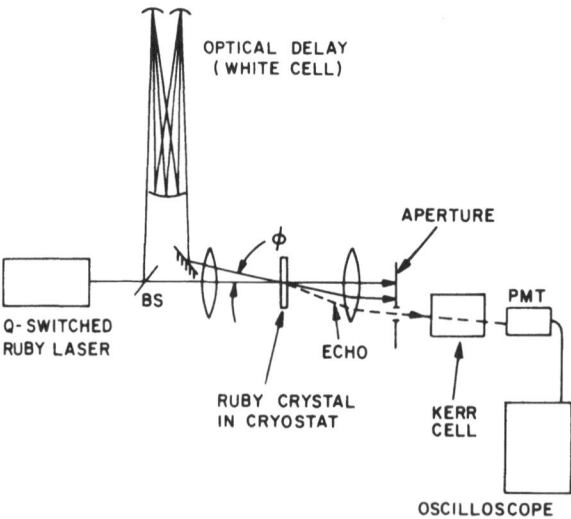

Fig. 4.6. A typical pulsed photon echo apparatus, used by *Hartmann* and coworkers to investigate echo phenomena in ruby. Q switching produces a 10-15 ns pulse of 200 kW peak power

substantially different decay rates. A discussion of this discrepancy is reserved for Chap. 5.

Advances in coherent transient spectroscopy, notably the recently developed techniques of Stark switching [4.63] and laser frequency switching [4.65], have made the observation of coherent transient decays a relatively simple and powerful method for directly measuring dephasing times in solids at low temperatures. Rather than gating the laser on and off, these techniques involve a rapid frequency shift of either the transition frequency of a particular set of ions (Stark switching) or the excitation frequency of the laser itself (laser frequency switching). In this way, two different frequencies are generated, and heterodyne detection may be used to observe the transient effects. A typical experimental setup, after [4.65] is shown in Fig. 4.7, where an intracavity E-O phase modulator is employed to shift the laser frequency.

It has been shown [4.65] that all of the coherent transients can be generated using Stark or frequency switching. For example, free induction decay (FID) occurs when the laser, after preparing a coherent state by remaining at a fixed frequency ν_1 for a time of at least T_2, is suddenly shifted to a new frequency ν_2 which differs from ν_1 by more than a homogeneous linewidth. The ions initially excited at ν_1 will continue to radiate in the forward direction at this frequency until they become dephased, and the ac output of the detector (typically a very fast photodiode) will exhibit a damped oscillation with a beat frequency of $\nu_2 - \nu_1$ and a decay envelope of $T_2/2$ [4.65]. If intracavity switching is used, the frequency shift is limited to approximately one-half of the laser cavity longitudinal mode spacing, or typically 100 to 150 MHz. External frequency switching is also possible [4.70], but in this case, the frequency shift occurs only while a voltage

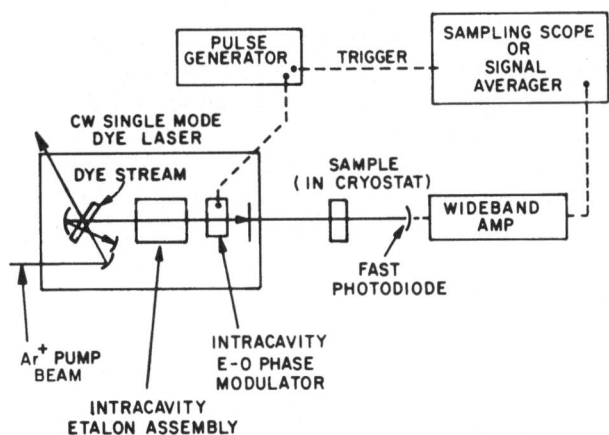

Fig. 4.7. Experimental apparatus for laser frequency shifting experiments. An X-cut AD*P crystal is used as an intracavity phase shifter, and a fast ($\sim$ 1 ns) p-i-n diode is used for detection

gradient is applied to the *E-O* crystal; consequently external switching may require much higher voltages to maintain the new frequency for a time greater than T_2.

Optical nutation occurs when a new set of ions, initially in the ground state, suddenly becomes resonant with the laser via either switching technique. A temporal oscillation in the transmitted signal amplitude occurs as the population of ions is repeatedly inverted and then stimulated back to the ground state. Once again, these oscillations damp out as the system becomes dephased. Although FID and nutation signals can superficially appear similar, there are obvious differences. The oscillations in FID represent the beat frequency between ν_1 and ν_2, independent of laser intensity, whereas the oscillations in the nutation signal are the amplitude fluctuations in the transmitted beam whose period is proportional to the square root of the laser intensity [4.64]. Typically in a laser frequency switching experiment, both FID and nutation are observed simultaneously (FID of the "old" set of ions and nutation of the "new" set), but the effects can be appropriately separated by adjusting the frequency difference and the laser intensity to enhance one or the other transient. For most accurate measurement of T_2, the laser intensity should be minimized to prevent power broadening, and consequently the FID is a more suitable measurement.

A proper sequence of frequency or Stark switching pulses can also produce photon echoes and other related phenomena such as "right angle echoes" [4.71] and edge echoes [4.72]. Measuring dephasing with photon echoes has the advantage that the laser linewidth, as long as it remains constant, does not influence the decay rate as it does in FID and optical nutation. This occurs because the enhanced decay due to the inhomogeneous distribution of ions excited by a finite laser width is reversed by the $\pi/2, \pi$ sequence of preparatory pulses [4.64].

Coherent transient spectroscopy is a very attractive technique for measuring the relatively long relaxation phenomena that predominate in many solids at low temperatures, i.e., on the order of or greater than one nanosecond. More recently, frequency switching [4.70] and pulsed photon echo measurements [4.73] have been extended to the sub-nanosecond regime, the latter by using a mode-locked laser providing picosecond pulses. However, since time-domain and frequency-domain experiments in principle supply the same information about relaxation, the expected relaxation rates will generally dictate which technique is favored.

In cases where results of time-domain and frequency-domain experiments have been directly compared [4.31], some discrepancies exist, as mentioned earlier. While FLN and hole burning suffer from the residual inhomogeneous broadening effects discussed in Sects. 4.1.2 and 4.2.1, coherent transient measurements appear to be limited by high-frequency laser jitter [4.70]. For the extremely narrow lines of a few hundred kilohertz found in some rare earths [4.32], the laser must be very carefully frequency stabilized [4.74,75]. Residual power broadening and heating effects may adversely affect both hole burning and coherent transient measurements.

4.4 Light Scattering

The technique of Raman scattering and more recently, the different variations of coherent Raman scattering (such as CARS [4.76, 77], RIKES [4.78], HORSES [4.79], etc.) have been invaluable tools in the study of vibrational modes in gases and condensed matter, but so far these methods have had only limited application in examining electronic transitions in solids. A review article concerning these coherent Raman methods has recently appeared [4.80].

Electronic Raman scattering will in general provide information about the widths of low-lying electronic transitions, but the typical cross sections are even weaker than conventional Raman scattering [4.81] and therefore difficult to observe. Nevertheless, such experiments have been pursued in rare earth and transition metal crystals with reasonable success [4.82]. For the purposes of studying relaxation phenomena via homogeneous line broadening, this technique is less suitable than others discussed above because the linewidths in the scattering experiments are limited by inhomogeneous broadening. However, for transitions which are homogeneously broadened, as in most solids at elevated temperature, scattering would allow the homogeneous linewidths of low-lying transitions to be obtained directly, whereas fluorescence measurements would require complicated deconvolution of several different widths. Alternate methods for direct determination of low-lying levels would require appropriate tunable infrared sources and detectors.

It has been recently shown that electronic scattering from excited metastable levels can also be observed [4.83], but the information obtained in this way is likely to have little advantage over more conventional spectroscopic methods.

4.5 Photoacoustic (or Optoacoustic) Spectroscopy

Photoacoustic spectroscopy (PAS) has gained recent popularity in the study of solids [4.84]. With this technique, the nonradiative decays of an excited optical center are directly observed by measuring the heat evolved after optical excitation. In a typical arrangement, the sample is sealed in a small chamber containing some exchange gas, and the heat given off after repetitive (sinusoidal or chopped) excitation produces pressure fluctuations in the gas which may be synchronously detected with a sensitive microphone coupled to the chamber [4.85]. Low temperature variations of this technique also exist, employing thermometers attached directly to the sample [4.86].

This method provides complementary information to that obtained by conventional absorption spectroscopy and has been shown to be extremely useful for determining the spectra of highly absorbing or scattering materials [4.87]. It has also proven successful in determining the absolute quantum efficiencies of certain rare earth transitions [4.88]. However, with this technique one is unable to either probe beneath an inhomogeneously broadened line or detect energy transport which does not involve a substantial nonradiative component. Thus, for the purpose of this discussion, its usefulness is somewhat limited.

4.6   Degenerate Four-Wave Mixing (Transient Grating Spectroscopy)

The last technique to be discussed in this chapter differs from all previous ones in that it provides no information about line-broadening mechanisms. It does, however, in principle allow the observation of spatial energy migration and hence, it is relevant to the tenor of this and previous chapters.

Referred to by different names, such as "phase conjugate wave generation," "degenerate four-wave mixing" [4.89], "transient grating" [4.90] or "holographic absorption grating spectroscopy" [4.91], or "forced Rayleigh scattering" [4.92], these techniques are all variations of essentially the same process, which is easy to understand with a simple picture: A pump beam at a wavelength absorbed by ions or molecules in the solid is split into two beams which overlap within the sample. The spatial overlap between the two coherent beams produces a stationary pattern of constructive and destructive interference whose spacing can be varied by changing the beams' angular separation. This interference produces a periodic modulation in the density of excited impurities and hence, in the index of refraction (which can then scatter light).   A third (probe) beam is directed onto the induced grating, and a fourth beam, that portion of the probe which is scattered according to the Bragg condition, reveals the status of the grating.

For purposes of detecting spatial migration, it is easier to visualize a pulsed experiment [4.90]. Once the grating is formed, with ions alternately in the ground or excited state according to the interference fringes, it will decay with the radiative lifetime of the excited state in question. However, if spatial migration is also occurring, the "valleys" of unexcited ions will start to fill in, and the decay of the grating will exceed the radiative decay rate. By observing the intensity of the scattered probe beam as a function of delay after the pump excitation, it is possible to monitor the grating decay and compare it with the measured radiative lifetime to determine whether any migration is occurring. For highest spatial resolution, the two interfering pump beams should be counterpropagating, in which case the interference peak separation will be $\lambda/2\bar{n}$. This will then allow observation of spatial migration over distances of less than $\lambda/4\bar{n}$, or typically about 0.1 micrometers.

Other variations of the basic experiment described above are possible. For example, pump and probe beams need not be of the same wavelength, since once formed, the grating will presumably Bragg scatter light of any chosen frequency. If the two pump beams are themselves of different wavelengths, a propagating rather than stationary grating will result, exhibiting a beat pattern. Note that these techniques are, of course, no longer "degenerate four-wave mixing." In a cw method, use of degenerate counter-propagating pump beams and a probe beam of equal intensity and nearly co-propagating with one of the pumps will produce two simultaneous stationary gratings, one of small spacing and one of much larger spacing. If spatial transport is occurring, the lifetime of the grating will clearly depend on the grating spacing. It can then be shown that with appropriate polarization of pumps and probe, the scattered wave will undergo a polarization rotation which depends on the relative lifetimes of the two gratings [4.93].

While these experiments show great promise for observing spatial migration of the optical excitation [4.94], the distances involved are very large when compared to the typical impurity spacings. Consequently, the energy transport must be quite rapid (compared to the radiative lifetime) to be observed in this way. Furthermore, any trapping of the excitation at vacancy or impurity trap sites may hinder this type of large scale diffusion. In addition, these experiments are extremely sensitive to any mechanical vibration or thermal expansion during the course of the measurement.

4.7 Conclusion

This concludes the discussion of recent experimental techniques in the laser spectroscopy of solids. Clearly, space limitations have prevented a thorough discussion of all of these methods, and other more conventional types of spectroscopy (such as modulated absorption techniques) have been ignored entirely. The emphasis here has been on the use of optical methods to extract relaxation and energy migration rates — the dynamical processes of optical centers in the excited state. The following three chapters deal with the specific experimental results obtained by employing many of the methods discussed herein — first for ionic and then for molecular solids.

References

4.1   Z. Genack, M. Mcfarlane, R.G. Brewer: Phys. Rev. Lett. 37, 1078 (1976)
4.2   See, e.g., K. Shimoda (ed.): *High-Resolution Laser Spectroscopy*, Topics in Applied Physics, Vol. 13 (Springer, Berlin, Heidelberg, New York 1976)
4.3   L.S. Letokhov: JETP Lett. 6, 101 (1967); P.M. Lee, M.L. Skolnick: Appl. Phys. Lett. 10, 303 (1967); V.N. Lisityn, V.P. Chebotayev: Sov. Phys. JETP 27, 227 (1968)
4.4   N.G. Basov, I.N. Kompanetz, O.N. Kompanetz, V.S. Letokhov, V.V. Nikitin: JETP Lett. 9, 345 (1969)
4.5   C.J. Borde: C.R. Ac. Sc. 271, 371 (1970)
4.6   T.W. Hänsch: M.D. Levenson, A.L. Schawlow: Phys. Rev. Lett. 26, 946 (1971)
4.7   C. Wieman, T.W. Hänsch: Phys. Rev. Lett. 36, 1170 (1976)
4.8   M.S. Sorem, A.L. Schawlow: Opt. Commun. 5, 148 (1972)
4.9   B. Couillaud, A. Ducasse: Opt. Commun. 13, 398 (1975)
4.10 D. Pritchard, J. Opt, T.W. Ducas: Phys. Rev. Lett. 32, 641 (1974); F. Biraben, B. Cagnac, G. Grynberg: Phys. Rev. Lett. 32, 643 (1974); M.D. Levenson, N. Bloembergen: Phys. Rev. Lett. 32, 645 (1974); T.W. Hänsch, K. Harvey, G. Meisel, A.L. Schawlow: Opt. Commun. 11, 50 (1974)

4.11   M.S. Feld, A. Javan: Phys. Rev. 177, 540 (1969)
4.12   A. Szabo: Phys. Rev. Lett. 25, 924 (1970)
4.13   R.I. Personov, E.I. Al'Shits, L.A. Bykovskaya: JETP Lett. 15, 431 (1974); also Opt. Commun. 6, 169 (1972)
4.14   A. Compaan: Phys. Rev. B5, 4450 (1972)
4.15   P.M. Selzer, D.L. Huber, B.B. Barnett, W.M. Yen: Phys. Rev. B17, 4979 (1978)
4.16   A.P. Marchetti, W.C. McColgin, J.H. Eberly: Phys. Rev. Lett. 35, 387 (1975)
4.17   P.M. Selzer, W.M. Yen: Opt. Lett. 1, 90 (1977)
4.18   N. Pelletier-Allard: Colloq. Int. CNRS 255, 289 (1977)
4.19   N. Motegi, S. Shionoya: J. Lumin. 8, 1 (1973)
4.20   R. Flach, D. S. Hamilton, P. M. Selzer, W. M. Yen: Phys. Rev. B15, 1248 (1977)
4.21   C. Brecher, L.A. Riseberg: Phys. Rev. B13, 81 (1976)
4.22   M.J. Weber, J.A. Paisner, S.S. Sussman, W.M. Yen, L.A. Riseberg, C. Brecher: J. Lumin. 12/13, 729 (1976)
4.23   L.E. Erickson: Phys. Rev. B11, 77 (1975)
4.24   A.P. Marchetti, M. Scozzafava, R.H. Young: Chem. Phys. Lett. 51, 424 (1977)
4.25   L.E. Erickson: Opt. Commun. 15, 246 (1975)
4.26   P.M. Selzer, D.L. Huber, D.S. Hamilton, W.M. Yen, M.J. Weber: Phys. Rev. Lett. 36, 813 (1976)
4.27   D.S. Hamilton, P.M. Selzer, W.M. Yen: Phys. Rev. B16, 1858 (1977)
4.28   R.S. Meltzer: Unpublished
4.29   See e.g., R. Loudon: *The Quantum Theory of Light* (Clarendon Press, Oxford 1975) Chap. 5
4.30   A.M. Stoneham: Rev. Mod. Phys. 41, 82 (1969)
4.31   R.M. Shelby, C.S. Yannoni, R.M. Macfarlane: Phys. Rev. Lett. 41, 1739 (1978)
4.32   L.E. Erickson: Phys. Rev. B16, 4731 (1977)
4.33   T. Kushida, E. Takushi: Phys. Rev. B12, 824 (1975)
4.34   B. Spencer: Unpublished
4.35   C. Delsart, N. Pelletier-Allard, R. Pelletier: Opt. Commun. 16, 114 (1976)
4.36   R.G. Bennett: J. Chem. Phys. 41, 3037 (1964)
4.37   W.B. Gandrud, H.W. Moos: J. Chem. Phys. 49, 2170 (1968)
4.38   M.J. Weber: Phys. Rev. B4, 2932 (1971)
4.39   R.K. Watts, M.J. Richter: Phys. Rev. B6, 1584 (1972)
4.40   N. Yamada, S. Shionoya, T. Kushida: J. Phys. Soc. Jpn. 32, 1577 (1972)
4.41   R.C. Powell, Z.G. Soos: J. Lumin. 11, 1 (1975) and references therein
4.42   D. Solson, S.M. George, T. Keyes, V. Vaida: J. Chem. Phys. 67, 4941 (1977) and references therein
4.43   P.M. Selzer, D.S. Hamilton, W.M. Yen: Phys. Rev. Lett. 38, 858 (1977)
4.44   D.L. Huber, D.S. Hamilton, B. Barnett: Phys. Rev. B16, 4642 (1977)
4.45   B.A. Wilson, J. Hegarty, W.M. Yen: Phys. Rev. Lett. 41, 268 (1978)
4.46   T. Holstein, S.K. Lyo, R. Orbach: Phys. Rev. Lett. 36, 891 (1976)
4.47   T. Holstein, S.K. Lyo, R. Orbach: Phys. Lett. 62A, 55 (1977)
4.48   P.A. Fleury, K.B. Lyons: Phys. Rev. Lett. 36, 1188 (1976)
4.49   J.E. Mack, D.P. McNutt, F.L. Roesler, R. Chabbal: Appl. Opt. 2, 873 (1963)
4.50   See e.g., R.B. Scott: *Cryogenic Engineering* (Van Nostrand, Princeton 1959)
4.51   See, e.g., W.A. Fitzsimmons, L.W. Anderson, C.E. Riedhauser, J.M. Vrtilek: I.E.E.E. J. QE 12, 624 (1976); M. Feldman, P. Lebow, F. Raab, H. Metcalf: Appl. Opt. 17, 774 (1978); M.G. Littman, H.J. Metcalf: Appl. Opt. 17, 2224 (1978;) I. Shoshan, U.P. Oppenheim: Opt. Commun. 25, 375 (1978)
4.52   T.W. Hänsch: Appl. Opt. 11, 895 (1972)
4.53   T.F. Gallagher: Unpublished
4.54   See, e.g., B. DiBartolo: *Optical Interactions in Solids* (John Wiley, New York 1968) p. 408

4.55 H. de Vries, D.A. Wiersma: Phys. Rev. Lett. **36**, 91 (1976)
4.56 S. Voelker, R.M. Macfarlane, A.Z. Genack, H.P. Trommsdorff, J.H. van der Waals: J. Chem. Phys. **67**, 1759 (1977)
4.57 A. Szabo: Phys. Rev. **B11**, 4512 (1975)
4.58 T. Muramoto, S. Nakanishi, T. Hashi: Opt. Commun. **21**, 139 (1977)
4.59 See, e.g., M. Ducloy, M.S. Feld: In *Laser Spectroscopy III*, ed. by J.L. Hall, J.L. Carlsten, Springer Series in Optical Sciences, Vol. 7 (Springer, Berlin, Heidelberg, New York 1977) p. 243; J.R.R. Leite, M. Ducloy, A. Sanchez, D. Seligson, M.S. Feld: Phys. Rev. Lett. **39**, 1469 (1977)
4.60 C. Delsart, N. Pelletier-Allard, R. Pelletier: Appl. Phys. Lett. **31**, 443 (1977)
4.61 J.J. Song, J.H. Lee, M.D. Levenson: Phys. Rev. **A17**, 1439 (1978)
4.62 N.A. Kurnit, I.D. Abella, S.R. Hartmann: Phys. Rev. Lett. **13**, 567 (1964)
4.63 R.G. Brewer, R.L. Shoemaker: Phys. Rev. **A6**, 2001 (1972)
4.64 R.L. Shoemaker: In *Physics of Quantum Electronic Series*, Vol. II, ed. by S.F. Jacobs, M. Sargent III, J.F. Scott, M.O. Scully (Addison-Wesley, Reading, Mass. 1974) p. 453
4.65 R.G. Brewer, A.Z. Genack: Phys. Rev. Lett. **36**, 959 (1976); A.Z. Genack, R.G. Brewer: Phys. Rev. **A17**, 1463 (1978) and references therein
4.66 A number of articles relating to coherent transient phenomena are found in *Laser Spectroscopy III*, ed. by J.L. Hall, J.L. Carlsten, Springer Series in Optical Sciences, Vol. 7 (Springer, Berlin, Heidelberg, New York 1977)
4.67 R.H. Dicke: Phys. Rev. **93**, 99 (1954)
4.68 E.L. Hahn: Phys. Rev. **77**, 297 (1950)
4.69 P.F. Liao, S.R. Hartmann: Opt. Commun. **8**, 310 (1973)
4.70 R.G. DeVoe, R.G. Brewer: Phys. Rev. Lett. **40**, 862 (1978)
4.71 A.H. Zewail: in *Laser Spectroscopy III*, ed. by J.L. Hall, J.L. Carlsten, Springer Series in Optical Sciences, Vol. 7 (Springer, Berlin, Heidelberg, New York 1977) p. 268
4.72 A. Szabo, M. Kroll: Opt. Lett. **2**, 10 (1978)
4.73 W.H. Hesselink, D.A. Wiersma: Chem. Phys. Lett. **56**, 227 (1978)
4.74 R.M. Macfarlane, A.Z. Genack, S. Kano, R.G. Brewer: J. Lumin. **18/19**, 933 (1979)
4.75 R.G. DeVoe, A. Szabo, S.C. Rand, R.G. Brewer: Phys. Rev. Lett. **23**, 1560 (1979)
4.76 P.D. Maker, R.W. Terhune: Phys. Rev. **137**, 801 (1965)
4.77 R.F. Begley, A.B. Harvey, R.L. Byer, B.S. Hudson: J. Chem. Phys. **61**, 2466 (1974); Appl. Phys. Lett. **25**, 387 (1974)
4.78 D. Heiman, R.W. Hellwarth, M.D. Levenson, G. Martin: Phys. Rev. Lett. **36**, 189 (1976)
4.79 I. Chabay, G.K. Klauminzer, B.S. Hudson: Appl. Phys. Lett. **28**, 27 (1975)
4.80 M.D. Levenson, J.J. Song: "Coherent Raman Spectroscopy", in *Coherent Nonlinear Optics*, ed. by M.S. Feld, V.S. Letokhov, Topics in Current Physics, Vol. 21 (Springer, Berlin, Heidelberg, New York 1980) p. 293–373
4.81 A. Kiel, S.P.S. Porto: J. Mol. Spectrosc. **32**, 458 (1969)
4.82 J.A. Koningstein, P. Grunberg: Can. J. Chem. **49**, 2336 (1971); also J.A. Koningstein: Colloq. Intern. C.N.R.S., **255**, 243 (1977)
4.83 B. Halperin, J.A. Koningstein: J. Chem. Phys. **69**, 3302 (1978)
4.84 For a review of this method, see Yoh-han Pao: *Optoacoustic Spectroscopy and Detection* (Academic Press, New York 1977)
4.85 For an efficient photoacoustic cell design, see J.F. McClelland, R.N. Kniseley: Appl. Opt. **15**, 2967 (1976)
4.86 M.B. Robin, N.A. Kuebler: J. Chem. Phys. **66**, 169 (1977)
4.87 A. Rosencwaig: Anal. Chem. **47**, 592A (1975)
4.88 R.S. Quimby, W.M. Yen: Opt. Lett. **3**, 181 (1978)
4.89 R.W. Hellwarth: J. Opt. Soc. Am. **67**, 1 (1977); S.M. Jensen, R.W. Hellwarth: Appl. Phys. Lett. **32**, 166 (1978); P.F. Liao, D.M. Bloom: Opt. Lett. **3**, 4 (1978)
4.90 D.W. Phillion, D.J. Kuisenga, A.E. Siegmann: Appl. Phys. Lett. **27**, 85 (1975)

4.91 R.G. Harrison, P. Key, V.I. Little, G. Magyar, J. Katzenstein: Appl. Phys. Lett. **13**, 253 (1968) T.A. Shankoff: Appl. Opt. **8**, 2282 (1969)
4.92 D.W. Pohl, S.E. Schwarz, V. Irniger: Phys. Rev. Lett. **31**, 32 (1973)
4.93 D.S. Hamilton, D. Heiman, Jack Feinberg, R.W. Hellwarth: Opt. Lett. **4**, 933 (1979)
4.94 J.R. Salcedo, A.E. Siegman, D.D. Diott, M.D. Fayer: Phys. Rev. Lett. **41**, 131 (1978)

5. High Resolution Laser Spectroscopy of Ions in Crystals

W. M. Yen and P. M. Selzer

With 25 Figures

It is the intent of this chapter to review specific results obtained when the laser spectroscopic methods presented in Chap. 4 are used in the investigation of ions in crystalline solids. Specifically, the majority of the work that has been done up to date and which will be surveyed here has been addressed to optically active ions in insulators. The general features of the spectra of these materials are well established and form the basis of the tutorial presented in Chap. 1. Of course, the application of new empirical technique invariably results in new findings, and this area of condensed matter Physics is no exception. Principally, the higher resolutions attainable in the frequency and temporal domains have given us a much more detailed view of the processes which affect the optically excited states of solids.

This chapter is divided into a part dealing with static and one dealing with dynamic features of the spectra of these crystalline systems. Section 5.1 surveys the status of studies of homogeneous line-broadening mechanisms in dilute crystals. The second section continues with a brief survey of high resolution laser measurements of other static features where the suppression of inhomogeneous statistical contributions has played an important role. Dynamic features are reviewed in Sect. 5.3 which deals with interline and intraline energy transfer processes arising from ion-ion interactions and generally occurring at higher dopant concentrations. These results are relevant to the theoretical treatments appearing in Chaps. 2 and 3. A review of the dynamical properties of concentrated materials is reserved for the final section; here various excitonic effects become prevalent and again we wish to illustrate that tunable laser spectroscopy can be used to good advantage.

In each category, we have selected particular systems for detailed review and have referred to the relevant literature which illustrates the theme of this volume.

5.1 Homogeneous Linewidth Studies

As has been noted in Chap. 4, the various laser techniques surveyed there can be used to suppress contributions arising from random sources and thus allow us to derive intrinsic properties of transitions. One such intrinsic property is the linewidth of the transition; for a metastable fluorescent state at zero temperature in the limit of no other interactions, this width should simply reflect the radiative lifetime of the state as is required by the uncertainty principle. Fluorescence line narrowing (FLN), optical hole burning and optical coherent transients have all been recently employed in the study of this problem in dilute systems with considerable success. Several hitherto unresolved features in the static spectra have also been obtained through these studies even as the homogeneous linewidths measured begin approximating their theoretical limits.

The theory of optical line broadening in solids is intimately connected with phonon-induced effects familiar in ESR spin relaxation; this has been reviewed briefly in Chap. 1 and more extensively elsewhere [5.1]. Through earlier studies, it has been found that the homogeneously broadened widths of transitions in solids may in general be described by a combination of single-phonon and multi-phonon processes interacting with the electronic states. Invariably in these earlier studies, which are referenced below, it was necessary to mathematically deconvolute the inhomogeneous contributions to arrive at the intrinsic width. This procedure complicated the analysis for one, and of course limited the accuracy and prevented the observation of any structure buried beneath the inhomogeneous envelope. This is true particularly in regions where the intrinsic width is small or comparable to the average inhomogeneous contribution, i.e., at low temperatures where phonon-induced relaxation is relatively slow. Subsequent to the development of various laser techniques alluded to above, direct intrinsic width or relaxation measurements have been made possible in these temperature regions, and it is indeed here that laser techniques have made impressive contributions.

In pursuing optical line-broadening mechanisms in ionic crystals, two systems, $Al_2O_3:Cr^{3+}$ (ruby) and $LaF_3:Pr^{3+}$, have been examined more extensively than any others. These crystals, exemplary of $3d$ and $4f$ systems respectively, have been studied with both frequency- and time-domain experiments in the low-temperature regime and serve generally to illustrate the historical progression of the application of increasingly sophisticated and maturing laser spectroscopic techniques. In each case we shall see that capabilities recently developed are allowing linewidth measurements approaching the theoretical intrinsic limit.

5.1.1 Ruby [$Al_2O_3:Cr^{3+}$]

The level scheme of Cr^{3+} $(3d)^2$ in Al_2O_3 appears in Figs. 1.2, 1.3. The transitions of interest in this case entail the trigonal components of the ^2E excited state and the ^4A ground state and comprise the well-known R_1 and R_2 lines. The fluorescence lifetime of the R_1 transition at low temperatures is of the order of a few ms ; thus its linewidth has an intrinsic lifetime limitation of a $\sim 10^2$ Hz.

The temperature dependence of the R_1 and R_2 lines in ruby was first measured by *Schawlow* [5.2] and later by *McCumber* and *Sturge* [5.3]. The latter determined that above 77 K, the homogeneous linewidth could be described by assuming dephasing in the excited ^2E state via Raman scattering of phonons. Below 77 K, however, strain broadening predominated and masked the homogeneous contribution. The magnitude of the strain broadening in a good crystal is of the order of a few tenths of a cm^{-1} (1 cm^{-1}=30 GHz). Later studies of Cr^{3+} in a cubic system (MgO) confirmed these results, and *Imbusch* and his co-workers [5.4] first carried out the mathematical deconvolution necessary to analyze the behavior of this single $^2E \rightarrow {}^4A$ transition at lower temperatures. During this same period, extensive ESR studies were conducted on ruby. Of particular interest to this discussion are the optically detected excited-state ESR studies by *Geschwind* et al. [5.5, 6] which provided information regarding relaxation processes occurring between the two ^2E excited-state levels, $2\bar{A}$ and $\bar{E}$, separated by 29 cm^{-1}. This direct-phonon process contributes to the temperature dependence of the R_1 line below 77 K; however, the additional broadening was not observed optically till much later by *Muramoto* et al. [5.7].

Historically and because of the early availability of high-power, temperature-tuned ruby lasers, the first experiments to measure line-broadening mechanisms at low temperatures were conducted on the R_1 line by *Kurnit* et al. [5.8] in 1964. This experiment has become a classic inasmuch as it first demonstrated the existence of the photon echo in general. *Abella* and his co-workers measured the decay of the echo intensity as a function of $\pi/2$ to π pulse separation and as a function of temperature [5.9]. While their data were not accurate enough to quantify results, they showed that the echo decay rate in time increased dramatically between 4 and 14 K (as would be expected from the exponential dependence of the Orbach process affecting the $\bar{E}$ and $2\bar{A}$ states) [5.5]. It was also established in this work that a magnetic field must be applied to the sample, along the c or optic axis in their case, in order for the echoes to be observed. The authors properly hypothesized that the presence of a sufficiently strong magnetic field inhibited an additional relaxation mechanism, to wit, dephasing due to local fields at Cr sites produced by the Al nuclear moments.

Subsequent work by *Compaan* et al. [5.10], *Lambert* et al. [5.11] and *Compaan* [5.12] provided more extensive photon echo data in ruby. Decay rates were measured as a function of Cr^{3+} concentration, magnetic field, laser intensity and field orientation. Particularly noteworthy is that considerable pulse separations, extending to 600 ns, were achieved. Echo envelopes were found to decay nonexponentially as

$$I = I_0 \exp(-K\tau^{1/2}) \qquad (5.1)$$

where τ is the $\pi/2-\pi$ pulse separation and K is a constant proportional to $n^{1/2}$ where n is the Cr concentration. These results by *Compaan* seemingly implied ion-ion dynamics reminiscent of the two pulse diffusion results of *Hahn* in NMR [5.13].

In 1973, *Liao* and *Hartmann* [5.14] reported new measurements in ruby obtained with a gated cw laser rather than using pulsed systems with optical delays. In direct contradiction with *Compaan*, these experimenters found strictly exponential echo delays over many e foldings and persisting some five times longer than had been previously measured. These results are shown in Fig. 5.1; there typically for a concentration of 0.05 at.% at 3 K, the decay time is 600 ns, implying an optical spin dephasing time, T_2, of 2.4 μs or a homogeneous width of 130 kHz. These longer, exponential decays have subsequently been reconfirmed by Stark switching experiments [5.15]. It is thus very likely that the earlier results which exclusively utilized pulsed flash lamp ruby sources suffered from experimental artifacts. A possible source of these technical difficulties might lie in the manner the delay time between pulses is generated; in a White cell optical delay line used by the Chicago group, increased delays are generated by increasing the number of reflections in the cell. By increasing the number of passes, deterioration of the beam quality and hence the focusability of the π pulse might be expected, thus influencing the spatial overlap of the two exciting beams. The net result would be in effect a reduction in the net rotation produced by the π pulse and hence a spuriously enhanced lost in coherence [5.16].

A recent work by *Samartsev* and co-workers [5.17] has extended the analysis of echo phenomena to include three driving pulses and these authors have experimentally observed stimulated and multiple echoes in ruby, as well as echo delays and advances. Unfortunately, their pulse delays did not extend beyond a few hundred nanoseconds and relaxation data were not obtained.

Despite a certain amount of discrepancy left in the absolute magnitude of the relaxation parameters, photon echoes, free induction decay (FID) and the rela-

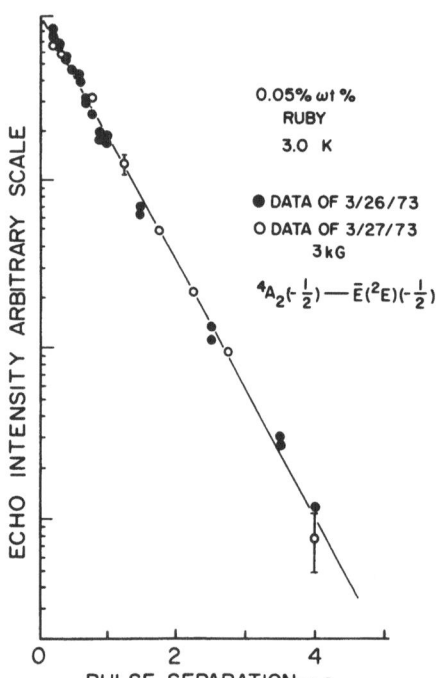

Fig. 5.1. Photon echo intensity vs pulse separation in the R_1 line of ruby showing exponential decay over several decades. The appropriate T_2 value is four times the measured decay in the figure. This arises from the position of the echo (2τ) and the definition of T_2 vis-a-vis the echo amplitude [5.14]

ted echo ENDOR (designated PENDOR) [5.18] double resonant experiments have provided a wealth of qualitative if not quantitative information about the nature of the interactions affecting the relaxation in the R_1 lines in ruby. For example, the Al-Cr nuclear hyperfine interactions in both the ground and the excited states have been obtained from PENDOR measurements [5.19, 20] and have been tabulated by *Meth* and *Hartmann* [5.21]. Echo modulations due to off axis magnetic field geometry have also been extensively investigated and have yielded good agreement with the above-mentioned PENDOR results. The variation of the decay rate, i.e., increasing as $n^{1/2}$ where n is the Cr^{3+} concentration, has also been verified in several different experiments, but as to this writing no adequate model has been proposed to account for this effect [5.12, 14].

A parallel effort to measure homogeneous linewidths of ruby in the frequency domain has been undertaken by *Szabo* and others. To this end, the technique of FLN was employed by *Szabo* for the first time in crystals[5.22]. The early measurements which used a cw laser succeeded in narrowing the resonant R_1 fluorescence to a width of approximately 120 MHz in $H = 0$ and 40 MHz when a field, $H \parallel c$, was applied for a sample with 0.1 at.% Cr^{3+} at 4.2 K [5.23]. These widths, while far narrower than any other previously measured optical width in a solid, were still considerably larger than the roughly 180 kHz predicted for this concentration [5.14]. The inhomogeneous character of the narrowed line in these experiments was further supported by its non-Lorentzian shape and the fact that the narrowed width scaled more strongly with concentration than the expected square root dependence found in echo measurements. Once again, superhyperfine interactions were proposed to account for the residual width.

In an effort to improve the spectral resolution, *Szabo* then used optical hole burning and employed an amplitude modulation scheme to generate both pump and probe beams from the same cw laser [5.24]. In a 0.03 at.% sample at 4.2 K, he obtained a deconvoluted width of roughly 70 MHz in zero field and an instrumental limited width of 5 MHz in an applied field of 400 Oe. A second, broader hole was observed as a background in the applied field experiments and a ground state cross relaxation mechanism was proposed to account for this effect. Unfortunately, the bandwidth limitations of the modulation scheme prevented the probe beam from scanning more than 35 MHz from the pump frequency.

A slightly different hole burning method was developed by *Muramoto* et al. [5.25] where instead of modulating the cw pump beam to obtain a sideband probe, these authors applied a sawtooth Stark field to the sample after burning a hole in the absorption with a cw laser. The Stark shift of the R_1 line then served to scan the hole past the fixed frequency pump as the transmission was monitored. The sawtooth was produced in a time scale long compared to T_2 and yet very short relative to the radiative lifetime of the R_1 level. The reported results were generally consistent with the earlier work, indicating a width of the order 100 MHz at zero field and narrowing to $\sim$ 5 MHz in a 220 Oe magnetic field. This experimental arrangement allowed larger probe scanning capability, and the authors were also able to detect two side holes arising from cross relaxation to other ground-state

Zeeman levels. The broad background hole observed by *Szabo* was not seen and may have been spurious. Results from this work are shown in Fig. 5.2. These results were later reconfirmed by an analysis of Stark switched coherent transients [5.26].

Recently cw FLN experiments have continued to be refined in their resolution. One such trace is illustrated in Fig. 5.3 derived from a recent experiment by *Jessop* [5.27]. Though in this figure and Fig. 5.2 the intrinsic limits have not been attained, they clearly illustrate the impressive resolutions which are almost now routinely achievable using laser spectroscopic techniques. It is perhaps worthwhile reminding those readers who are used to spectroscopic units that 10 MHz is equivalent to $1/3$ of a $m\,cm^{-1}$.

In another recent FLN experiment in ruby by *Selzer* et al. [5.28], other interesting features of the low-temperature R_1 widths have been investigated. With no applied field, they found that with the laser excitation near line center of the inhomogeneously broadened line, the resulting FLN widths were consistent with the 100 MHz value of *Szabo* and *Muramoto*. However, as the laser was tuned into the wings of the inhomogeneous absorption, particularly on the low-energy side, dramatic increases in the resulting fluorescence linewidths were observed.

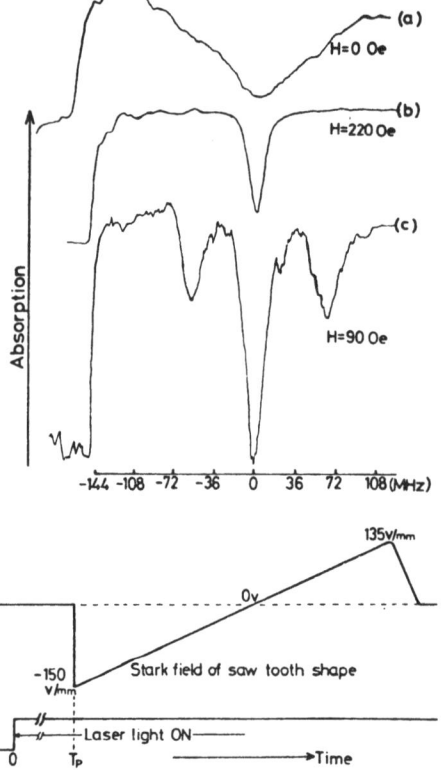

Fig. 5.2. Traces of the observed hole burned into the R_1 transition of ruby at low temperature using a Stark switching-scanning technique. The shape of the applied Stark field and the laser on-off sequence are shown in the inserts below. Note the narrowing of the hole as a function of applied field and the appearance of the two side holes in *c*. The shape of the holes is Lorentzian [5.25]

Both the detuning wavelength from line center for the onset of this effect and the linewidth values showed strong concentration dependences. This is shown in Fig. 5.4. From an analysis of the data it was concluded that these lines could not be homogeneously broadened, but most likely exhibited unresolved splitting, or lift-

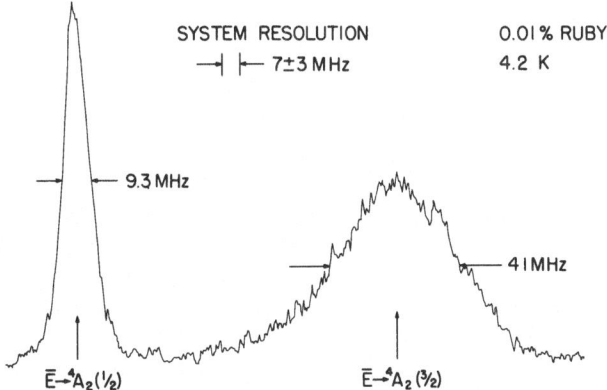

Fig. 5.3. Fluorescence line narrowing in the R_1 transition of ruby. The two lines correspond to transitions to the ground-state component split in a H field. The trace illustrates the resolutions attainable using cw FLN techniques and still contains inhomogeneous components. System resolution is 9 MHz with a 400 Oe field applied $\parallel c$. [5.27]

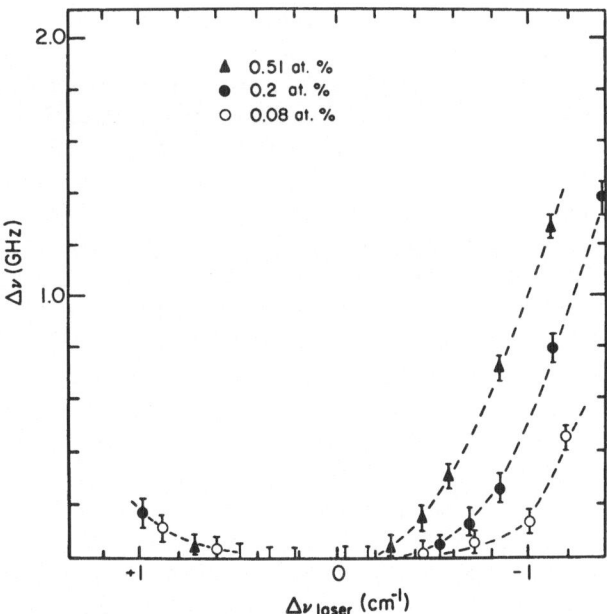

Fig. 5.4. Measured FLN resonant linewidths in the R_1 line of ruby shown as a function of concentration and as a function of laser pump frequency within the inhomogeneous $^4A \to \bar{E}$ transition. Sample temperature was 10 K [5.28]

ing of the degeneracy of the relevant levels due to weak pairing or clustering inter-actions between distant Cr ions. It might be that effects attributable to such pair-ing have also been observed in photon echo measurements [5.12], and in ESR of the ground state [5.29]. It would therefore appear that the inhomogeneous absorp-tion or emission lines of ruby have a more complex structure than expected from merely a statistical distribution of energy levels in a random strain field.

Our knowledge of the line-broadening mechanisms in ruby to date may thus be summarized as follows: At liquid helium temperatures, Cr-Al superhyperfine interactions determine both the homogeneous and the residual inhomogeneous linewidths of the ruby R_1 line. Experiments utilizing coherence would indicate that the homogeneous width is less than 100 kHz at line center, for low concen-trations (0.01 at.%) and in sufficiently large applied fields ($\sim$ 1 kOe). FLN and hole burning experiments are at least consistent with this value if it is assumed that the observed widths are instrumental resolution limited; however, even a lar-ger spectral width would not contradict the coherent transients results because of residual inhomogeneous broadening effects as has been noted in Chap. 4.

With no applied fields, the random alignment of the Cr and Al spin produces echo decays which are too rapid to measure. Assuming an instrumental response lower limit of 20-50 ns for a typical echo experiment, a homogeneous width $>$ 2 MHz would be implied. Extrapolation of the applied magnetic field data of [5.14] to zero field would be consistent with this value. The spectroscopically, i.e., FLN etc., determined width of approximately 100 MHz is therefore likely to contain residual inhomogeneous components — a speculation which is supported by the Voight profile of the lines [5.23].

The linewidth of R_1 then increases as a function of increasing temperature and manifests the direct and Raman terms arising from phonon-ion interactions (1.18, 19). Excited state ESR measurements [5.5], direct FIR laser measure-ments of the $R_1 \leftrightarrow R_2$ ($2\bar{A} \leftrightarrow \bar{E}$) relaxation [5.30] and the R_2 lifetime mea-surements of *Rives* and *Meltzer* [5.31] have helped to determine the value of the direct-process coefficient. It would appear that the spin-flip and non spin-flip relaxation times at low temperatures are 16 ns and 1 ns respectively; the coefficient of broadening from the resonant process is a combination of these two quantities and appears to be of the order of 0.013 cm^{-1} [5.7]. This term makes a compara-ble contribution to Raman scattering in the 15-30 K region; then the latter becomes dominant as the temperature is further raised. Thermal line-shift measurements of the R_1 line [5.32] and theoretical calculations [5.33] have also appeared and are consistent with our general understanding of these effects.

The concentration dependence of the echo decay implies that Cr-Cr inter-actions may enhance the dephasing process. At the present time, it is not known whether this effect is related to optical energy transfer, involves ground-state cross relaxation, or arises because of Cr clustering. Further discussion of energy transfer in ruby and its possible relationship to homogeneous line broadening is reserved for Sect. 5.3.

5.1.2 Trivalent Praseodymium in Lanthanum Fluoride [$LaF_3 : Pr^{3+}$]

$LaF_3 : Pr^{3+}$ is a lanthanide or rare earth crystal system which has been most thoroughly studied with many of the laser techniques discussed in Chap. 4. The LaF_3 crystalline host is an easily grown nonhygroscopic crystal of low symmetry in which substantial quantities of Pr^{3+} may be substitutionally introduced. Pr^{3+} has two active 4f electrons (Fig. 1.1) which give rise to a number of transitions at convenient visible tunable laser wavelengths. Though the local symmetry of the active ion sites in LaF_3 remains somewhat controversial [5.34], it is sufficiently low (C_{2v} at most) to lift all degeneracies in all the $4f$ Stark levels, except the Kramer's degeneracy when that applies.

Two levels have been studied principally, namely the $^3P_0$, with a ground state transition at 478 nm and having a radiative lifetime of ~ 40 μs, and the $^1D_2$ located at 592 nm and having a lifetime of 0.5 ms because of spin selection rules.

Temperature-dependent linewidth and line shift measurements in low-concentration samples were conducted by *Yen* et al. [5.35] on the $^3P_0$ state of Pr^{3+} in this compound. Because of the more complex $4f$ energy level scheme vis-a-vis ruby, the temperature dependence of the linewidths was fitted with various coefficients required by phonon-relaxation theory. In these earlier studies, the effects due to resonant phonons (direct process) were first demonstrated. Similar measurements were made later in a Kramer's system to make direct contact with ground state ESR T_1 studies [5.36]. Again as in ruby, the low-temperature linewidths were found to contain strain-broadened contributions which required mathematical deconvolution.

Erickson [5.37] first applied laser spectroscopic techniques to this sytem. He used a cw dye laser to study the FLN of the nonresonant $^1D_2 \rightarrow (^3H_5)_1$ [1] transition and found that phonon-relaxation theory accounted adequately for the linewidth behavior above 20 K; below this temperature, an unexplained residual linewidth of approximately 1 GHz persisted. This width was later shown to be the residual inhomogeneous broadening of nonresonant transitions arising from crystalline field parameters, and is generally termed the "accidental coincidence" effect in FLN [5.38,39] (see Chap. 4).

Somewhat later, *Erickson* [5.40] also studied the $^1D_2 \rightarrow (^3H_4)_1$ resonant transition where he employed a double-chopper technique to avoid laser saturation of the PMT. With these measurements, he obtained a good fit to phonon-induced relaxation theory above 10 K, but a small residual width of ~ 10 MHz, somewhat larger than the instrumental resolution, persisted; his results are shown in Fig. 5.5. Nuclear hyperfine interactions were hypothesized as a cause of this residue, and it is also certainly possible that other mechanisms including laser jitter may have contributed to these results.

---

[1] For $4f$ ions we use the notation $(X)_i$, where the subscript denotes the ith component of the X Stark manifold with $i = 1$ being the lowest in energy.

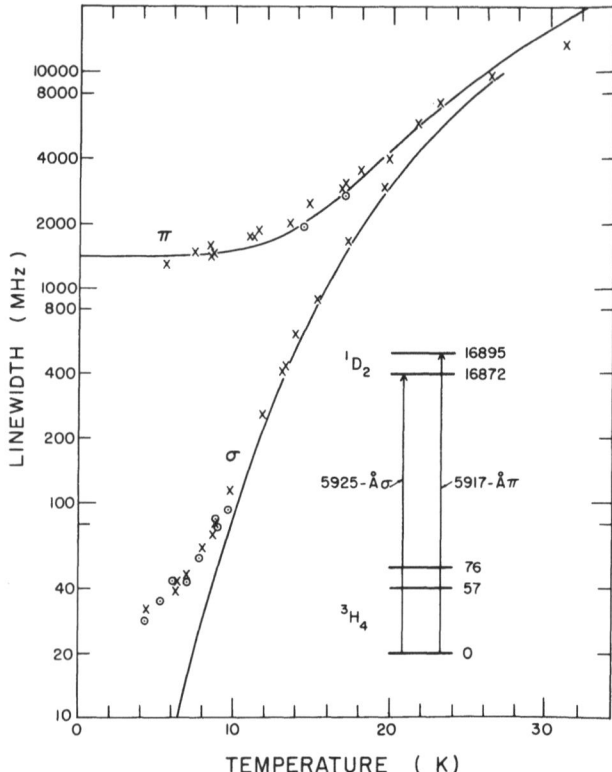

Fig. 5.5. Temperature dependence of the linewidths of the $^1D_2 \rightarrow (^3H_4)_1$ transition in LaF$_3$: Pr^{3+}. The limiting width π transition reflects the direct-phonon relaxation probability of the 16895 state. The value of the σ transition at low temperatures reflects the presence of hyperfine components. Solid lines are theoretical curves from phonon-relaxation theory [5.37]

A similar FLN study was conducted by *Flach* et al. [5.39] on the resonant $^3P_0 \rightarrow (^3H_4)_1$ fluorescence. The pulsed nature of this experiment limited the resolution such that only an upper limit of approximately 20 MHz could be placed on the 1.5 K width of this transition. Again and not surprisingly, the total temperature dependence of the linewidth agreed well with predictions. Consistency with the earlier results was attained in these two studies. The determination of the direct-process coefficients was much more accurate owing, of course, to the higher resolutions and suppression of the inhomogeneous contributions. The source of the temperature variation of linewidths in 4f systems is thus well accounted for by phonon relaxation theory.

Erickson and others then concentrated their efforts on the residual broadening remaining at low temperatures. Following a similar path as that taken in ruby, *Erickson* turned from FLN to hole burning in an effort to improve resolutions achievable [5.41]. A more or less parallel effort in the time domain was undertaken by *Takeuchi* [5.42, 43], *Genack* [5.44], *Yamagishi* [5.45] and *Chen* [5.46] with their respective co-workers, using coherent transient techniques. Historically, the first echoes in a rare earth system were observed in $LaF_3:Nd^{3+}$ in 1972 by *Chandra* et al. [5.47], but, by and large, coherent effects have been pursued almost exclusively in the Pr^{3+} system.

Erickson's measurements using hole burning [5.41] and optical-rf double resonance [5.48] revealed a limiting linewidth of 200 kHz for the $(^3H_4)_1 \rightarrow (^1D_2)_1$ absorption. This is the narrowest optical line in a solid measured to date using direct spectroscopic techniques. Note, however, that a Lorentzian line shape was assumed in the deconvolution in order to obtain this value. A Gaussian shape, arising from dipolar or Van Vleck broadening, would result in a deconvoluted linewidth more than twice as large. Nevertheless, *Erickson's* linewidth value represented a milestone and was at least consistent with the initial values derived from FID [5.44] and echo measurements [5.45] [the latter made on the $(^3H_4)_1 \rightarrow {}^3P_0$ transition].

Optically detected spin-transient experiments by *Shelby* et al. [5.49], on the $^1D_2$ transition demonstrated that the ground-state widths of *Erickson*, obtained with double resonance, were still inhomogeneously broadened. An implication of this work was that the optical linewidth measurement would have a similar strenuous component. This suggestion was substantiated by the refined optical FID measurements of *Macfarlane* et al. [5.50], in which a measured T_2 of 480 ns ($\Delta\nu$=660 kHz) was shown to be primarily due to laser jitter. Similarly, more careful photon echo measurements by *Chen* et al. [5.46, 51] on the $^3P_0$ state revealed a modulation in the echo decay which for various reasons was not detected in [5.43]. Nearly complete rephasing of the echo was observed to occur in zero field at a pulse spacing of 240 ns, suggesting that the real T_2 was considerably longer than the 490 ns quoted in the earlier work.

More recently, *DeVoe* et al. [5.52] using a highly stabilized dye laser substantially reduced laser jitter and obtained a FID decay as long as 16 μs in the $^3H_4 - {}^1D_2$ transitions in $LaF_3:Pr^{3+}$. Perhaps we are approaching the last word on this problem: *Macfarlane* and co-workers [5.53] eliminated the effects of laser jitter by using a gated high resolution (1 MHz) dye laser to produce photon echoes in Pr^{3+} doped $YAlO_3$ and LaF_3. They obtained linewidths of 2.0 kHz and 5.0 kHz in the $(^1D_2)_1$ state respectively with an applied field of 80 Oe. These high resolving powers are now comparable with Mössbauer resolutions. Similarly, by using rf-laser double resonance techniques, *Rand* et al. [5.54] irradiated the ^{19}F NMR thus reducing dipolar fluctuations at the Pr^{3+} sites in $LaF_3:Pr^{3+}$, the $(^1D_2)_1$ FID was observed to narrow to $\sim$ 2 kHz as well. These measurements are now close to their respective T_1 limits. Some of the recent FID results are shown in Fig. 5.6. *Chen* and his co-workers [5.55] have also extended their photon echo measurements of the $^3P_0$ state in $LaF_3:Pr^{3+}$ to longer pulse separations (up to 10 μs). Though the echo modulation alluded to earlier persists, the envelope decays exponentially with a linewidth of 40 kHz for a 0.01 at.% Pr^{3+} sample (see Figs. 5.8, 9).

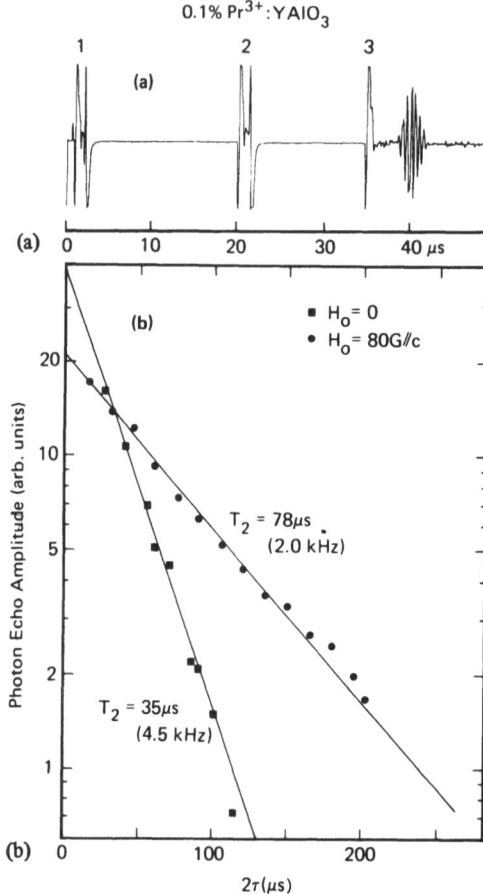

Fig. 5.6a, b. Heterodyne detected photon echoes in the $^1D_2$ state of Pr^{3+} in $YAlO_3$. (a) Shows the echo observed by this technique: 1 is the $\pi/2$ pulse, 2 the π pulse and 3 marks the beginning of the heterodyne pulse. (b) Echo delays recorded at 1.9 K in no field and in 80 G field; the line widths are HWHM and represent the narrowest optical widths observed to date. For details see [5.53]

An additional concentration dependent dephasing likely arising from ion-ion interactions is also reported in this work.

These experimental methods have been applied to Eu^{3+} most recently and the linewidths have been found to be considerably narrower [5.56].

5.1.3 Actinides and Other Centers

Generally the study of the optical properties of the trivalent actinides ($5f$) ions in solids has been limited to the identification of spectral features. Only recently has a comprehensive survey of the actinides in hosts such as $LaBr_3$ and $LaCl_3$ appeared and theoretical attempts in dealing with crystalline field effects intermediate between rare earths and transition metals have been undertaken.

Currently attention has been turned to phonon-ion interactions in these systems. *Hessler* et al. [5.57] have measured the linewidth behavior of the FLN fluorescence of the $D_1 \rightarrow Z_1$ transition of $LaCl_3:Np^{3+}$ shown in Fig. 5.7a. The

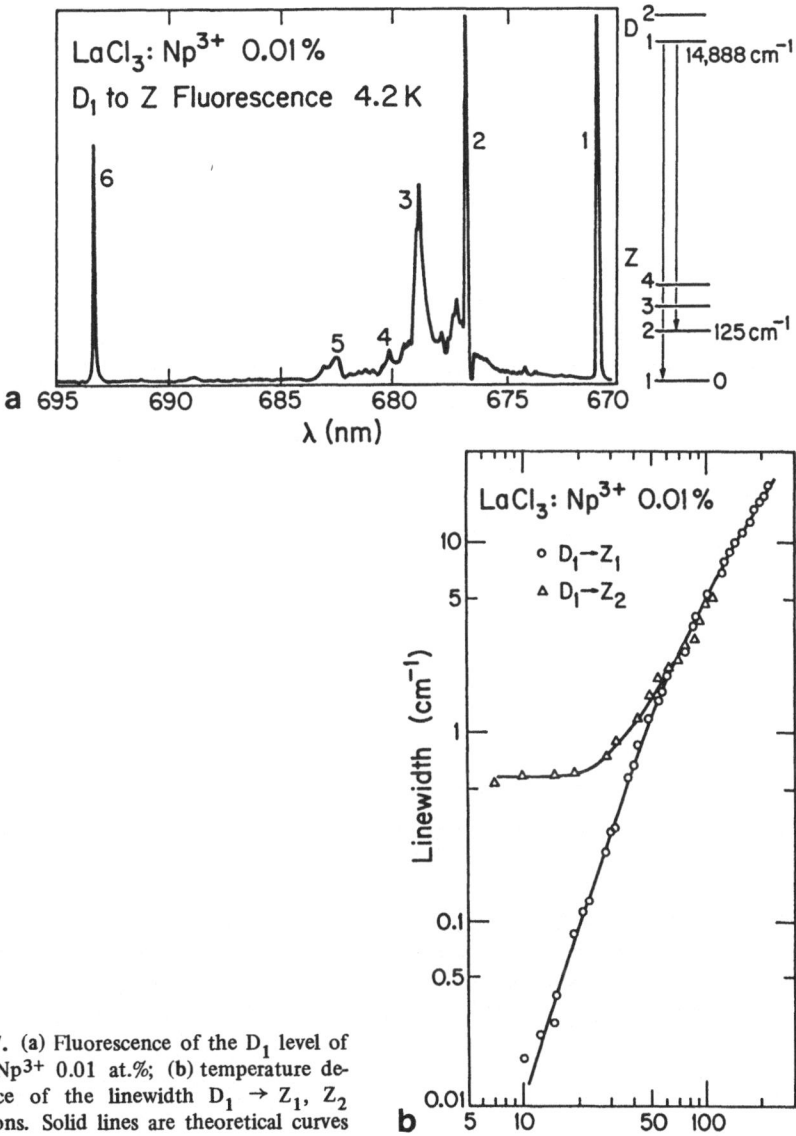

Fig. 5.7. (a) Fluorescence of the D_1 level of LaCl$_3$:Np^{3+} 0.01 at.%; (b) temperature dependence of the linewidth $D_1 \rightarrow Z_1$, Z_2 transitions. Solid lines are theoretical curves derived from phonon-relaxation theory [5.57]

increased lattice coupling relative to its rare earth counterpart, Pm^{3+}, is evidenced by the presence of well-developed phonon sidebands accompanying the transition and by the possible occurrence of Fano type resonances in the first excited ground-state level. The temperature dependence of the linewidth is shown in Fig. 5.7b. The low-temperature value is limited by the pulsed laser resolution and is of the order of 100 MHz. Again, the widths are well described by direct process relaxation in the ground and excited states and by a nonresonant Raman scattering term. The coefficients of these processes may be commensurate with larger phonon-ion interaction strengths.

Macfarlane et al. have used both coherent transients [5.58] and hole burning techniques [5.59] to study low-temperature properties of aggregated color centers in alkali halide crystals. It has generally been understood that these systems exhibit sharp, zero phonon lines which are inhomogeneously broadened at low temperatures with typical widths of the order of 0.1 nm [5.60]. In particular, *Macfarlane* studied the $^1A_1 \rightarrow {}^1E$ transition of the F_3^+ center in NaF and obtained consistent results between both experiments, giving a dephasing time of 18 ns and a width of 17 MHz, respectively. It was determined that the broadening of the line resulted from population decay from the excited state of 10 ns time constant, corresponding to an instance where $T_2 = 2T_1$.

Interestingly, in their hole burning experiment the above authors observed a two-phase recovery with time constants of 2.5 s and 70 min. These two components were determined, respectively, as recovery from population hole burning due to a metastable triplet state reservoir and from photochemical hole burning in which active electrons tunnel to an adjacent center. The specificity in frequency ($\sim$ 1 MHz) and in position within the sample and the duration of the hole hold obvious implications for storage technology.

Finally, we have already mentioned the photon echoes observed in LaF_3: Nd^{3+}. Similar work exists on Nd^{3+} in $CaWO_4$ and YAG [5.61].

5.2 Static Spectral Features

Clearly the possibility of suppressing inhomogeneous features in the spectra of ions in solids through the use of laser excitation increases our ability to derive intrinsic single-ion properties. This is true regardless of the nature of the inhomogeneous fluctuations. In this section, we will survey experiments in which the primary emphasis has been the elucidation of spectral features rather than the determination of line-broadening phenomena.

5.2.1 Hyperfine and Other Structures

As has been alluded to in the previous section, work towards the resolution of the line broadening problem has also elucidated various mechanisms which lead to hyperfine and more refined interactions. Generally, the dopant material has been Pr^{3+}; in this case it is not only because of its desirable optical properties but also because Pr has only one stable isotope with a nuclear spin $I = 5/2$.

For the case of $LaF_3:Pr^{3+}$ the symmetry of the ion sites lifts all degeneracy, and hence all its levels are singlets. In the absence of magnetic fields there is no magnetic hyperfine interaction. The nucleus, however, has a small but finite quadrupole moment which when allowed to interact with the electronic functions,

creates 3 doubly degenerate levels. Operative selection rules for transitions between excited- and ground-state levels allow for population changes through optical pumping; its subsequent readjustment by applied rf fields leads to detection of
the rf-optical double resonance [5.48]. Similarly, the simultaneous coherent excitation of the multiplets in the hyperfine structure is responsible for the beat modulation of the photon echoes in the time domain. The structure in the frequency
domain may be recovered by Fourier transformation of the echo envelope [5.62].
Using these methods, the quadrupole–induced hyperfine structure obeys Hamiltonians of the type

$$\mathcal{G} = P[I_z^2 + \eta(I_x^2 - I_y^2)/3] \tag{5.2}$$

where the quadrupole interaction parameter P and the quadrupole asymmetry
parameter η are specific to the electronic states in question. In this notation the
measured parameters were $P = 4.185$ MHz and $\eta = 0.105$ for the $^3H_4$ ground state
[5.48, 62], -1.10 MHz and -0.48 for the $^1D_2$ and 0.293 MHz and 0.516 for the
$^3P_0$ excited states. The ground-state and excited–states principal axes for their
respective Hamiltonians differ. Recent results of *Chen* et al. [5.62] in the Fourier
transformation of photon echo modulation are shown in Figs. 5.8, 5.9.

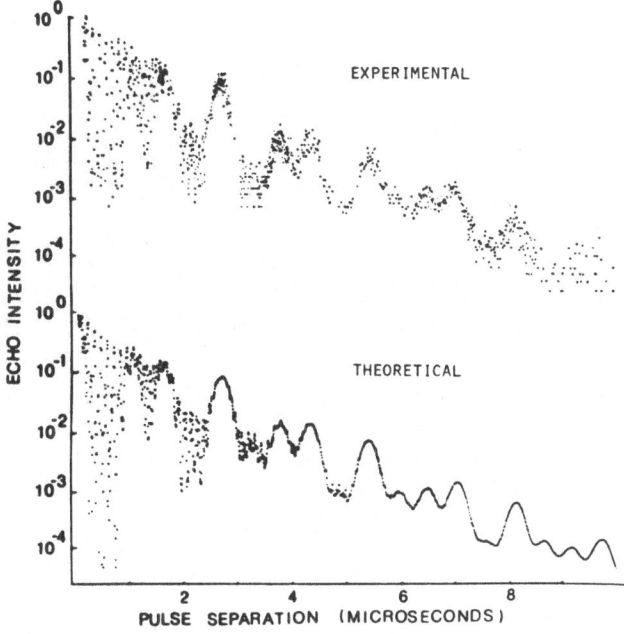

Fig. 5.8. Photon echo intensity shown as a function of laser pulse separation in the $^3P_0 \rightarrow$
$(^3H_4)_1$ transition of $LaF_3:Pr^{3+}$ (0.03 at.%) at low temperatures. The modulation arises from
coherent interference of the various quadrupolar levels. Results of theoretical modeling are
shown in the lower trace. [5.62]

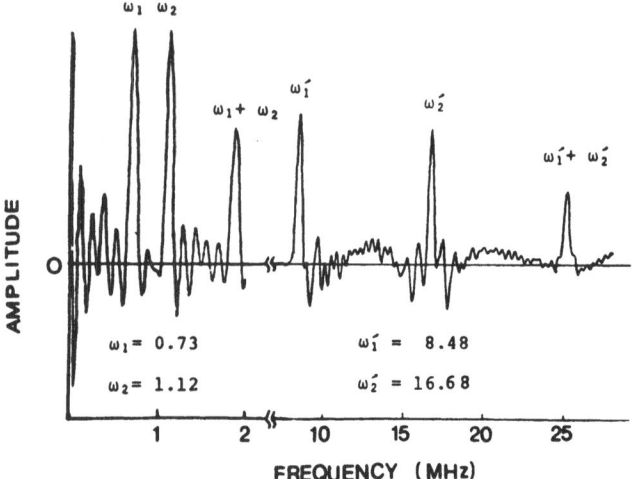

Fig. 5.9. Fourier transform of results shown in Fig. 5.9 for LaF_3:Pr^{3+}. The various peaks correspond to ground $(^3H_4)_1$ and excited $(^3P_0)$ state quadrupolar splittings. The frequencies ω listed in Fig. 5.9 are the fundamental frequencies of these splitting with primes referring to the excited state. [5.62]

Delsart et al. in a sequence of papers have successfully investigated the magnetic hyperfine interaction in other crystalline systems containing Pr^{3+}. Matrices such as $LaAlO_3$, $LaCl_3$ and $LaBr_3$ have higher symmetries than LaF_3, resulting in a doublet ground state characterized by the magnetic quantum number $\mu = \pm 2$. The magnetic interaction between the electronic crystalline field states and the nuclear spin results in six equally spaced, doubly degenerated states in the $(^3H_4)_1$ state. Using cw FLN, these authors found in their work that the minimum fluorescence width obtainable in $LaAlO_3$:Pr^{3+} was 2.4 GHz — too broad to resolve any hyperfine structures [5.63]. They ascribed this width to a rapid relaxation between the hyperfine sublevels, but it is more likely that inhomogeneous sources may have been the cause. In later work, on $LaCl_3$ and $LaBr_3$ doped with Pr^{3+}, the $(^1D_2)_1 \rightarrow$ $(^3H_4)_1$ transition was observed to have an instrumental limited width at low temperatures, and the magnetic hyperfine structure could be extracted [5.64]. High-resolution absorption and fluorescence measurements on these crystals have allowed the determination of the electronic Zeeman g factor, the ground-state second-order Zeeman hyperfine interaction and the ground-state dipole magnetic hyperfine constant. Small distortions from the C_{3h} point symmetry have also been detected in these compounds [5.65]. Figure 5.10 shows a representative FLN trace of the hyperfine structure for $LaCl_3$:Pr^{3+}. In all of this, quadrupole interactions are small compared to the magnetic interactions and hence are neglected. In a more recent work, these same workers considered temperature changes of the hyperfine spectra at temperatures higher than 4.2 K and were able to estimate a value of T_1 for the $^1D_2$ states [5.66].

Erickson has also investigated the hyperfine interactions of the $^1D_2$ and $(^3H_4)_1$ states of $YAlO_3:Pr^{3+}$ at low temperatures. The site symmetry in this case is even lower than that of LaF_3, and the hyperfine interactions consist of a second-order magnetic interaction and the nuclear quadrupolar interaction. Fig. 5.11 shows recent results of *Erickson* in this compound obtained through the use of the rf-laser double-resonance technique [5.67].

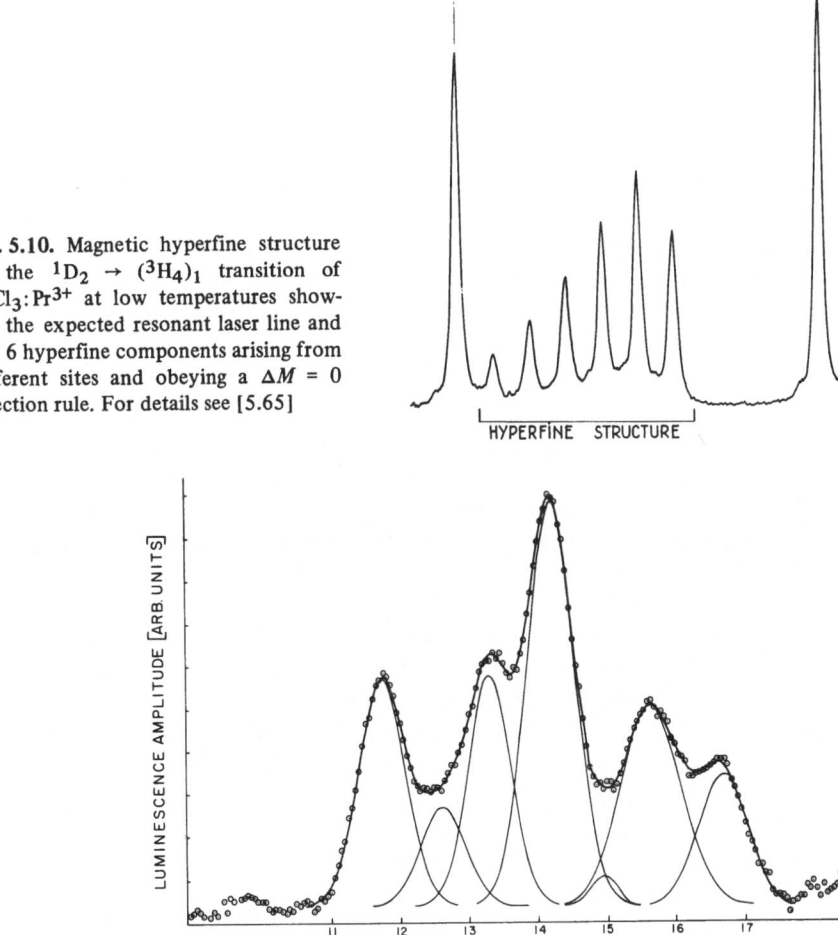

Fig. 5.10. Magnetic hyperfine structure of the $^1D_2 \rightarrow (^3H_4)_1$ transition of $LaCl_3:Pr^{3+}$ at low temperatures showing the expected resonant laser line and the 6 hyperfine components arising from different sites and obeying a $\Delta M = 0$ selection rule. For details see [5.65]

Fig. 5.11. Hyperfine splitting of the $^1D_2$ state of $YAlO_3$. The trace is a measure of the fluorescence amplitude plotted against the difference between saturating and probing optical beams (see [5.41]). The enhanced spectrum consists of seven lines appearing at various combinations of ground-state and excited-state frequencies. In this case, the ground-state splitting is 14.124 ± 0.024 MHz and the $^1D_2$ splittings are 1.461 ± 0.025 and 1.065 ± 0.026 MHz. The central peak has a width of 603 kHz [5.67]

Generally, of course, the ground-state hyperfine splittings are attainable through conventional ESR measurements. The laser techniques here described have advantages in their simplicity and potentially have their greatest benefit in determining excited-state parameters. It must be noted, however, that a transfer of excitation between excited levels must occur to measure these splittings because only the state in resonance with the laser is initially populated in FLN and in absorptive hole burning. It is also to be noted that use of low resolution lasers, as was the case in the work by *Chen* et al. [5.62] in photon echoes alleviates this problem by "covering the line," i.e., exciting all the hyperfine state simultaneously. There the limitation becomes the response of the system to frequencies, hence to splittings, that are too large.

5.2.2 Site and Impurity Selectivity

Inhomogeneous effects, as has been mentioned in Chap. 1, are of course not limited to the relatively small contributions affecting the linewidths of transitions. In some host materials, for example, the dopant ions can substitute into different and distinct crystallographic sites each describable by its own parameters. The resulting spectrum is a composite of these sites and can be highly complex.

Wright and his co-workers have used high resolution selective laser excitation in order to unravel such complex spectra and determine the symmetries and compositions of the individual ion sites. Their initial work was conducted in CaF_2: Er^{3+} [5.68]. CaF_2 is a cubic system which is widely used as a host for trivalent rare earth ions; the latter enter substitutionally into Ca^{2+} sites, requiring charge compensation which occurs through relocation of fluoride ions at interstitial sites and by the creation of a number of other centers. Each of these additional defects can be located in many different positions relative to the *Er* impurity thus producing variations in the otherwise cubic *Er* site symmetry. It might be noted that because of this, the transitions from Er sites which are unperturbed are electric dipole forbidden and hence are expected to be weak. *Tallant* and *Wright* used a high-resolution tunable dye laser to selectively populate only one *Er* site at a time, and the greatly simplified fluorescence from each site was then measured in turn. From the number and the splitting of the lines, information about the point symmetry was thus derived. Specifically, monitoring of single fluorescence identifiably belonging to a particular site while the laser was tuned provided complementary information about the energy level schemes correlated to that site. Fig. 5.12 shows an example of this type of study, CaF_2:Ho^{3+} in this case [5.69]. It is to be noted that unperturbed Ho^{3+} cubic sites contribute little if at all to the normal broadband fluorescence.[2] These studies have allowed to consider in detail the equilibrium dynamics of defects in these crystals [5.70, 71].

---

[2]CaF_2 doped with 4f ions are extensively used as thermoluminescence (TL) devices. Much of the modeling there assumes the luminescence to originate from cubic sites, see [5.155].

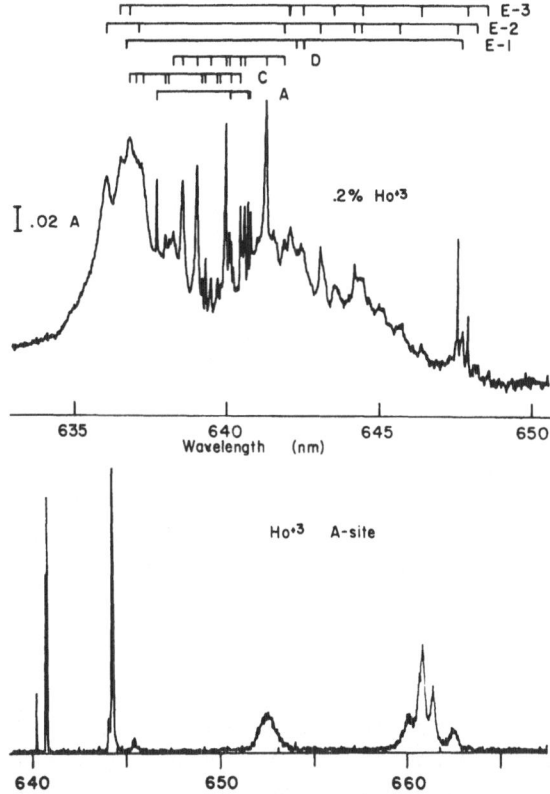

Fig. 5.12. Absorption spectrum of the $^5I_8 \rightarrow {}^5F_5$ transition of 0.2 mol.% Ho^{3+} in CaF_2. Lines above represent various sites contributing to this complex spectrum. Site A corresponds to trigonal single ion sites which, when selectively excited by a laser, produce the fluorescence shown in the lower trace. C, D, and various E sites arise from clusters of Ho^{3+} ions and their charge compensation. No evidence is seen of cubic sites. [5.69]

Since these experimental methods rely on photon detection schemes and hence can potentially be extremely sensitive, *Miller* et al. [5.72] have proposed that rare-earth ions in these materials can also serve as fluorescent probes for determining trace amounts of other foreign ions present in the condensed host. This is because the spectrum of the probe is highly sensitive to its surroundings; thus a rare earth in the vicinity of another type of impurity will generally have a sufficiently unique spectrum as to distinguish it from ions in the "normal" sample. Measurements of the absorption strength and other parameters such as the radiative lifetime of the newly perturbed sites as compared to the normal ones can then provide a quantitative analysis of the impurity.

Wright [5.73] and *Gustafson* [5.74] have employed this idea to study a sample of $BaSO_4:Eu^{3+}$ with traces of PO_4^{3-} impurity and found that a Eu ion with a nearby PO_4 was clearly distinguishable from one in a distant, unperturbed site and thus could be used as a tool for quantitative analytical chemistry. Various other systems have also been explored by these authors [5.74]; with the addition of time resolution, they have demonstrated that chemical reactions in solids may also be followed using these laser excitation techniques. Generally, the work of this group represents an excellent example of high-resolution laser spectroscopy in crystals used as a probe to examine microscopic crystalline structure, trace impurity distributions and, with the additional dimension of time resolution, chemical reactions and processes in the condensed phases.

Similar selective-site techniques have been used by *Dubicki* et al. [5.75] and by *Ferguson* [5.76] and his co-workers in the study of transition metal systems: the spectra of Mn^{2+}, singly and doubly doped Mn^{2+}, and Cu^{2+} in $KMgF_3$ and $KZnF_3$ hosts. By coupling site excitation with uniaxial stress and time resolution, these authors have been able to isolate single-ion spectra as well as dimers and trimers of various species.

Laser selective excitation has also found uses in the study of actinide spectra in solids. Various features of uranium in its different valency states have been investigated in this manner [5.77]. It is, however, in the heavier tripositive actinides where these techniques have been most useful. Because of their nuclear activity, samples containing the heavier actinides [5.78], Es^{3+} for example, tend to be contaminated by the sequential daughter products, in this case Bk, Cf, and Cm. Standard spectroscopic studies attempting to unravel the spectra of trivalent $5f$ ions all were hampered to some degree by these effects. Recently *Hessler* et al. [5.79] have investigated $LaCl_3$ doped with Es^{3+} and hence Cf^{3+} and Bk^{3+} using laser techniques. This has allowed them to uniquely determine the levels belonging to each of these ions.

Finally, the origin of the inhomogeneous contribution may arise from structural disorder introduced by the host. The ultimate example of such a host would be an amorphous solid or a glass and these problems will be discussed in a later chapter. Intermediate cases where the host remain crystalline but contain local disorder such as mixed crystals, discussed in Chap. 1, are clearly also susceptible to attack using the techniques described above. Some studies of Nd^{3+} doped $Y_3Al_{5(1-x)}Ga_{5x}O_{12}$ and in $CaF_2:YF_3$ mixed systems have recently appeared in the literature [5.80].

5.2.3 Assisted Transitions

High-resolution laser spectroscopy will not only simplify complex, multiple site spectra arising from pure electronic transitions but may also be useful in deciphering sideband structure in their accompanying phonon or magnon structure if the latter are site specific. *Aoki* and *Abelia* [5.81] first employed this strategy to study the vibronic sideband emission of Cr^{3+} in $LaAlO_3$.[3] They found that by exciting the R_2 transition directly with a tunable source, a vibronic spectrum free of complications from pair lines was obtained.

Similarly, it is generally accepted that fluorescence magnon sidebands of Mn^{2+} in antiferromagnetic fluorides are obscured by much stronger transitions

---

[3] Historically, *Dietz* and co-workers first obtained the intrinsic magnon sideband $\sigma_1{}^*$ of MnF_2 using nontunable laser excitation, see [5.143].

from perturbed sites of Mn^{2+} which serve as efficient excitation traps. Time-resolved resonant laser spectroscopy has been used by *Strauss* et al. [5.82] to observe the intrinsic magnon sideband of the $^4T_{1g}$ transition of $KMnF_3$ prior to the rise of the impurity contributions. Similar techniques were employed by *Chiang* et al. [5.83] to observe the magnon sideband structure in MnF_2.

Wilson has attempted to carry this idea yet one step further by using FLN techniques to probe impurity effects on the magnon sidebands. As nonmagnetic Zn impurities are doped into MnF_2, both the pure transitions and their associated intrinsic structure broadens due to the additional strains induced by Zn [5.84]. If the laser excites only one subset of Mn ions within the broadened line, a side-band specific to that subset of ions should be obtained in the absence of any ion-ion interactions which disperse the optical energy. These experiments done on the $^4T_{1g}$ so-called E_1 line are difficult because the transitions are extremely weak and complications arise because of the introduction of intense trap fluorescence. Pre-liminary results, shown in Fig. 5.13, indicated that narrowing of the sidebands does indeed occur even in the sidebands accompanying traps. Hence, in principle this technique may be useful for studying the density of magnon states as magnetic or nonmagnetic impurities are added. This is because the position within the in-homogeneous profile has a one-to-one correlation with the local distribution of near neighbors of the active ion and the magnetic excitations are specific to these neighbors.

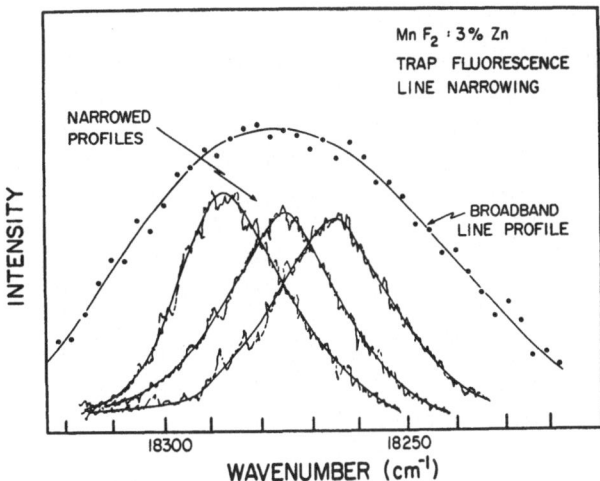

Fig. 5.13. FLN of disorder broadened trap magnon ($^4T_{1g}$) sideband of MnF_2 doped with 3% Zn^{2+}, showing narrowing of the sidebands corresponding to pumping of various sites. The broad envelope is the inhomogeneously broadened trap sideband. The phonon sideband was pumped in these experiments reducing the site selectivity but improving the signal. [5.85]

5.3 Single-Ion and Interion Dynamics

We turn now from static spectral studies to a discussion of the dynamics of the optically excited states — first to the dynamics of a single ion accessible to observation by use of high-resolution and/or pulsed lasers and second to the interion dynamics which occur as the concentration is raised. These latter processes arise from ion-ion interactions, either nonradiative (exchange or multipolar) or radiative. Nonradiative energy transfer is for the purpose of discussion divided into nonresonant, near resonant and resonant, respectively, to be found in Sects. 5.3.2a-c. Radiative transfer or trapping [5.86] was discussed in the context of phonon assistance in Chap. 2.

It is in the area of these ion dynamics that, in our opinion, laser spectroscopy in solids has made its most significant and noticeable impact, inasmuch as the results have led to the reexamination of some of the fundamental concepts which underlie the propagation of energy in condensed phases. It is also very likely that these particular areas will develop rapidly as even more refined techniques are brought to bear on these problems.

Much of the theoretical background and mathematical formalism related to energy transfer along with illustrative examples have been covered in Chaps. 2 and 3, particularly concerning the near-resonant case. Thus some of these areas will only be touched upon briefly. Once again, ruby and $LaF_3 Pr^{3+}$ have been most extensively studied probably because of the availability of good quality samples.

5.3.1 Single-Ion Dynamics

Before addressing the topic of energy transfer, we will survey a few experiments where modern laser techniques have made possible measurements of dynamical parameters not readily accessible to standard optical spectroscopic methods. The simplest process affecting single ions is the thermalization of their population through ion-lattice interactions, i.e., T_1 processes.

One such study illustrating a purely laser spectroscopic optical measure of T_1 was done by Broer et al. [5.87] on the $^4S_{3/2}$ state of LaF_3:Er^{3+}. The Stark manifold here is comprised of two Kramers doublets separated by 29 cm^{-1}. The Kramers degeneracy was lifted by an applied field and narrow band, pulsed laser excitation was employed to initially populate one level of the lowest $^4S_{3/2}$ doublet. Redistribution of the population within these states resulted in the rise of fluorescence originating from the other component of the doublet, the time constant of the rise is just T_1. Figure 5.14 summaries these results; it is seen that the lowest doublet of the $^4S_{3/2}$ manifold manifests Orbach relaxation via the higher component of the $^4S_{3/2}$ states at the higher temperatures. The lower temperature behavior has a T dependence appropriate for a direct process between the doublet but was found to be field independent and thus probably indicates cross relaxation to other paramagnetic impurities. The time scale involved here is in the vicinity of 10^{-5} s.

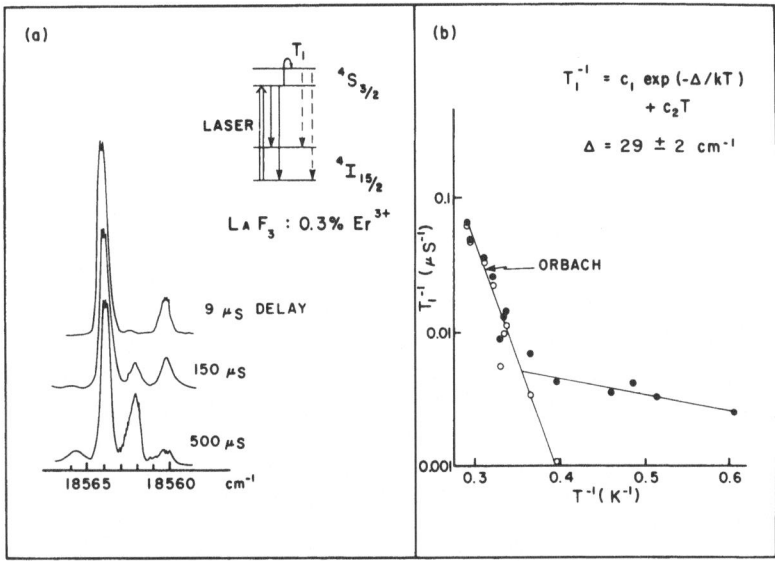

Fig. 5.14a,b. Spin lattice relaxation measurement in the $^4S_{3/2}$ of $LaF_3:Er^{3+}$. (a) Time resolved luminescence spectra after initial laser excitation with insert showing the principal thermalization path; (b) Full circles show the measured values of $1/T_1$ as a function of temperature. The open circles which are values after subtracting the pseudo "direct process" contribution show a clear Orbach process in the excited state. [5.87]

Similar dynamics can also be determined in the nanosecond scale. *Meltzer* and *Wood* [5.88] have used a short pulse-nitrogen pumped dye laser and single-photon correlation techniques to measure the direct-process relaxation of a state in Tb^{3+} to a state located some 5 cm^{-1} below it. The basis of their detection system is a time-to-amplitude converter which converts the time between the laser (start) pulse and detection of a photon (stop) pulse into a proportional voltage for further processing by a multichannel analyzer. This type of detection is useful for low-count rates such as occur when the emission is highly quenched.

Using the above system, *Meltzer* and *Rives* have measured excited-state-spin lattice relaxation in $Y(OH)_3:Tb^{3+}$ [5.89] and in ruby [5.31]. These experiments also form a basis for the laser generation of narrowband phonons through the direct relaxation between levels. Experiments in the ^2E level of ruby have not only provided accurate values of the no-spin-flip relaxation process but also have revealed interesting phonon dynamics which unfortunately are outside the scope of this review [5.90].

By using polarization labelling spectroscopy, *Lee* et al. [5.91] have extended these purely optical methods to the picosecond time domain. Their measurement entails the use of two independently driven dye lasers, one of which is used to populate the level to be studied and the second is used as a polarization probe

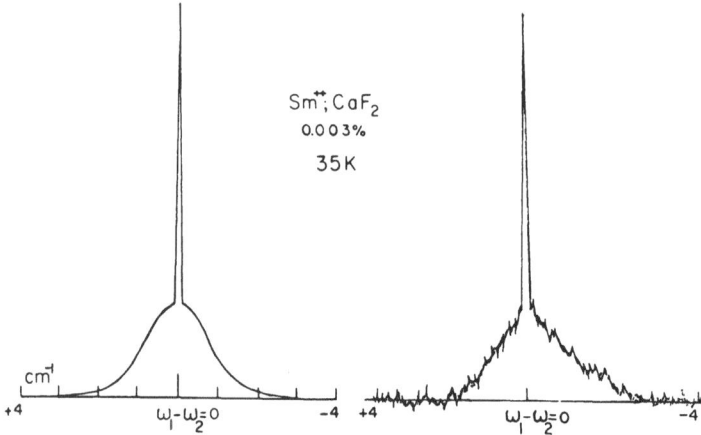

Fig. 5.15. The theoretical and experimental polarization spectroscopy line shapes in 0.003% Sm^{2+}:CaF_2. The theoretical curve was calculated using a density matrix formalism. The best fit was found with a homogeneous width of 1.75 cm^{-1} (T_2 = 6 ± 0.6 ps). The uncertainty arises mostly from laser resolutional problems. The level is centered at 14497 cm^{-1}. [5.91]

[5.92]. These workers have investigated the relaxation properties of one of the $4f \rightarrow 5d$ transitions in CaF_2:Sm^{2+} and have obtained T_1 and T_2 values of 8 ns and 82 ps, respectively. Some of their results are illustrated in Fig. 5.15. Undoubtedly this represents the first of many such measurements in this new temporal region even as picosecond pulsed lasers become more readily available.

Optical and rf or microwave double-resonance experiments have of course been applied to the study of relaxation phenomena in excited and ground states for some time [5.6]. Tunable dye laser excitation, however, has added an additional dimension to these studies inasmuch as all transitions of interest may now be studied and the resolution allows extreme site selectivity. Recent studies of laser − ESR in ruby [5.93] and in *F* centers in CaO [5.94] by *Boccara* and co-workers are but two examples of this.

As discussed in Sect. 5.1.2, *Erickson* also noted the substantial optical pumping which occurred when he investigated the $^1D_2$ state of LaF_3:Pr^{3+} with a single mode cw laser[5.40,41]. The population changes induced by the laser were found to be consistent with a recovery time of 0.5 s at 4.5 K, a value arising from nuclear-spin relaxation between the ground-state components. He therefore developed an optical rf double-resonance technique which allowed him to investigate various static spectral features of the Pr system (Sects. 5.1.2, 5.2.1). The fluorescence amplitude of the $^1D_2 - {}^3H_4$ transition was monitored as a weak rf field applied to the sample was scanned from 0.05 MHz to 50 MHz. As various resonances occurred, population shifted from one hyperfine level to another thus producing absorptive or emissive changes. This was the first experimental demonstration of optically detected NMR in solids in the absence of an applied dc magnetic field. Similar consideration led to the estimate of T_1 in LaBr$_3$ by *Pelletier-Allard* also alluded to in Sect. 5.2.1 [5.66].

These few examples by no means exhaust the literature on the subject of single-ion dynamics but serve only to demonstrate the power of modern laser techniques in these investigations.

5.3.2 Energy Transfer

Another type of dynamics involving the excited state of ions in crystals arises from ion-ion interactions which often occur as the dopant concentration is increased, and which result in energy transfer and/or diffusion within the active system. There are two distinct aspects to this problem: The first entails the understanding of the microscopic interactions which allow the ions to communicate while the second requires the translation of these microscopic interactions to the macroscopic quantities which are observed in the laboratory. The role of laser spectroscopy, particularly FLN, in advancing the knowledge of these areas formed the basis of Chaps. 2 and 3 and thus these specific areas will only be briefly reviewed in this section, with emphasis on the experimental literature.

There are, for the purposes of discussion, three types of transfer which need to be dealt with. The lines of demarcation are necessarily somewhat fuzzy inasmuch as several processes may occur simultaneously in a given system. Various types of transfer processes are illustrated in Fig. 5.16a-c.

a) Nonresonant Energy Transfer (Interline)

We define nonresonant energy transfer as occurring when the energy mismatch between an optically excited "donor" (D) ion and an unexcited "acceptor" (A) to which the excitation will be transferred is much greater than kT and/or the inhomogeneous linewidth. This criteria is applicable to crystalline solids at low temperature in general but has to be applied with care in the case of glasses where inhomogeneous widths may be orders of magnitude larger than kT in the liquid helium range. These cases include the situation where A may itself be an optically excited ion, in which case related upconversion processes will be observed [5.95]. The likely mode of excitation transfer is a multipolar electrostatic or an exchange ion-ion cross relaxation process coupled to the emission or absorption of single or multiple phonons. Such processes have been observed in many rare earth and transition metal ion systems, and the literature is replete with examples [5.96]. It was in fact to deal with this type of problem that the phenomenological theories of *Dexter* and *Förster* [5.97] and their subsequent refinements were developed. In the dilute D and A limit, one need only consider a D → A transfer process, which when macroscopically averaged gives rise to *Inokuti* and *Hirayama* type decays in the donor system [5.98]. As either concentration or temperature of the donors is raised, D → D transfer processes must also be considered as has been discussed in Chap. 3. An illustration of the macroscopic dynamics involved here

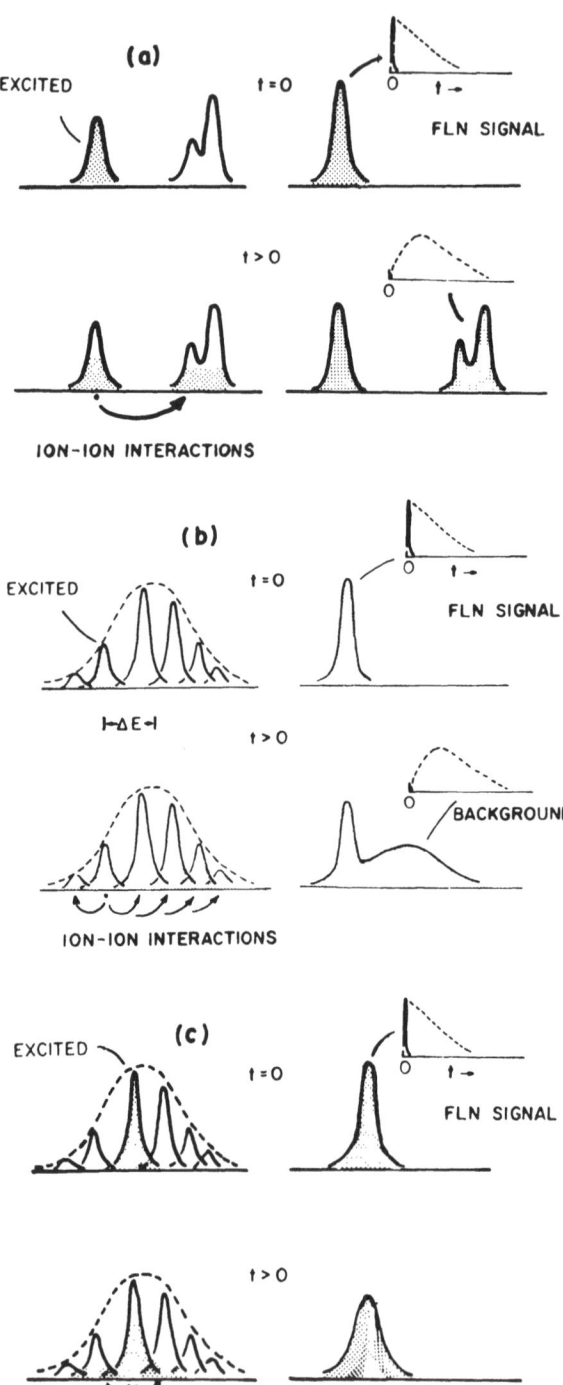

Fig. 5.16a-c. Diagramatic representation of various transfer processes studied with laser spectroscopic techniques, showing (a) nonresonant; (b) near resonant; and (c) resonant processes. The FLN plays the role of donor, D, while various spectral features play the role of acceptors, A

has recently been completed by *Hegarty* [5.99] on $PrF_3:Nd^{3+}$ and is discussed in Chap. 3.

In our definition of nonresonant transfer, then, we deal with interline transfer and we include the following cases: i) transfer between like ion species in drastically different crystallographic sites (again amorphous solids are excluded), ii) transfer between intermediate states of like ions followed by multiphonon relaxation, iii) transfer betweeen single ions and multiple-ion complexes such as pairs, dimers and trimers, iv) transfer between two distinct ion species, and v) transfer between excitations of the lattice and impurity centers.

All of the above processes have, by and large, been established through the use of conventional means, and various features of the measurements have been brought to agreement with theory. Laser spectroscopy studies in these systems, because of their resolution and hence selectivity, can considerably simplify the interpretation of results and, hence, yield additional details to our understanding of these problems.

The work of *Wright* and his co-workers serve as illustration of various types of interline processes of the type i) − iii) mentioned above [5.100, 101]. In work on charge compensated $CaF_2:Er^{3+}$ and $BaF_2:Er^{3+}$ they have demonstrated cross relaxation, upconversion, and transfer to ion complexes, all of which are mediated by phonons and are preceded and/or followed by multiple phonon emission. Studies of this type have helped these workers in deciphering the complexities of defect equilibrium in solids [5.69], an important contribution.

Less comprehensive work exhibiting transfer between distinct crystallographic sites has dealt with Pr^{3+} in PrF_3 [5.102] and CaF_2 [5.103]. In the case of PrF_3, transfer was observed between normal sites and various imperfection sites but the transfer dynamics between the irregular sites was found to be limited, thus implying preferential rather than random positioning of these imperfections within the crystal. More recent work has considered divalent and trivalent Eu in various hosts [5.104] and Nd^{3+} in mixed crystals [5.80].

The case of transfer between two distinct ions is somewhat less well documented inasmuch as laser spectroscopic studies are concerned. This is principally because various aspects of sensitization have in the past been so thoroughly studied [5.105]. Again, here the dynamics of transfer between Pr^{3+} to Nd^{3+} by *Hamilton* et al. [5.106] and by *Hegarty* [5.99] have helped elucidate aspects of macroscopic averaging and strain effects.

Transfer from intrinsic excitation of the host to specific impurity sites has been investigated by *Hsu* and *Powell* [5.108] wherein they have observed transfer from $CaWO_4$ excitations to Sm^{2+} placed as an acceptor.

By joint laser and heat pulse excitation and with application of magnetic fields, *Heber* and his co-workers [5.107] have investigated transfer of energy between single Cr^{3+} ion and various types of pairs in ruby and in $YaAlO_3$. The results, particularly in the former case [5.28], are relevant to the discussion in Sect. 5.3. 2b.

As has also been discussed in Chap. 1, energy transfer in ordered materials such as antiferromagnetic MnF_2 manifests itself as a spatial migration to trapping centers formed by highly perturbed Mn^{2+} ions adjacent to impurities [5.109] and to various impurities (Er^{3+}, Eu^{3+}) [5.110] which are intentionally introduced into the lattice. Recently *Wilson* et al. [5.111] have conducted a comprehensive laser spectroscopic study of nominally pure MnF_2 and MnF_2:Er^{3+} and Eu^{3+} and have measured various transfer rates from the intrinsic $^4T_{1g}$ excitons to the variety of trap centers. These studies in conjunction with earlier temperature dependent studies have allowed some conclusions to be drawn in regard to the structure of the traps and the nature of the migration and dynamic equilibrium existing between traps and intrinsic excitations [5.110].

The rare-earth platinates of the type $RE_2[Pt(CN)_4]_3 \cdot mH_2O$ have also been studied using laser spectroscopic methods [5.112]. In these compounds which are largely one dimensional, excitation may reside in the $[Pt(CN)_4]^{2-}$ stacks and eventually find its way to a rare-earth activator. An interesting feature of these systems is that energy levels of the platinate exciton may be readily tuned by pressure or by rare-earth concentration changes. Certain nonlinear effects also appear in these sytems.

In all prior conventional studies and in a number of the presently cited ones, the energy transfer process has been identified by studying temperature and concentration dependence of the donor lifetime. Parameters quantifying these interactions have then been derived assuming specific macroscopic averaging models. Some caution should be exercised in accepting all such results. *Birgeneau* [5.113], in his analysis of transfer in ruby, has pointed out, for example, that a straightforward concentration dependent analysis ignores the increasing nonlocalized nature of the Cr^{3+} excitation. Similarly, as Chap. 3 emphasizes, the Inokuti and Hirayama results often used in analysis have ignored back transfer which may be important in many of the cases studied.

b) Near-Resonant Energy Transfer (Intraline)

Chapters 2 and 3 have considered near resonant energy transfer in detail, and in an effort to avoid redundancy, we shall only reemphasize some of the important aspects of this process. By near-resonant energy transfer we mean the interaction of ions which are in every respect analogous except for a slight difference in the coordination and hence in their energy as might be induced by small statistical fluctuation in the crystalline host. This is precisely the case that occurs in the strain broadening observed in transitions of ions in crystals at low temperature and it is here that time resolved FLN has played its most important role. This type of behavior is diagramatically illustrated in Fig. 5.16b.

Perhaps one of the most interesting features of all time-resolved FLN studies conducted on near resonant systems, i.e., narrowed to inhomogeneous line transfer, is that the spectral dynamics which result are quite similar in different materials even when the mechanisms responsible for the transfer are different. Figure 3.7 shows these spectral dynamics for the case of LaF_3:Pr^{3+}; similar dynamics for ruby are shown in Fig. 5.17. It is to be noted that the whole inhomogeneous background fluorescence begins rising simultaneously, implying interactions to ions in all parts of the distribution. In addition, the FLN component is observed to decay without appreciable broadening. These general results of FLN experiments made it clear that the earlier phenomenological treatment of the transfer process was altogether too simplistic [5.97]. In that approach spectral diffusion was assumed to occur resonantly through step-wise linewidth overlap and would have resulted in a gradual broadening of the narrowed line to fill the inhomogeneous background, as is illustrated in Fig. 5.16c.

This recent work is then principally responsible for the reexamination of the theory; the developments in this area discussed in Chap. 2 are more fundamental in that they deal directly with the total ion-phonon system, and the role of phonons becomes explicit in the conservation of the energy mismatch required of the transfer [5.114]. This later treatment reduces to that of *Förster-Dexter* under the appropriate conditions.

Again, $LaF_2:Pr^{3+}$ has played an important role in these developments. Early time-evolved FLN studies on this system by *Flach* et al. [5.115] and *Selzer* et al. [5.116] showed the donor to acceptor transfer rates to be asymmetric as a function of excitation position within the $^3P_0$ profile; the rates were found to be higher when the high-energy wing was pumped. This asymmetry was found to persist until the lattice temperature became comparable to the inhomogeneous linewidth and illustrated the intrinsic involvement of the phonon population. The transfer rate was also found to increase with a T^3 dependence at low Pr^{3+} concentration

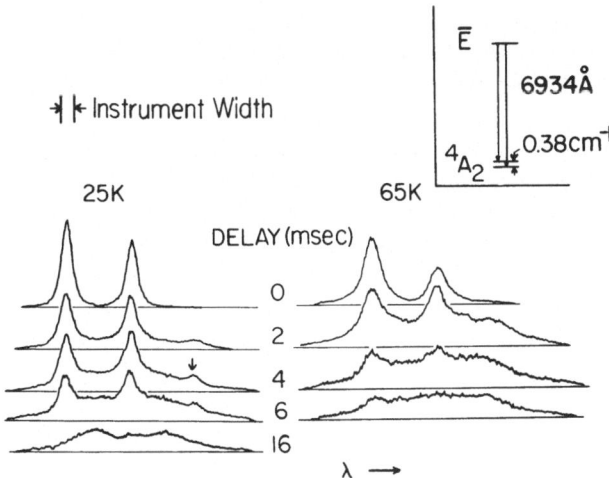

Fig. 5.17. Time evolution of the FLN signal of the R_1 transition in ruby as a function of temperature. The two lines seen at $t=0$ correspond to the 0.38 cm^{-1} ground-state splitting. The trace at 25 K with 16 ms delay represents the inhomogeneous width of the R_1 line under broadband pumping conditions. Note the rise of this component simultaneously at all delays. FLN components show no broadening as a function of time. The third peak in the 25 K trace results from radiative trapping [5.86]. [5.118]

and allowed the identification of a "one-phonon, second-order" mediated multipolar process as the responsible interaction [5.114]. It is to be noted that this particular process is very weakly dependent on the energy mismatch ΔE between ions, and that this property considerably simplifies the spatial averaging discussed in Chap. 3 [5.117].

Similar studies have been conducted in ruby [5.118]. There transfer between near resonant ions is found to increase linearly with temperature when kT exceeds ΔE and is thus consistent with a "one-phonon" mediated exchange interaction.

In considering these processes, we have inherently assumed the conditions implicit in "microscopic strain broadening," i.e., that the interacting neighbor ions of an excited ion are representative of the whole inhomogeneous distribution and thus on the average these neighbors are mismatched in their energy by quantities of the order of the inhomogeneous linewidth. Generally, this appears to be the case in the rare-earth systems studied to date where the ion-ion coupling is weak and hence the excitation is localized. However, in viewing the ruby system in toto and taking into consideration the $R \rightarrow$ pair line (N) transfer in this system, the validity of this specific assumption has been challenged by *Imbusch* [5.119] and others [5.28], and is discussed below.

c) Resonant Energy Transfer

As the dopant concentration and, consequently, the coupling strength between ions, J, increases, the possibility exists for resonant energy transfer to occur. If one of two identical ions coupled by the interaction J is placed in an excited state, the occupation probability of the excitation will coherently oscillate between the two ions with a transfer rate of J/h [5.120]. An analogous classical system would be two identical coupled pendula. If there is an energy mismatch of ΔE between the two upper levels, resonant transfer can still occur as long as $\Delta E \ll J$. This coherent process must be contrasted to the incoherent hopping of the excitation from one ion to another, i.e., phonon-assisted transfer, the rate of which is proportional to J^2 (see Chap. 2). Furthermore, the former mechanism is temperature independent, occurring even at absolute zero, since no phonons or line-broadening processes are required.

In a stoichiometric crystal, the coherent coupling of resonant ions leads to a Frenkel exciton, which can have a substantial optical bandwidth and a well defined wave vector, k. Coherent propagation may also occur in impurity-doped systems, but translational symmetry is no longer conserved and k is no longer a very meaningful entity. As dopant ion concentration is increased from a very low value, the migration of optical excitation should therefore progress from no transfer to (temperature-dependent) incoherent hopping to coherent, excitonic propagation. At very low temperatures where the phonon-assisted processes are frozen out, *Lyo* proposed [5.121] that the transition from no propagation to rapid coherent propagation should under appropriate circumstances occur abruptly at a critical dopant concentration, c_0. This is the phenomenon of so-called "Anderson localization" [5.122]. Unfortunately, the existence of coherent, resonant excitation transfer in impurity ion crystals has been extremely difficult to verify and is presently the subject of considerable controversy.

With the introduction of codopant acceptor ions as "traps," it is presumably possible to determine whether rapid excitation transfer between donor ions is occurring by studying the fluorescence decay of the donors and acceptors. However, these experiments merely monitor spatial migration and in general cannot distinguish between resonant and near-resonant donor interactions. The use of time-resolved FLN has added another dimension by allowing direct observation of near-resonant processes: these incoherent mechanisms should give rise to the spectral behavior described in the previous section, whereas resonant transfer should not produce any spectral evolution. Hence, by combining donor-acceptor studies with FLN, if rapid donor-donor transfer is occurring with no spectral dynamics, it may be inferred that the transfer mechanism must be truly resonant. Such reasoning would be further substantiated if the rapid donor-donor transfer was observed even at the lowest temperatures.

This type of argument has been recently used by *Selzer* et al. [5.28] to propose the existence of distinct, simultaneous resonant and near-resonant excitation transfer in ruby. They observed an explicit temperature-dependent spectral transfer which could be frozen out at 2 K and an implicit temperature-independent spatial migration of the excitation from isolated Cr ions to Cr pair traps. A critical element in their reasoning was the detection of strictly exponential R_1 (donor) decays with a time constant shorter than the radiative lifetime measured in low concentration samples, and N_2 (acceptor) decays with an exponential tail at this same R_1 decay rate, as shown in Fig. 5.18.

In a related experiment, *Koo* et al. [5.123] claim to have seen a "mobility edge" in ruby around a critical concentration of 0.3 at .% Cr. If true, this would

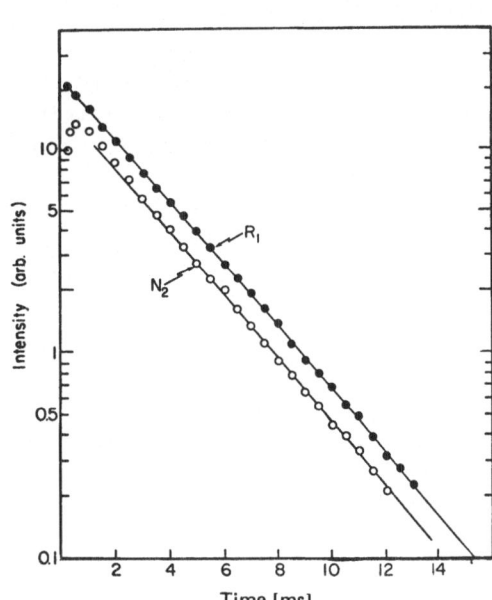

Fig. 5.18. R_1 and N_2 lifetime decays for a 0.51 at.% powdered sample of ruby at 5 K, showing exponential tracking and implying the presence of fast resonant R_1 transfer, see [5.28, 119]. [5.28]

not only be the first time that such an effect was directly observed, but it would also verify the predicted Anderson localization in this system [5.121]. Yet, because of demonstrated complexities associated with inhomogeneous line broadening in ruby and the ambiguities arising therefrom [5.28], this experiment remains inconclusive without further verification.

How then might it be possible to unequivocally demonstrate the existence of resonant transfer in these dilute systems (where clustering effects can presumably be ignored)? One way may be through coherent transient experiments. If resonant transfer is associated with a dephasing of the coherently prepared system, there would be a decrease in T_2 as the concentration is raised. As mentioned in Sect. 5.1.1, such a dependence has been seen in ruby, but it has not been determined whether this arises from optical energy transfer or strictly ground-state Cr-Cr interactions.

Yet, a lack of any additional concentration-dependent dephasing would not refute the existence of resonant transfer. Since the transfer is itself presumably coherent, the coherence induced by the laser would not necessarily be lost through ion-ion interactions. Alternatively, one can picture the resonant interaction as forming an "extended state" of coupled ions. It is this extended state, an eigenstate of the system, which is excited by the laser, and its dephasing via phonon or other processes may closely approximate that of a single ion, or localized state. An experimental approach might be to search for any subtle yet abrupt changes in dephasing as the concentration is increased, indicating a transition from localized to nonlocalized behavior.

Another possible way to detect coherent transfer is to utilize a two-step excitation procedure in conjunction with FLN. Due to accidental coincidence, which was discussed in Chap. 4, resonant transfer can occur between ions which have identical ground and upper levels (the only levels involved in the excitation exchange) and yet whose intermediate levels are different by the observed residual inhomogeneous width. By exciting the system with high resolution, first to the intermediate level and then to the upper level, it is possible to preselect ions which have all three levels in exact resonance. Under this excitation condition, fluorescence from the uppermost state to the intermediate levels will not exhibit the accidental coincidence residual width and will simply have a linewidth determined by the convolution of the two laser photons. Resonant energy transfer might then show up in one of two ways: either a spreading of the nonresonant fluorescence between upper and intermediate levels with time until the residual width is "filled," or else the immediate appearance of the full residual width, even though two-step, high-resolution excitation was used. The latter would occur if the resonant transfer were very rapid on the time scale of the experiment, or if all the different ions with different intermediate levels were coupled into an extended state. Using the extended-state picture, it could be argued that the system may undergo a fluorescent transition to any of its numerous intermediate levels regardless of excitation procedure. Since the transition from localized to delocalized behavior is predicted to be abrupt, one might observe the sudden appearance of the residual linewidth as the concentration is increased. The residual width and the two-step excitation process are schematically illustrated in Figs. 5.19a, b.

An alternative way to observe the same phenomenon would be to burn a narrow hole in the nonresonant fluorescence by stimulating a downward transition, and observing the time evolution of the hole as a function of dopant concentration, as shown schematically in Fig. 5.19c. A sudden inability to burn a hole as the concentration increases would imply the onset of delocalized states. This latter experiment has been recently attempted in $LaF_3:Pr^{3+}$ and a typical hole in the nonresonant fluorescence is shown in Fig. 5.20. As yet, no evidence of rapid filling has been seen in concentrations up to 70 at.% Pr [5.124].

All of the above methods differ from previous efforts to measure resonant transfer in that no attempt is made to observe spatial migration of the excitation (as in donor-acceptor experiments). In contrast, the transient grating method des-

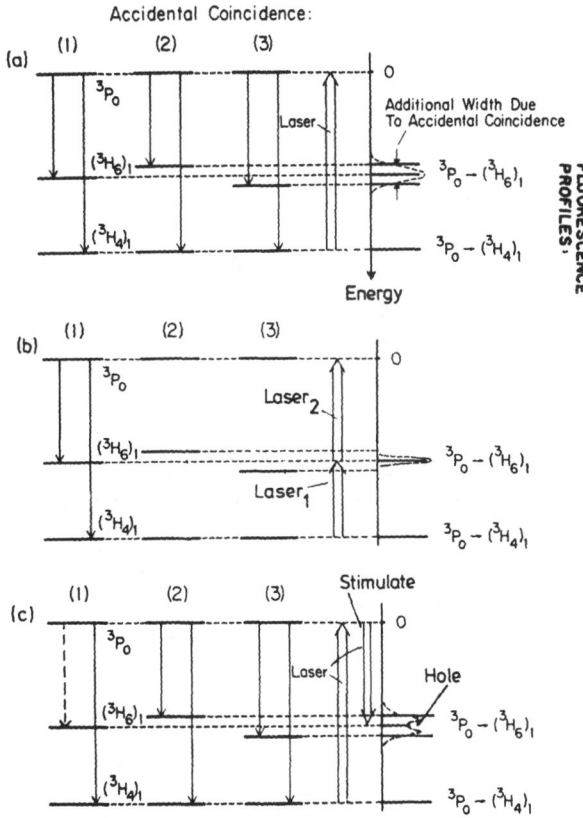

Fig. 5.19. (a) Schematic representation of the "accidental degeneracy" effect and its uses through two-photon laser spectroscopy allowing the study of resonant transfer processes; (b) entails the use of two absorptive photons; (c) one photon is used to stimulate one subset of ions in the excited state burning a hole in the intermediate fluorescence.

TWO PHOTON HOLE BURNING

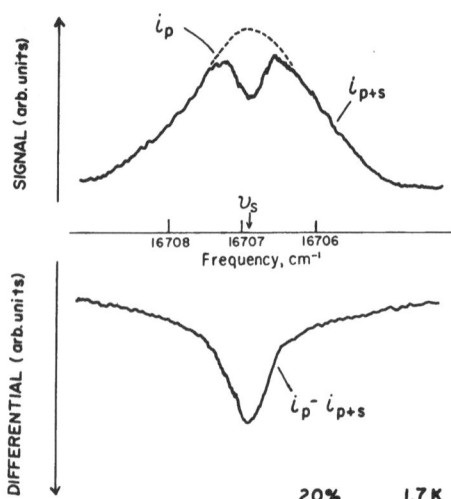

Fig. 5.20. Hole burned in the intermediate $^3P_0 \to {}^3H_6$ fluorescence of LaF_2:Pr^{3+} using method (c) of Fig. 5.19. By employing a differential method with pump on (I_{p+s}) and off (I_p) the hole may be investigated without the background. Results are shown in the lower trace. Any broadening of the hole as a function of time would denote presence of resonant transfer. No dynamics were observed in this sample with 70 at.% Pr^{3+}. [5.124]

cribed in Chap. 4 relies solely on spatial effects and provides no pertinent spectral information. Yet, this technique might help determine whether resonant transfer, if it exists, results in migration over long distances because it provides a way of easily and accurately determining excitation migration over distances variable from 500 Å to several microns. Some recent transient grating and degenerate 4-wave mixing experiments [5.125] in ruby have yielded null results, indicating that in this system spatial transport does not occur over such large distances.

The problem of resonant energy transfer in a disordered system, particularly at finite temperature, is a highly complex one, and any model must consider the energy and spatial distribution of the dopant ions. While calculations have so far assumed a strictly random distribution of both parameters, real systems may exhibit clustering and macroscopic strains which could severely complicate any concentration-dependent studies. Possible evidence of these nonrandom effects has been seen in ruby [5.28, 29], and at the high concentrations necessary to obtain sufficiently large interaction strengths in many rare-earth systems, clustering may be prevalent. Thus, an unequivocal demonstration of resonant optical energy transfer in ionic systems may require the development or discovery of a new, more nearly ideal system for investigation – one in which the coupling energy between optical centers exceeds ΔE at a low enough dopant concentration such that any clustering or macroscopic strain effects are avoided.

5.4 Concentrated Materials and Excitonic Effects

In this section recent results in the laser spectroscopy of concentrated and ordered ionic crystals are surveyed. Once again, a few exemplary systems have been chosen for detailed discussion and the general literature in this area will be cited.

5.4.1 PrF_3 and $PrCl_3$

As a natural consequence of the work on Pr^{3+} cited in the earlier sections, the stoichiometric compounds PrF_3 and $PrCl_3$, have also been subjected to comprehensive investigations. These two systems are good candidates for investigation because good optical quality crystals are also available and comparison between these two similar systems is of some interest.

PrF_3 has a structure which is similar to that of LaF_3 and can be formed stoichiometrically as $La_{1-x}Pr_x F_3$ when x is increased to 1. It follows that the levels of Pr^{3+} are once again nondegenerate because of the very low local symmetry [5.35]. The same argument is valid for $PrCl_3$, but there as we have also mentioned (Sect. 5.2.1), Pr^{3+} ions have a C_{3h} symmetry and some of the degeneracy remains. The ground state is thus, for example, a doublet with $\mu = \pm 2$ [5.126]. Because of this, $PrCl_3$ orders magnetically and/or electrically [5.127]. In addition, as compared to the low-concentration crystals, considerable cross relaxation and hence fluorescence quenching occurs in the concentrated compounds. The $^3P_0$ state lifetime decreases from 40 μs to ~ 1 μs at low temperatures in increasing Pr^{3+} in the fluoride to 100% and is completely quenched at higher temperatures. The same state in $PrCl_3$ displays little concentration quenching, presumably owing to its less energetic phonons or less favorable cross-relaxation channels. Various transitions in both compounds have been made to lase [5.128, 129].

Many authors have chosen to discuss concentrated materials in terms of an excitonic description of the collectively excited electronic states. Bearing our discussion on resonant transfer in mind, however, for weakly coupled systems, whence $J < \Delta\nu_{inh}$, a localized or single-ion description will frequently suffice. It appears that this is the case in the fluoride.

As the concentration of Pr in LaF_3 is increased, the inhomogeneous widths of the $(^3H_4)_1 - {}^3P_0$ transition increase and then decrease due to the mixed crystal effects discussed earlier. In the pure PrF_3, the $^3P_0 \rightarrow (^3H_4)_1$, transition is only about 0.5 cm^{-1} in width. *Hamilton* et al. [5.102] have shown that this width remains inhomogeneous by demonstrating FLN in this transition and obtaining in time resolved fashion the spectral diffusion dynamics illustrated in Fig. 5.21. They also obtained the surprising result that PrF_3 behaves in all qualitative aspects as low concentration $LaF_3:Pr^{3+}$. Thus *Hamilton* was able to show that the same "one-phonon, second-order" [5.114] mediated dipole-dipole interaction was responsible for the dynamics observed in this excited system. The transfer rate was found to increase as a function of temperature with a power dependence T^n, with $n \simeq 5$ which was ascribable to phonon phase structure. The spec-

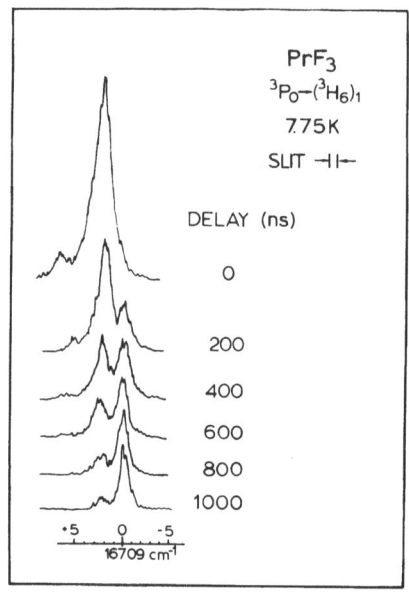

Fig. **5.21.** Time-resolved FLN spectrum of the $^3P_0 \rightarrow (^3H_6)_1$ transition of PrF$_3$ at 7.75 K. The spectrum suffers from accidental degeneracy but nevertheless can be used to monitor the $^3P_0$ population dynamics as in Fig. 3.7. The presence of dynamics indicates that the $^3P_0$ absorption is inhomogeneously broadened. The 1000 ns trace approximates the broadband excited fluorescence of PrF$_3$. [5.102]

tral evolution of the fluorescence in PrF$_3$ under FLN conditions is then what one would expect from microscopic strain broadening, i.e., a narrowed transition feeding the full inhomogeneous background. The precise interpretation of these results remains somewhat ambiguous for in order to have truly random strains associated with neighbors to the excited Pr, very short-range randomized perturbations ($a \simeq 2.5$ Å) would be required. An alternative which has not been eliminated is the possibility that lattice strains vary slowly with respect to the lattice parameter, a, and that the observed dynamics occur initially between groups of truly resonant ions (hence showing no spectral dynamics) eventually propagating macroscopic distances to other groups, and thus fortuitously mimicking the low-concentration results. Attempts to observe this resonant transfer using the hole burning techniques of Fig. 5.19c have so far failed in PrF$_3$ because of experimental difficulties connected with stimulation [5.129].

In contrast to PrF$_3$, no resolvable narrowing nor spectral dynamics were observed in PrCl$_3$. *German* and *Kiel* [5.130] have reported that fluorescence linewidths in certain Pr transitions in PrCl$_3$, specifically those involving doublet terminal states, $(^3F_2)_1$, $(^3H_4)_1$, exhibited considerable broadening as the temperature was lowered below 100 K. This anomalous behavior was interpreted as arising from ordering fluctuations and was substantiated by the observation that the linewidth decreases with applied magnetic fields. A number of other transitions in this material involve singlet terminal states, and no anomalous broadening is apparent in these. The low-temperature width of the $^3P_0 \rightarrow (^3H_6)_1$ transition at low temperature is again of the order of 0.5 cm^{-1}, where phonon relaxation effects should be negligible at 2 K, in this case. If the width of these later transitions reflects inhomogeneous effects, then the very fact that no FLN was observed in the *Hamilton* experiments implies very rapid energy migration, i.e., faster than their temporal resolution 20 ns, and hence orders of magnitude faster than PrF_3. Alternatively, this residual linewidth could be caused by exciton dispersion within the interacting levels. In such a case, the bands would have to be thermalized to some extent, and this is entirely plausible in this case. Once again, complete elucidation of this system must await further experimentation with greater temporal resolution.

5.4.2 Tb(OH)$_3$

In the ferromagnetic insulator Tb(OH)$_3$, *Cone* and *Meltzer* [5.131] have identified an unusually shaped, double-peaked emission line to be an intrinsic band-to-band exciton emission from the $^5D_4$ to the $^7F_5$ Tb^{3+} levels. The emission-line shape could be well fit assuming quasithermodynamic equilibrium across the zone and nearly one-dimensional ion-ion coupling, as shown in Fig. 5.22.

While the splitting of the peaks gives a measure of the exciton dispersion, the sign of this dispersion was unobtainable from static measurements. These authors [5.132] therefore excited a nonthermal distribution of k states near $k=0$ with a pulsed laser and watched the time evolution of the system. Time-resolved fluores-

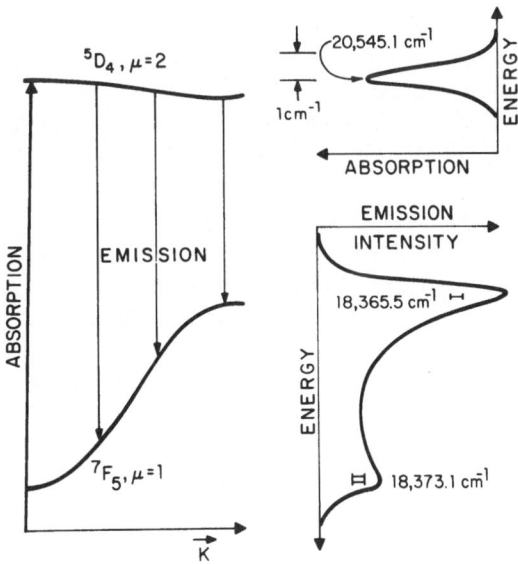

Fig. 5.22. Shape of the exciton band $^5D_4 \rightarrow {}^7F_5$ in Tb(OH)$_3$ at 1.3 K. The absorption shown in upper insert corresponds to the creation of $k = 0$ excitons leading to emission in peak II of lower insert. The time evolution of the emission from peak II to peak I allows a direct measure of thermalization times across the Brillion zone. Relaxation times have been estimated at 1 μs. [5.133]

cence from the low- and high-energy peaks of the emission line were separately measured using the single-photon time correlation techniques mentioned in Sect. 5.3.1. From these results it was easy to determine that zone-center emission gave the higher energy peak, and zone edge the lower energy peak, thus revealing the sign of the dispersion.

Time resolved measurements also permitted the authors to obtain insights into the scattering mechanism across the zone in the $^5D_4$ state [5.133]. Although the results are still preliminary at this time, several interesting features of the scattering process have been observed. The process appears to depend strongly on laser excitation wavelength within the inhomogeneous absorption as well as on magnetic field; the typical scattering times is on the order of 1 μs in a 30 kOe field at 1.3 K. At these relatively high fields, no temperature dependence is observed below roughly 10 K, but above this temperature the scattering rate starts to increase. As the field is reduced, the scattering rate also increases and the onset of temperature dependence is lower. All of these results are consistent with a model [5.134] in which scattering at the lowest temperatures involves exciton-impurity interactions, whereas at slightly higher temperatures exciton-magnon interactions predominate. The relative influence of the latter process may be altered by varying the strength of the applied magnetic field.

5.4.3 Magnetic Excitons

Among all antiferromagnetic crystals, MnF_2 has served a role comparable to that of ruby or $LaF_3:Pr^{3+}$ in impurity ion systems. Magnon sidebands were first identified in this compound [5.135], and since that time various static spectral features have been examined with increasing detail so that the optical properties are generally well understood. Relevant energy levels have been touched upon in Chap. 1 and reviews of the spectroscopy in ordered materials have also appeared [5.136]. In an earlier section, we have also mentioned that intrinsic sideband structure of the $^4T_{1g}$ state excitons had been studied using laser spectroscopy. For example, Chiang et al. [5.83] successfully obtained the π_1 sideband of E_1 using these techniques. It is also interesting to note that line broadening of optical transitions in the magnetic ordered phase was identified as occurring via magnon Raman scattering [5.137]. In this section we shall concentrate our attention on some interesting dynamics of the intrinsic magnetic excitons which have been derived exclusively through the use of tunable lasers. Various processes in this system are illustrated schematically in Fig. 5.23.

Dietz and his co-workers [5.138] first investigated the intrinsic exciton dynamics using a pulsed argon monofrequency laser in conjunction with time-resolved spectroscopy. In these experiments, they were able to observe the decay dynamics of the intrinsic exciton, E_1, and its accompanying sideband σ_1^* in fluorescence and to study their behavior as a function of applied stress and temperature. From these studies these authors concluded that zone-center and zone-boundary excitons, as derived from the E_1 and σ_1^* behavior, respectively, were

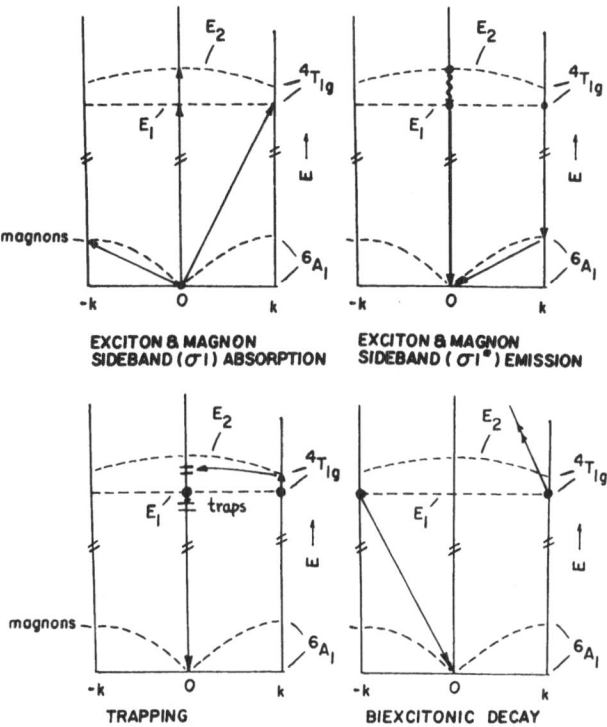

Fig. 5.23a–d. Schematic representation of diverse optical processes in antiferromagnetic MnF$_2$. (a) Exciton E_1 and E_2 and magnon sideband σ_1; (b) fluorescence E_1 and sideband σ_1^*; (c) trapping and activation via E_2; (d) biexcitonic decay of two E_1 excitons

distinguishable and that the scattering time across the Brillouin zone was relatively slow at low temperatures. They also concluded that the E_1 exciton had a negligible dispersion and that the σ_2 sideband shape could only be fitted if the E_2 exciton had a similarly small dispersion. The time resolution in these studies, because of various reasons, was relatively slow and some of the conclusions drawn were on the basis of the long time behavior ($\sim$ ms) of the E_1 and σ_1^* decays. In this regime, the appropriate model for energy diffusion might be extremely complex and has as yet not been fully explored.

Macfarlane and *Luntz* [5.139] using tunable pumping measured the scattering time across the zone boundary directly by creating $k \cong 0$ or zone boundary excitons (E_1 or σ_1 pumping) and monitoring fluorescence from the opposite portion of the zone (σ_1^* or E_1). These authors concluded that the scattering time across the zone was of the order of 1 μs rather than the $\sim$200 μs earlier reported by *Dietz*. A linear dependence of the scattering rate on pump power was observed and mechanisms for the thermalization of the excitons through exciton-exciton interactions were suggested.

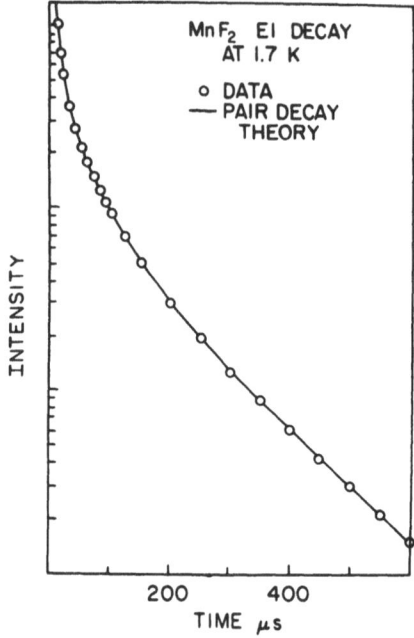

Fig. 5.24. Biexcitonic decay of the E_1 excitons in MnF_2 at 1.7 K. The solid line is the theoretical fit assuming this process; the pair annihilation cross section is of the order of 10^{-13} cm^2. [5.140]

Wilson et al. [5.140] in later work and using techniques similar to the above but with better response time, detected at 1.7 K a nonexponential decay of the E_1 exciton when the system was pumped resonantly with an intense laser pulse. This decay was found to be independent of impurity concentration and was found to be a function of the pumping intensity. To explain this nonlinear effect, the authors proposed a model of biexcitonic decay in which two E_1 excitons interact in an inverted cross relaxation, producing a ground state and a yet more excited Mn^{2+} ion. This simple model of pair decay yielded an excellent fit to the observed behavior as is shown in Fig. 5.24. As the temperature was raised, thermally activated transfer masked this decay mode by providing additional deexcitation channels, and the nonexponential decay vanished. The states involved in the latter were placed some 10 cm^{-1} above E_1, somewhat lower than the E_1-E_2 splitting at k=0. The nature of these states is controversial at this time inasmuch as they could imply a $\sim$5 cm^{-1} negative E_2 dispersion, as has been suggested in recent Raman scattering work [5.141]. On the other hand, these activating states could arise from weak, imperfection perturbations on the E_2 state producing shallow traps. Once again, further work remains to be done in this area addressed to the dispersion and to the long time behavior questions.

When created, excitons and magnons will produce magnetization changes in an antiferromagnetic lattice if localized to a specific sublattice. *Holzrichter* et al. [5.142] were able to observe these changes in a laser-induced photomagnetism experiment conducted in MnF_2. By applying a stress to the crystal [5.143] they selectively excited excitons on only one sublattice and thereby were able to observe a magnetic induction signal in a coil wound around the sample. The decay of the signal was found to be considerably faster than any trapping or radiative time and the decay rate was observed to increase as temperature and pump power were increased. This rapid change in sample magnetization was construed as evidence of a spin-allowed scattering process in which excitons and extant magnons exchange sublattices, with the latter subsequently decaying in a time less than 50 ns. This resulted in one exciton switching sublattices thus decreasing the net magnetization. It might also be that following this process, nonlinear biexcitonic effects contributed to the further deterioration of the signal.

The compounds $KMnF_3$ and $RbMnF_3$ have also been studied using laser techniques. These antiferromagnets are perovskites in structure, and the Mn^{2+} ions have cubic local symmetry. The excited states because of the remaining degeneracy exhibit Jahn-Teller distortions which may be driven by the applications of external perturbations. The dynamics of the $^4T_{1g}$ state excitons have been investigated by *Strauss* and his co-workers [5.82, 144] who in a recent work confirmed the nonlinear behavior of the excitonic fluorescence decay in these two materials. In order to conserve *k*-vectors, the biexcitonic decay occurs pairwise between *k* and −*k* excitons. Depending upon thermalization within the exciton band and the details of the excitonic density of states, the time evolution of the spectra will be related to the dispersion of the band or to exciton-exciton interactions [5.132]. Fig. 5.25 shows a shift in the peak of the emission in $KMnF_3$ which has been identified as arising from exciton-exciton interactions [5.144]. The particular shift shown in this figure, it is noteworthy to mention, occurred only during the period in which the nonexponential biexcitonic decay contributed significantly. At later times, corresponding to reduced excitonic concentrations, other shifts were observed, seemingly supporting the conjecture of the existence of very shallow traps. It would appear again that dispersion is small in this compound. It is also interesting to note that the application of uniaxial stresses in these cubic magnetic systems enhances the intrinsic emission of the unperturbed Mn^{2+} ions by reducing

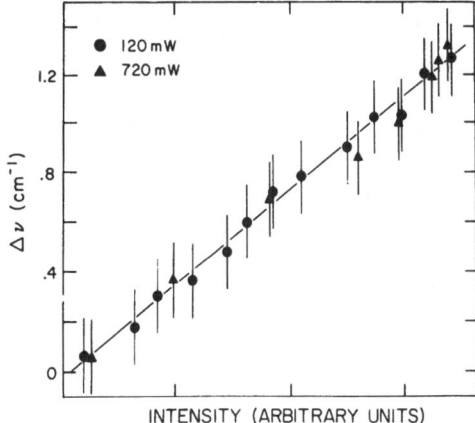

Fig. 5.25. Lineshifts observed in $KMnF_3$ as a function of time following pulsed laser excitation into the $^4T_{1g}$ exciton states. The shift results from strong exciton interactions as the biexcitonic decay is in progress. [5.144]

the trapping probabilities through the lowering of the dimensionality. It is conceivable that considerable populations can be built up in the excited states through this artifice and that more dramatic nonlinear effects are observable in these systems.

Other magnetically ordered insulators which have been investigated using laser related techniques include $TbPO_4$ [5.145], the rare earth orthochromates [5.146] and $GdCl_3$ [5.147]. Historically, the biexcitonic decay process was observed first in the case of $TbPO_4$ which because of its shielded $4f$ electrons represented a highly localized excitonic case. Various photomagnetic experiments have been conducted in the orthochromates by *Tsushima* and co-workers. In this class of materials, the easy axis of magnetization for the rare earths is considerably different in the ground and various excited states. Thus optical excitation induces changes not only through spin changes but also through readjustments of magnetization direction. $GdCl_3$ is a ferromagnetic compound which only recently has been observed to emit fluorescence, and studies are still in a very preliminary stage.

Various laser techniques have been applied to the study of phase transitions in magnetic materials. For these purposes a tunable modulated laser and detection system was developed by *Egbert* et al. [5.148]. A number of investigations have centered on metamagnetic $FeCl_2$ and have yielded information on the tricritical point in this material [5.149]. A detailed discussion of phase transitions as studied by optical methods has appeared elsewhere [5.150].

5.4.4 Other Concentrated Materials

Recently there has also been considerable interest in another class of stoichiometric concentrated materials, NdP_5O_{14} (NPP) or neodynium pentaphosphate being an example [5.151]. It was determined that these materials were efficient fluorescers and some very unusual properties were soon after derived. In NPP the radiative lifetime of the $^4F_{3/2}$ (~ 1.06 μm transition) is found to decrease only by about a factor of three from the lifetime measured in the dilute YP_5O_{14}:Nd^{3+} and the decrease is found to be a linear rather than quadratic function of concentration. The decay of the $^4F_{3/2}$ state is also invariably exponential over several e foldings at all concentrations and is only weakly temperature dependent. Various models have been proposed to account for these properties but no one can explain all the observations in their totality. In recent work, *Flaherty* and *Powell* [5.152] have conducted a laser spectroscopic study in this material. These workers pumped the $^4G_{5/2}$, $^4G_{7/2}$ states of Nd^{3+}, observing the $^4F_{3/2} \rightarrow {}^4I_{9/2}$ transition. They were able to identify a number of nonequivalent sites of NPP. No dynamics were observed in the regular NPP sites and the authors concluded that no spectral diffusion was occuring in this case. Unfortunately, the widths encountered at low temperatures in NPP were of the same order of magnitude as the resolution of their pump source, and their nonresonant pumping scheme added complications to their interpretation. No interline dynamics were observed between regular and nonequivalent sites, indicating that no trapping occurs in NPP. This latter observation led these authors to conclude that quenching occurs at the sample surfaces even though their concentration studies are somewhat ambiguous. It is more likely that the answer to this behavior can be found in the macroscopic spatial averaging, taking the appropriate donor and acceptor ion dynamics into account. As it has been shown in Chap. 3 generally, and by *Fay* [5.153] specifically for this instance, rapid donor-donor transfer leads to this type of behavior readily.

Compounds of the type $K_5NdLi_2F_{10}$ and $K_5PrLi_3F_{10}$ have been synthesized recently, and with small variations, e.g., the radiative lifetime is found to be temperature dependent, they behave similarly in a generic sense [5.154].

It is clear that additional experimentation and spectroscopic studies are needed to fully elucidate the microscopic and macroscopic properties of these interesting compounds. It is also possible that these compounds may also provide us with the means to consider the transition region between dilute localized excitations and concentrated delocalized one.

5.5 Conclusion

The purpose of this chapter has been to survey what we believe to be some of the exciting areas of topical interest in the spectroscopy of ions in crystals. As these systems are examined with recently developed instruments and techniques, new details are often discovered which increase our understanding of the intricate processes occurring in these materials.

At this point, the general interactions of dilute ions with the host crystal are fairly well understood due to recent hole burning and coherent transient studies. It should be evident from the foregoing discussions, however, that the transition region between isolated single-ion behavior and exciton-like behavior is presently the most complicated, the least understood, and consequently the most challenging area of current investigation. The various ion-ion interaction mechanisms have been elucidated, but the interplay of these processes in a real strain-broadened crystal may lead to highly complex behavior as the concentration is raised.

Understanding optical interactions of ions in crystals is not only an interesting and formidable problem in itself but it could also result in the development of new and superior solid-state laser materials. Furthermore, the dynamics of these systems are somewhat analogous to other forms of energy transport in solids and may therefore lead to the better understanding of seemingly unrelated problems in solid-state physics, such as those existing in disordered systems.

In the following chapter a similar discussion is extended to the investigation of ions in amorphous hosts. Many of the processes involved are very similar to those in crystal systems, but the large inhomogeneous broadening due to the numerous different local environments and the additional decay modes of the amorphous state result in some significantly different and interesting optical properties.

There are some areas in which clearly new results will soon be appearing and they are principally connected with new experimental developments. We can forecast that the areas of investigation will likely expand to the near uv and IR and that the temporal domain of the subnanosecond spectroscopy will also become more or less routine.

Finally, we add that laser uses in other phases of spectroscopy, Raman, Brillouin, etc., have needless to say been extremely active and have often provided complementary results. Similarly, equivalent techniques have found their uses in semi-conductor studies. As usual, excellent reviews of these important areas exist and hence these topics were not covered in this chapter.

References

5.1   B. DiBartolo: *Optical Interactions in Solids* (Wiley, New York 1968), Chap. 15; also S. Hüfner: *Optical Spectroscopy of Transparent Rare-Earth Compounds* (Academic Press, New York 1978)
5.2   A.L. Schawlow: In *Advances in Quantum Electronics*, ed. by J.R. Singer, (Columbia University Press, New York 1962) p. 50
5.3   D.E. McCumber, M.D. Sturge: J. Appl. Phys. **34**, 1682 (1963)
5.4   G.F. Imbusch, W.M. Yen, A.L. Schawlow, D.E. McCumber, M.D. Sturge: Phys. Rev. **133**, A1029 (1964)
5.5   S. Geschwind, G.E. Devlin, R.L. Cohen, S.R. Chinn: Phys. Rev. **137**, A1087 (1965)
5.6   S. Geschwind: In *Electron Paramagnetic Resonance*, ed. by S. Geschwind (Plenum, New York 1972)
5.7   T. Muramoto, Y. Fukada, T. Hashi, Phys. Lett. **48A**, 181 (1974)
5.8   N.A. Kurnit, I.D. Abella, S.R. Hartmann, Phys. Rev. Lett. **13**, 567 (1964)
5.9   I.D. Abella, N.A. Kurnit, S.R. Hartmann: Phys. Rev. **141**, 391 (1966)
5.10 A. Compaan, L.Q. Lambert, I.D. Abella: Phys. Rev. Lett. **20**, 1089 (1968)
5.11 L.Q. Lambert, A. Compaan, I.D. Abella: Phys. Lett. **30A**, 153 (1969)
5.12 A. Compaan, Phys. Rev. **B5**, 4450 (1972)
5.13 E.L. Hahn: Phys. Rev. **80**, 580 (1950)
5.14 P.F. Liao, S.R. Hartmann: Opt. Comm. **8**, 310 (1973)
5.15 A. Szabo, M. Kroll: Opt. Lett. **2**, 10 (1978)
5.16 A. Szabo: Private communication
5.17 V. V. Samartsev, R. G. Usmanov, G. M. Ershov, B. Sh. Khamidullin: Sov. Phys. JETP **47**, 1030 (1978); translated from Zh. Eksp. Teor. Fiz. **74**, 1979 (1978)
5.18 P.F. Liao, R. Leigh, P. Hu, S.R. Hartmann: Phys. Lett. **41A**, 285 (1972)
5.19 P.F. Liao, S.R. Hartmann: Phys. Rev. **B8**, 69 (1973)
5.20 P.F. Liao, P. Hu, R. Leigh, S.R. Hartmann: Phys. Rev. **A9**, 332 (1974)
5.21 S. Meth, S.R. Hartmann: Opt. Comm. **24**, 100 (1978)
5.22 A. Szabo: Phys. Rev. Lett. **25**, 924 (1970)
5.23 A. Szabo: Phys. Rev. Lett. **27**, 323 (1971)
5.24 A. Szabo: Phys. Rev. **B11**, 4512 (1975)
5.25 T. Muramoto, S. Nakamishi, T. Hashi: Opt. Comm. **21**, 139 (1977)
5.26 T. Muramoto, S. Nakamishi, T. Hashi: Opt. Comm. **24**, 316 (1978)
5.27 P.E. Jessop, T. Muramoto, A. Szabo: Phys. Rev. **B21**, 926 (1980)
5.28 P.M. Selzer, D.L. Huber, B.B. Barnett, W. M. Yen: Phys. Rev. **B17**, 4979 (1978)
5.29 J.C. Murphy: Private communication
5.30 K.F. Renk, J. Deisenhofer: Phys. Rev. Lett. **26**, 764 (1971)
5.31 J.E. Rives, R.S. Meltzer: Phys. Rev. **B16**, 1808 (1977)
5.32 T. Muramoto, T. Hashi: Phys. Lett. **51A**, 423 (1975)

5.33   T. Kushida, M. Kikuchi: J. Phys. Soc. Japan **23**, 1333 (1967); E.I. Penepelitsa, Sov. Phys. Solid State **17**, 1660 (1976); translated from Fiz. Tverd. Tela **17**, 2490 (1975)

5.34   A. Zalkin, D.H. Templeton, T.E. Hopkins: Inorg. Chem. **5**, 1466 (1966) and references therein

5.35   W.M. Yen, W.C. Scott, A.L. Schawlow: Phys. Rev. **136**, A271 (1964)

5.36   W.M. Yen, W.C. Scott, P.L. Scott: Phys. Rev. **137**, A1109 (1965)

5.37   L.E. Erickson: Phys. Rev. **B11**, 77 (1975)

5.38   T. Kushida, E. Takushi: Phys. Rev. **B12**, 824 (1975); M.J. Weber, J.A. Paisner, S.S. Sussman, W.M. Yen, L.A. Riseberg, C. Brecher: J. Lumin. **12/13**, 729 (1976)

5.39   R. Flach, D.S. Hamilton, P.M. Selzer, W.M. Yen: Phys. Rev. **B15**, 1248 (1977)

5.40   L.E. Erickson: Opt. Comm. **15**, 246 (1975)

5.41   L.E. Erickson: Phys. Rev. **B16**, 4731 (1977)

5.42   N. Takeuchi, A. Szabo: Phys. Lett. **50A**, 316 (1974)

5.43   N. Takeuchi: J. Lumin. **12/13**, 743 (1976); A. Szabo, N. Takeuchi: Opt. Comm. **15**, 250 (1975)

5.44   A.Z. Genack, R.M. Macfarlane, R.G. Brewer: Phys. Rev. Lett. **37**, 1078 (1976)

5.45   A. Yamagishi, A. Szabo: Opt. Lett. **2**, 160 (1978)

5.46   Y.C. Chen, S.R. Hartmann: Phys. Lett. **58A**, 201 (1976)

5.47   S. Chandra, N. Takeuchi, S.R. Hartmann: Phys. Lett. **41A**, 91 (1972)

5.48   L.E. Erickson: Opt. Commun. **21**, 147 (1977)

5.49   R.M. Shelby, C.S. Yannoni, R.M. Macfarlane: Phys. Rev. Lett. **41**, 1739 (1978)

5.50   R.M. Macfarlane, A.Z. Genack, S. Kano, R.G. Brewer: J. Lumin. **18/19**, 933 (1979)

5.51   Y.C. Chen, K.P. Chiang, S.R. Hartmann: Opt. Commun. **26**, 269 (1978)

5.52   R.G. DeVoe, A. Szabo, S.C. Rand, R.G. Brewer: Phys. Rev. Lett. **43**, 1560(1979)

5.53   R.M. Macfarlane, R.M. Shelby, R.L. Shoemaker: Phys. Rev. Lett. **43**, 1726(1979)

5.54   S.C. Rand, A. Wokaum, R.G. DeVoe, R.G. Brewer: Phys. Rev. Lett. **43**, 1868   (1979)

5.55   Y.C. Chen, K.P. Chiang, S.R. Hartmann: Opt. Commun. **28**, 181 (1979)

5.56   R.M. Shelby, R.M. Macfarlane: Phys. Rev. Lett. **45**, 1098 (1980)

5.57   J.P. Hessler, R. Brundage, J. Hegarty, W.M. Yen: Opt. Lett. **5**, 348 (1980)

5.58   R.M. Macfarlane, A.Z. Genack, R.G. Brewer: Phys. Rev. **B17**, 2821 (1978)

5.59   R.M. Macfarlane, R.M. Shelby: Phys. Rev. Lett. **42**, 788 (1979)

5.60   D.B. Fitchen, R.H. Silsbee, T.A. Fulton, E.L. Wolf: Phys. Rev. Lett. **11**, 275 (1963)

5.61   N. Takeuchi, S. Chandra, Y.C. Chen, S.R. Hartmann: Phys. Lett. **46A**, 97(1973)

5.62   Y.C. Chen, K.P. Chiang, S.R. Hartmann: Phys. Rev. **B21**, 40 (1980)

5.63   C. Delsart, N. Pelletier-Allard, R. Pelletier: Opt. Commun. **11**, 84 (1974)

5.64   C. Delsart, N. Pelletier-Allard, R. Pelletier: J. Phys. B: Atom. Mol. Phys. **8**, 2771 (1975); C. Delsart, N. Pelletier-Allard, and R. Pelletier: Phys. Rev. **B16**, 154 (1977)

5.65   N. Pelletier-Allard: Colloq. Int. *CNRS* **255**, 289 (1977)

5.66   N. Pelletier-Allard, R. Pelletier, C. Delsart: J. Phys. C. Solid State Phys. **10**, 2005 (1977)

5.67   L.E. Erickson: Phys. Rev. **B19**, 4412 (1979)

5.68   D. R. Tallant, J.C. Wright: J. Chem. Phys. **63**, 2074 (1975)

5.69   M.E. Seelbinder, J.C. Wright: Phys. Rev. **B20**, 4308 (1979)

5.70   D.R. Tallant, D.S. Moore, J.C. Wright: J. Chem. Phys. **67**, 2897 (1977)

5.71   D.S. Moore, J.C. Wright: Chem. Phys. Lett. **66**, 173 (1979)

5.72   M.P. Miller, D.R. Tallant, F.J. Gustafson, J.C. Wright: Anal. Chem. **49**, 1474 (1977)

5.73   J.C. Wright: Anal. Chem. **49**, 1690 (1977)

5.74   J.C. Wright, F.J. Gustafson: Anal. Chem. **50**, 1147A (1978)

5.75   L. Dubicki, J. Ferguson, G.A. Osborne, I. Tabjerg: Colloq. Int. *CNRS* **255**, 217 (1977)

5.76   J. Ferguson, H.U. Gudel, E.R. Kransz, H.J. Guggenheim: Ncl. Phys. **28**, 879 (1974)

5.77   J.P. Hessler, J. Hegarty, C.G. Levey, G.F. Imbusch, W.M. Yen: J. Lumin. **18/19**, 73 (1979)

5.78   J.P. Hessler, W.T. Carnall: In *Chemistry of Lanthanide and Actinides*, ed. by N. Edelstein (American Chemical Society, Washington, D.C. 1980)

5.79   J.P. Hessler, J.A. Caird, W.T. Carnall, H.M. Crosswhite, R.J. Sjoblom, F. Wagner, Jr.: *The Rare Earths in Modern Science and Technology (1978)*, ed. by G.J. McCarthy and J.J. Rhyne (Plenum Press, New York 1979) p. 507

5.80   T.T. Basiev, Yu. K. Voron'ko, A. Ya. Karasik, V.V. Osiko, I.A. Shcherbakov: Sov. Phys. JETP **48**, 32 (1978); trans. Zh. Eksp. Teor. Fiz. **75**, 66 (1978); M. Zokai, R.C. Powell, G.F. Imbusch, B. DiBartolo: J. Appl. Phys. **50**, 5930 (1979)

5.81   S. Aoki, I.D. Abella: Appl. Phys. Lett. **26**, 653 (1975)

5.82   E. Strauss, V. Gerhardt, H. Riederer: J. Lumin. **12/13**, 239 (1976)

5.83   T.C. Chiang, P. Salvi, J. Davies, Y.R. Shen: Solid State Commun. **26**, 217 (1978); 527 (1978)

5.84   J. Hegarty, B.A. Wilson, W.M. Yen, T.J. Glynn, G.F. Imbusch: Phys. Rev. **B18**, 5812 (1978)

5.85   B.A. Wilson, J. Hegarty, W.M. Yen: Phys. Rev. **B24**, 6725 (1981)

5.86   P.M. Selzer, W.M. Yen: Opt. Lett. **1**, 90 (1977); T. Holstein, S.K. Lyo, R. Orbach: Phys. Rev. **B16**, 934 (1977)

5.87   M.M. Broer, J. Hegarty, G.F. Imbusch, W.M. Yen: Opt. Lett. **3**, 175 (1978)

5.88   R.S. Meltzer, R.M. Wood: Appl. Opt. **16**, 1432 (1977)

5.89   R.S. Meltzer, J.E. Rives: Phys. Rev. **B16**, 2442 (1977)

5.90   See, e.g., *Proceedings of the International Conference on Lattice Dynamics, Paris 1977* (Flammanion Sciences, Paris 1978)

5.91   J.H. Lee, J.J. Song, M.A.F. Scarparo, M.D. Levenson: Bull. Am. Phys. Soc. **24**, 901 (1979)

5.92   J. J. Song, J. H. Lee, M. D. Levenson: In *Laser Spectroscopy III*, ed. by J. L. Hall and J.L. Carlsten, Springer Series in Optical Sciences, Vol. 7 (Springer, Berlin, Heidelberg, New York 1977) p. 450

5.93   A.C. Boccara: Colloq. Int. *CNRS* **255**, 299 (1977)

5.94   N. Bontemps-Moreau, A.C. Boccara, P. Thibault: Phys. Rev. **B16**, 1822 (1977)

5.95   See, e.g., F.E. Auzel: Proc. IEEE **61**, 758 (1973)

5.96   J.C. Wright: In *Radiationless Processes in Molecules and Condensed Phases,* ed. by. F.K. Fong, Topics in Applied Physics, Vol. 15, (Springer, Berlin, Heidelberg, New York 1976) p. 239

5.97   D.L. Dexter: J. Chem. Phys. **21**, 836 (1953); T. Förster: Ann. Physik. **2**, 55 (1968)

5.98   M. Inokuti, F. Hirayama: J. Chem. Phys. **43**, 1978 (1965)

5.99   J. Hegarty, D.L. Huber, W.M. Yen: Phys. Rev. **B23**, 6271 (1981)

5.100   D.R. Tallant, M.P. Miller, J.C. Wright: J. Chem. Phys. **65**, 510 (1976)

5.101   M.P. Miller, J.C. Wright: J. Chem. Phys. **68**, 1548 (1978)

5.102   D.S. Hamilton, P.M. Selzer, W.M. Yen: Phys. Rev. **B16**, 1858 (1977)

5.103   P. Evesque, J. Kliava, J. Duran: J. Lumin. **18/19**, 646 (1979)

5.104   L.D. Merkle, R.C. Powell: J. Chem. Phys. **67**, 371 (1977); L.D. Merkle, R.C. Powell T.M. Wilson: J. Phys. C **11**, 3103 (1978); G.C. Venikoas, R.C. Powell: Phys. Rev. **B17**, 3456 (1978)

5.105   N. Yamada, S. Shionoya, T. Kushida: J. Phys. Soc. Jpn. **32**, 1577 (1972)

5.106   D.S. Hamilton, P.M. Selzer, D.L. Huber, W.M. Yen: Phys. Rev. **B14**, 2183 (1976)

5.107   J. Heber, H. Murman: Z. Phys. **B26**, 145 (1977); H. Siebold, J. Heber: J. Lumin. **22**, 297 (1981)

5.108   C. Hsu, R.C. Powell: J. Lumin. **10**, 273 (1975)

5.109   R.L. Greene, D.D. Sell, R.S. Feigelson, G.F. Imbusch, H.J. Guggenheim: Phys. Rev. **171**, 600 (1968)

5.110   J.M. Flaherty, B. DiBartolo: Phys. Rev. **38**, 5232 (1973); J. Hegarty, G.F. Imbusch: Colloq. Int. *CNRS* **255**, 199 (1977)

5.111   B.A. Wilson, W.M. Yen, J. Hegarty, G.F. Imbusch: Phys. Rev. **B19**, 4238 (1978)

5.112   H. Yersin, W.V. Ammon, M. Stock, G. Gliemann: J. Lumin. **18/19**, 774 (1979) and references therein

5.113   R.J. Birgeneau: J. Chem. Phys. **50**, 4282 (1969)

5.114   T. Holstein, S.K. Lyo, R. Orbach: Phys. Rev. Lett. **36**, 891 (1976)

5.115   R. Flach, D.S. Hamilton, P.M. Selzer, W.M. Yen: Phys. Rev. Lett. **35**, 1034 (1975)

5.116   P.M. Selzer, D.S. Hamilton, R. Flach, W.M. Yen: J. Lumin. **12/13**, 737 (1976)
5.117   D.L. Huber, D.S. Hamilton, B. Barnett: Phys. Rev. **B16**, 4642 (1977)
5.118   P.M. Selzer, D.S. Hamilton, W.M. Yen: Phys. Rev. Lett. **38**, 858 (1977)
5.119   G.F. Imbusch: Phys. Rev. **153**, 326 (1967)
5.120   See, e.g., V.M. Kenkre, R.S. Knox: Phys. Rev. Lett. **33**, 803 (1974)
5.121   S.K. Lyo: Phys. Rev. **B3**, 331 (1971)
5.122   P.W. Anderson: Phys. Rev. **109**, 1492 (1958)
5.123   J. Koo, L.R. Walker, S. Geschwind: Phys. Rev. Lett. **35**, 1669 (1975); also S. Chu, H.M. Gibbs, A. Passner, S. Geschwind: Bull. Am. Phys. Soc. **24**, 894 (1979)
5.124   E. Strauss, W.J. Miniscalco, J. Hegarty, W.M. Yen: J. Phys. C. Solid State Phys. **C14**, 2229 (1981)
5.125   H. Eichler: Opt. Acta. **24**, 631 (1977); P. F. Liao, D. M. Bloom: Opt. Lett. **3**, 4 (1978); D.S. Hamilton, D. Heiman, J. Feinberg, R.W. Hellwarth, Opt. Lett. **4**, 933 (1979)
5.126   G.H. Dieke: *Spectra and Energy Levels of Rare-Earth Ions in Crystals* (Interscience, New York 1968)
5.127   J.P. Harrison, J.P. Hessler, D.R. Taylor: Phys. Rev. **B14**, 2979 (1976)
5.128   F. Varsanyi, Appl. Phys. Lett. **19**, 169 (1971); K.R. German, A. Kiel, H.J. Guggenheim: Appl. Phys. Lett. **22**, 87 (1973)
5.129   J. Hegarty, W.M. Yen: J. Appl. Phys. **51**, 3545 (1980)
5.130   K.R. German, A. Kiel: Phys. Rev. Lett. **33**, 1039 (1974)
5.131   R.L. Cone, R.S. Meltzer: J. Chem. Phys. **62**, 357 (1975)
5.132   R.S. Meltzer: Solid State Comm. **20**, 553 (1976)
5.133   R.S. Meltzer, R.L. Cone: J. Lumin. **12/13**, 247 (1976)
5.134   H.T. Chen, R.S. Meltzer: Phys. Rev. Lett. **44**, 599 (1980)
5.135   R.L. Greene, D.D. Sell, W.M. Yen, A.L. Schawlow, R.M. White: Phys. Rev. Lett. **15**, 656 (1965)
5.136   R. Loudon: Adv. In Physics **17**, 243 (1968); V.V. Eremenko, E.G. Petrov: Adv. in Physics **26**, 31 (1977)
5.137   M.W. Passow, D.L. Huber, W.M. Yen: Phys. Rev. Lett. **23**, 477 (1969)
5.138   R.E. Dietz, A.E. Meixner, H.J. Guggenheim, A. Misetich: Phys. Rev. Lett. **21**, 1067 (1968); J. Lumin. **1/2**, 279 (1970)
5.139   R.M. Macfarlane, A.C. Luntz: Phys. Rev. Lett. **31**, 832 (1973)
5.140   B.A. Wilson, J. Hegarty, W.M. Yen: Phys. Rev. Lett. **41**, 268 (1978)
5.141   N.M. Amer, Tai-chang Chiang, Y.R. Shen: Phys. Rev. Lett. **34**, 1454 (1975); **36**, 1102 (1976); D.L. Rousseau, R.E. Dietz, P.F. William, H.J. Guggenheim: Phys. Rev. Lett. **36**, 1098 (1976)
5.142   J.F. Holzrichter, R.M. Macfarlane, A.L. Schawlow: Phys. Rev. Lett. **26**, 652 (1971)
5.143   R.E. Dietz, A. Misetich, H.J. Guggenheim: Phys. Rev. Lett. **16**, 841 (1966)
5.144   E. Strauss, W.J. Miniscalco, W.M. Yen, U. Kellner, V. Gerhardt: Phys. Rev. Lett. **44**, 824 (1980)
5.145   P.C. Diggle, K.A. Gehring, R.M. Macfarlane: Solid State Comm. **18**, 391 (1976)
5.146   K. Toyokawa, S. Kurita, K. Tsushima: Phys. Rev. **B19**, 274 (1976)
5.147   S. Yokono, W.J. Miniscalco, E. Strauss, W.M. Yen: Unpublished; R. Mahiou, B. Jacquier, R. Moncorgé: In *Rare Earths Spectroscopy*, ed. by B. Jezowska-Trzebiatowska, J. Legendziewicz, W. Strek (World Scientific, Singapore, 1985) p. 579
5.148   W.C. Egbert, P.M. Selzer, W.M. Yen: Appl. Opt. **15**, 1158 (1976)
5.149   J.F. Dillon, Jr., E. Yi Chen, H.J. Guggenheim: Phys. Rev. **B18**, 377 (1978) and references therein
5.150   Y. Farge: Colloq. Int. *CRNS* **255**, 243 (1977)
5.151   H.G. Danielmeyer: *Festkorperprobleme, XV*, ed. by J.H. Queisser (Pergamon, Braunschweig 1975) p. 253
5.152   J.M. Flaherty, R.C. Powell: Phys. Rev. **B19**, 32 (1979)
5.153   D. Fay, G. Huber, W. Lenth: Opt. Commun. **28**, 117 (1979)
5.154   B.C. McCollum, A. Lempicki: Mater. Res. Bull. **13**, 833 (1978); A. Lempicki, B.C. McCollum: J. Lumin. **20**, 291 (1979)
5.155   J.L. Merz, P.S. Pershan: Phys. Rev. **162**, 217 (1967)

Additional References with Titles

5.156   S. Chu, H.M. Gibbs, S.L. McCall, A. Passner: Energy transfer and Anderson localization in ruby. Phys. Rev. Lett. **45**, 1715 (1980)

5.157   A.Z. Genack, D.A. Weitz, R.M. Macfarlane, R.M. Shelby, A. Schenzer: Coherent transients by optical phase switching: Dephasing in $LaCl_3 : Pr^{3+}$. Phys. Rev. Lett. **45**, 438 (1980); **45**, 1044 (E) (1980)

5.158   M. Glasheck, R. Hond, A.H. Zewail: Observation of dipolar-induced spin dephasing in ionic solids using coherent optical-microwave spectroscopy. Phys. Rev. Lett. **45**, 744 (1980)

5.159   J. Hegarty, R.T. Brundage, W.M. Yen: Line-shape deconvolution in fluorescence line narrowing. App. Opt. **19**, 1889 (1980)

5.160   P.E. Jessop, A. Szabo: Visual observation of macroscopic inhomogeneous broadening of R_1 lines in ruby. Appl. Phys. Lett. **37**, 510 (1980)

5.161   P.E. Jessop, A. Szabo: High resolution measurements of the ruby R_1 line at room temperature. Opt. Commun. **33**, 301 (1980)

5.162   P.E. Jessop, A. Szabo: Resonant optical energy transfer in ruby. Phys. Rev. Lett. **45**, 1712 (1980)

5.163   J.H. Lee, J.J. Song, M.A.F. Scarparo, M.D. Levenson: Coherent population oscillations and hole burning in $Sm^{2+}:CaF_2$ using polarization spectroscopy. Opt. Lett. **5**, 196 (1980)

5.164   M.D. Levenson, R.M. Macfarlane, R.M. Shelby: Polarization spectroscopy measurements of homogeneously broadened color-center bands. Phys. Rev. **B22**, 4915 (1980)

5.165   P.F. Liao, A.M. Glass, L.M. Humphrey: Optically generated pseudo-Stark effect in ruby. Phys. Rev. **B22**, 2276 (1980)

5.166   R.M. Macfarlane, R.M. Shelby, A.Z. Genack, D.A. Weitz: Nuclear quardupole optical hole burning in the stoichiometric material EuP_5O_{14}. Opt. Lett. **5**, 462 (1980)

5.167   A. Monteil, E. Duval: Energy transfer between different emitters and spatial diffusion in ruby. J. Phys. C. Solid State Phys. **13**, 4565 (1980)

5.168   N. Pelletier-Allard, R. Pelletier: Laser induced fluorescence in $Nd^{3+}:LaCl_3$ I and II. J. Phys. (France) **41**, 855 (1980); **44**, 861 (1980)

5.169   R.C. Powell, D.P. Neikirk, J.M. Flaherty: Lifetime measurements, infrared and photoacoustic spectroscopy of NdP_5O_{14}. J. Phys. Chem. Solids **41**, 345 (1980)

5.170   K.K. Sharma, L.E. Erickson: NMR measurement of the hyperfine constant of an excited state of an impurity ion in a solid. Phys. Rev. Lett. **45**, 294 (1980)

5.171   R.M. Shelby, R.M. Macfarlane, C.S. Yannomi: Optical measurements of spin lattice relaxation in dilute nuclei: $LaF_3 : Pr^{3+}$. Phys. Rev. **B21**, 5004 (1980)

5.172   R.M. Shelby, R.M. Macfarlane: Frequency dependent optical dephasing in the stoichiometric material EuP_5O_{14} Phys. Rev. Lett. **45**, 1098 (1980)

5.173   H. Wolfrum, K. Lanzinger, K.F. Renk: Spin lattice relaxation by the direct process in optically excited Er^{3+} in LaF_3. Opt. Lett. **5**, 294 (1980)

Note: The work by *Chu* et al. [5.156] and by *Jessop* [5.162] reports on new results in the study of energy transfer in ruby; these results disagree with those reported by *Koo* [5.123]. They also report the observation of resonant transfer within the R_1 line which is slow compared to that postulated by *Imbusch* [5.119].

References [5.116, 169, 171] report on recent measurements including coherent transients in stoichiometric systems discussed in Sect. 5.4.4.

We are also aware of a number of recent preprints on spatial migration in NPP and on multiphoton and/or multi ion transitions which could not be incorporated because of various limitations.

6. Laser Excited Fluorescence Spectroscopy in Glass

M. J. Weber

With 22 Figures

6.1 Background

Glass, on a microscopic scale, epitomizes an inhomogeneous system. Being an in-inherently disordered medium, the environment of each ion in a glass is not identical as in a crystal. In addition, because of differences in the bonding to nearest-neighbor ions and, in multicomponent glass compositions, in the types and statistical distribution of more distant neighbor ions, the local fields at individual ion sites vary. This results in site-to-site differences in the energy levels and the radiative and nonradiative transition probabilities of paramagnetic ions in glasses. Broadband-excited optical absorption and emission spectra and excited-state decays consist of a superposition of contributions from individual ions distributed among the entire ensemble of local environments. The spectra exhibit inhomogeneous broadening and the decays do not have a single exponential time dependence.

When a narrowband source is used for excitation, only those ions resonant with the excitation quanta to within the homogeneous linewidth are excited. This site-selective excitation effectively reduces the inhomogeneous broadening and a line-narrowed fluorescence spectrum is obtained. In comparison to crystals where local strains and other imperfections produce inhomogeneous broadening of narrow-line transitions of $\gtrsim 1$ cm^{-1}, in glasses the same transition may be broadened by $\gtrsim 100$ cm^{-1}. Since the radiative lifetimes of some rare-earth ion transitions correspond to homogeneous linewidths as small as $\approx 10^{-8}$ cm^{-1}, line narrowing of ten orders of magnitude in principle is possible.

Fluorescence line narrowing (FLN) in glass is not new. In a pioneering study published in 1967, *Denisov* and *Kizel* [6.1] reported FLN and time-resolved spectra of Eu^{3+} emission in a borate glass. Narrow lines from a mercury lamp that matched the inhomogeneously broadened Eu^{3+} absorption bands were used for excitation. While the *Denisov-Kizel* report was brief and qualitative, it demonstrated both line narrowing and spectral diffusion and the dependence of these properties on paramagnetic ion concentration. This was done several years before *Szabo*'s report of laser-induced FLN in a ruby crystal [6.2].

Although any monochromatic source can be used for FLN in glass, a laser — especially a tunable laser — is the ideal spectroscopic tool. Yet another two years passed after *Szabo*'s paper before the first report of laser-induced FLN in glass appeared. Using a fixed-frequency cw laser, *Riseberg* excited a Nd^{3+} absorption line in a silicate glass, observed nonresonant line-narrowed fluorescence from a lower excited state, and studied the dependence of the line narrowing on Nd concentration [6.3]. In the same year, *Personov* et al. [6.4] showed that laser-induced FLN was also possible in organic glasses. Using a He-Cd laser, narrowed zero-phonon lines and vibronic structure normally obscured by inhomogeneous broadening were observed from fluorescing molecules in a low-temperature glassy alcohol.

The work of *Motegi* and *Shionoya* [6.5] in the following year revealed more fully the potential usefulness of laser-excited fluorescence for investigating the spectroscopic properties of ions in glasses. Their interest was in energy transfer and migration among europium ions in an inhomogeneously broadened system. The study was made using a tunable pulsed dye laser for excitation and observing time-resolved nonresonant fluorescence. It was evident from this work that large, site-dependent variations in the energy–level structure were observable from the resulting line-narrowed fluorescence. Thus, tunable FLN techniques provided a unique probe of the local fields and interactions at paramagnetic ion sites in glass.

Line-narrowed fluorescence and optical site selection spectroscopy of organic compounds continued and proved to be a sensitive probe of the local solvent environment [6.6]. In 1974, photochemical hole burning, the decrease in the line intensity of optical spectra due to "burning out" of impurity centers in the course of laser excitation, was observed in organic glasses [6.7, 8].

In the past five years, the techniques and applications of laser-induced fluorescence line narrowing have proliferated. Site-to-site variations in local fields, electron-phonon coupling, and ion-ion interactions and the effects of these interactions on energy levels and relaxation processes have been explored in both organic and inorganic glass. This chapter reviews the experimental data and interpretation of these studies. The fluorescence of paramagnetic ions in inorganic glasses has received the most extensive investigation and, therefore, will be the principal subject of this review. Investigations of fluorescence of molecules in organic glasses will be noted and references cited; however, it is beyond the intended length of this chapter to discuss these results in detail.

The general spectroscopic properties of ions and molecules in solids and the experimental techniques of laser spectroscopy are well covered in Chaps. 1, 4 and 5, and hence will not be repeated here. We concentrate instead on those spectroscopic features and experimental techniques specific to glasses, but applicable to other amorphous or disordered systems. Because the spectroscopic inhomogeneities in glass are large, FLN experiments are relatively simple; the interpretation and relationship of the spectroscopic features to structural details, however, have proved to be more difficult and subtle.

We begin with a brief review of the structure of inorganic glasses, a survey of paramagnetic ions and molecules useful for FLN studies, and experimental techniques and apparatus. This is followed by a summary of investigations of energy levels, radiative and nonradiative transition probabilities, homogeneous line broadening, and ion-ion energy transfer in glasses as studied by laser-excited fluorescence spectroscopy.

Fluorescence line narrowing in glass, in addition to being an important spectroscopic tool, may be of greatest importance as a microscopic probe of the local environments in glass. We therefore conclude with discussions of the applications of FLN results to the analysis of local structure in glasses, to studies of structural changes in glass, and the effects of inhomogeneities on energy extraction in laser glass.

6.2 Glass Properties

6.2.1 Structure

Structurally glass is a continuous random network lacking both symmetry and periodicity. The basic structural units which make up the glass network, for example SiO_4 tetrahedra or BO_3 triangles, have a definite geometry but are connected at corners to form a random three-dimensional network [6.9]. Several compounds exist in an amorphous phase singly (e.g., SiO_2, GeO_2, B_2O_3, P_2O_5) and constitute simple glass formers. Conditional glass formers (e.g., TeO_2, Al_2O_3, WO_3) require the presence of one or more additional compounds to form a glass. Other compounds added to the glass modify the network. Anions which normally bridge two network former cations may become nonbridging. Network modifier cations, such as alkali, alkaline earth, and higher valence state ions, are accommodated randomly in the network in close proximity to nonbridging anions. A simple two-dimensional representation of a sodium silicate glass structure illustrating both bridging and nonbridging oxygens is shown in Fig. 6.1.

Paramagnetic ions, depending upon their size and valence state, enter a glass as a network modifier cation or substitutionally for a network former cation. The former is the predominant case for rare-earth ions. It is frequently difficult to add ≈ 1 cationic % of rare earths to simple glass formers such as SiO_2, GeO_2, and B_2O_3. The networks of these single-component glasses are tightly bonded by

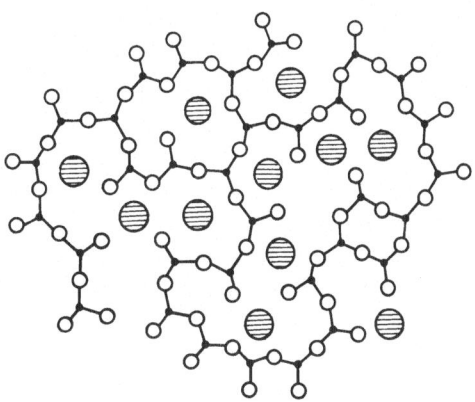

Fig. 6.1. Two-dimensional representation of a sodium silicate glass. The small black dots, open circles, and large shaded circles are Si^{4+}, O^{2-}, and Na^+ ions, respectively

bridging oxygens. Trivalent rare earths cannot easily enter this structure substitutionally because of size. If, however, a network modifier such as Na^+ is present, it disrupts the network and nonbridging oxygens occur. Rare earths may now be incorporated as modifier ions in the vicinity of the looser structure.

The variability in the sizes and valence states of known glass forming and glass modifying cations and anions [6.10] produces a very large range of possible local fields at an activator site. A multicomponent glass may contain several different modifier cations, each constituent of which is assumed to be evenly distributed throughout the whole structure. Glasses may also be composed of a combination of network formers as in borosilicate, aluminosilicate, and phosphotellurite glasses. Examples of mixed anion glasses include fluorophosphate, borofluoride, and chlorophosphate glasses. Because the nearest-neighbor anions affect the activator ion most strongly, mixed anion coordination introduces an additional degree of disorder into the glass structure at an activator site.

Phase separation is observed in a variety of glass forming systems including, for example, silicates, borates, chalcogenides, and fused salts [6.11]. Therefore, instead of a random structure extending isotropically and indefinitely in all directions, structural heterogeneities on a scale of several hundred Å may be present. These may occur as interconnected structures and occupy a large volume fraction.

A number of organic compounds, such as simple alcohols (ethanol, methanol), form glass-like structures at low temperatures. These can serve as hosts for fluorescing organic molecules and dyes and provide a distribution of different sites. Thin, amorphous polyvinylbutryal films have also been used as hosts.

6.2.2 Activator Ions and Molecules

Ions from several transition-metal groups can be used for laser-excited spectroscopy in inorganic glasses. The fluorescing transitions may be either intraconfigurational $(d \rightarrow d, f \rightarrow f)$ or interconfigurational $(d \rightarrow f, d^{10} \rightarrow d^9 p, s^2 \rightarrow sp)$. The former include both narrow and broad homogeneous linewidths; the latter are characterized by broad or very broad linewidths. Ions and excitation and fluorescence transitions that have been used for FLN studies are tabulated in Table 6.1. They include transition metal ions (Cr^{3+}, Mo^{3+}), rare-earth ions (Pr^{3+}, Nd^{3+}, Sm^{3+}, Eu^{3+}, Yb^{3+}), and one post-transition-group ion (Bi^{3+}) [6.12-33].

Trivalent lanthanide ions have been the most extensively used activator ions. They are attractive for several reasons. First, there are many fluorescing states and wavelengths to choose from among the $4f$ electronic configurations. Second, the $f\text{-}f$ transitions have small homogeneous linewidths—in the order of or, at low temperatures, very much smaller than the inhomogeneous linewidths. And third, the local fields in glass can be treated as small perturbations on the free-ion energy levels.

Of the lanthanide ions, trivalent europium is particularly attractive because of the small homogeneous linewidths of the $^7F - ^5D$ transitions and the relatively sim-

Table 6.1. Ions and transitions used for laser-excited fluorescence studies in inorganic glasses

Ion	Excitation	Fluorescence	Glass[a] [Ref.]
$Cr^{3+}(3d^3)$	$^4A_2 \rightarrow ^4T_2, ^4T_1$	$^4T_2 \rightarrow ^4A_2$	S [6.12]
$Mo^{3+}(4d^3)$	$^4A_2 \rightarrow ^2E$	$^2E \rightarrow ^4A_2$	P [6.13, 14]
$Pr^{3+}(4f^2)$	$^3H_4 \rightarrow ^3P_0$	$^3P_0 \rightarrow ^3H_4, ^3H_6$	G, FB [6.15]
$Nd^{3+}(4f^3)$	$^4I_{9/2} \rightarrow (^4G_{5/2}, ^2G_{7/2})$	$^4F_{3/2} \rightarrow ^4I_{9/2}, ^4I_{11/2}$	S [6.16, 17], P [6.17], G [6.17], T [6.17]
	$^4I_{9/2} \rightarrow ^2P_{1/2}$	$^4F_{3/2} \rightarrow ^4I_{9/2}, ^4I_{11/2}$	S, P, B, FB, FP [6.18]
	$^4I_{9/2} \rightarrow ^4G_{11/2}$	$^4F_{3/2} \rightarrow ^4I_{9/2}, ^4I_{11/2}$	S [6.3]
	$^4I_{9/2} \rightarrow ^4F_{3/2}$	$^4F_{3/2} \rightarrow ^4I_{9/2}, ^4I_{11/2}$	S [6.19, 20]
	$^4I_{11/2} \rightarrow ^4F_{3/2}$	$^4F_{3/2} \rightarrow ^4I_{11/2}$	S [6.19]
$Sm^{3+}(4f^5)$	$^6H_{5/2} \rightarrow ^4G_{5/2}$	$^4G_{5/2} \rightarrow ^6H_{5/2, 7/2, 9/2}$	S [6.21]
$Eu^{3+}(4f^6)$	$^7F_0 \rightarrow ^5D_{0,1,2,3}$	$^5D_0 \rightarrow ^7F_{0,1,2...}$	S [6.22, 23], P [6.5,24-26] B [6.27], FB [6.28], FP [6.29]
	$^7F_1 \rightarrow ^5D_{0,1}$	$^5D_0 \rightarrow ^7F_{0,1,2...}$	S [6.23], P [6.30, 31]
	$^7F_2 \rightarrow ^5D_0$	$^5D_0 \rightarrow ^7F_{0,1,2...}$	P [6.23]
$Yb^{3+}(4f^{13})$	$^2F_{7/2} \rightarrow ^2F_{5/2}$	$^2F_{5/2} \rightarrow ^2F_{7/2}$	S [6.32]
$Bi^{3+}(6s^2)$	$^1S_0 \rightarrow ^3P_1$	$^3P_0 \rightarrow ^1S_0$	G [6.33]

[a] S – Silicate, P – Phosphate, B – Borate, G – Germanate, T – Tellurite, FB – Fluoroberyllate, FP – Fluorophosphate

ple Stark structure of the ^7F_J and ^5D_J states for $J = 0, 1, 2$. Excitation of Eu^{3+} fluorescence is possible via a singlet state $^5D_0$, a magnetic-dipole transition $^7F_0 \rightarrow$ $^5D_1$ or $^7F_1 \rightarrow ^5D_0$, or electric-dipole transitions from $^7F_0$ to $^5D_2$ and other levels. Fluorescence occurs by both electric- and magnetic-dipole transitions. The $^2F_{5/2}$ and $^2F_{7/2}$ states of Yb^{3+} also provide a relatively simple energy-level scheme with moderately strong electric-dipole transitions, but in an experimentally less accessible wavelength range for tunable laser excitation. Neodymium, because of its use in glass lasers and the large base of existing spectroscopic data, has been studied in many hosts. The $^2P_{1/2}$ state provides a single level for nonresonant excitation and the $^4F_{3/2}$ state a doublet for resonant excitation studies. Trivalent praseodymium also has a singlet state for excitation ($^3P_0$) and has been well studied in crystals (see Chap. 5). Another potentially useful lanthanide ion is Gd^{3+} because of the very small splitting of the ground state ($^8S_{7/2}$), but a tunable near-ultra-violet laser is needed for excitation of the $^6P_{7/2}$ and higher excited states.

Several divalent lanthanide ions (Sm^{2+}, Eu^{2+}, Tm^{2+}) and trivalent Ce exhibit intense broadband, Stokes-shifted $5d \rightarrow 4f$ emission in various hosts. Thus far there are no reports of laser-excited fluorescence studies of these ions.

Actinide ions have a $5d^n$ ground configuration and are possible candidates for FLN studies. The $5f$ electrons of the trivalent actinides interact more strongly with the local static and dynamic fields than do the $4f$ electrons of the lanthanides, therefore, there are probably comparatively few fluorescing levels in glasses with good quantum efficiency [6.34]. In addition, most of the fluorescence transitions are between J states of high multiplicity which makes the interpretation and analysis of laser-excited studies more difficult. Potentially useful ions include Am^{3+} (transitions to $^7F_{0,1,2}$ states) and Cm^{3+} ($^6P_{7/2}-^8S_{7/2}$).

Trivalent chromium, which has been used extensively for spectroscopic studies in crystals because of its simple energy-level structure and narrow $^2E-^4A_2$ transition, is much less useful in glasses. The lowest excited state is usually the bottom of the $^4T_2$ band and broadband, Stokes-shifted $^4T_2 \rightarrow ^4A_2$ emission rather than $^2E \rightarrow ^4A_2$ emission is observed. This feature arises from the smaller value of the crystal-field parameter Dq in glasses and the resulting energy-level structure shown in Fig. 6.2. Also indicated in Fig. 6.2 is the value of Dq/B for the $4d^3$ ion Mo^{3+}. The crystal-field interaction is stronger (in phosphate glass $Dq \approx$ 2000 cm^{-1} for Mo^{3+} vs ≈ 1600 cm^{-1} for Cr^{3+}) and $^2E \rightarrow ^4A_2$ fluorescence is observed. But, as discussed later, the electron-phonon coupling is also stronger. Hence, the emission is broader and predominantly of vibronic nature and exhibits only small line narrowing effects for temperatures $\geqslant 77$ K.

Line emission from other transition-metal ions in glass is rare. Pairs of levels must have energy separations independent of the crystal field to zeroth order. Examples of ground electronic configurations, ions, and d–d transitions having this characteristic include d^2 (V^{3+}): $^3T_1 - ^1T_2$; d^3 (Cr^{3+}, Mo^{3+}): $^4A_2 - ^2E$ and $^4A_2 - ^2T_2$; d^5 (Mn^{2+}, Fe^{3+}): $^6S - ^4G$; and d^8 (Ni^{2+}): $^3A_2 - ^1E$. Examples of broadband d-d fluorescence transitions include d^5 (Mn^{2+}, Fe^{3+}): $^4T_1 \rightarrow ^6A_1$.

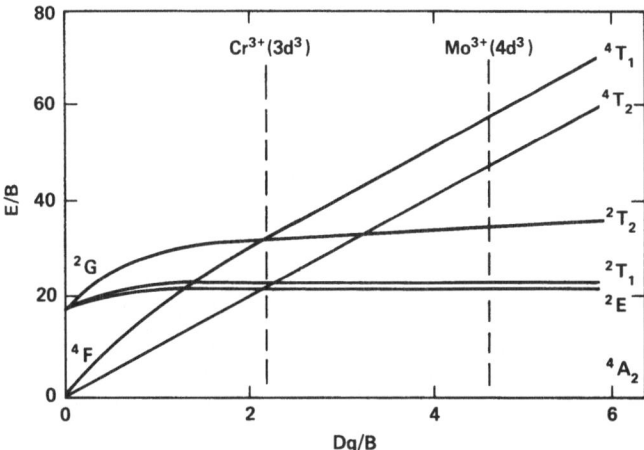

Fig. 6.2. Energy levels of the d^3 electronic configuration in an octahedral field. Typical values of Dq/B for trivalent chromium and molybdenium in glass are indicated

Table 6.2. Molecules and solvents used for laser-excited fluorescence studies in organic glasses. In all cases, the glasses were at liquid helium temperatures

Molecule	Solvent	Ref.
Perylene	Ethanol	[6.4, 7, 35]
Perylene	Polyvinylbutyral	[6.36]
9-aminoacridine	Ethanol	[6.7]
Naphthacene	2-methyltetrahydrofuran	[6.6]
Phenanthrene	Methylcyclohexane	[6.6]
Naphthalene	Ethanol-methanol (4:1)	[6.37]
Anthracene	Ethylether-isopropanol (5:2)	[6.37]
Tetracene	Ethylether-isopropanol (5:2)	[6.37]
Tetracene	Ethanol-methanol (4:1)	[6.38]
Quinizarin	Ethanol-methanol (3:1)	[6.39]

Ions characterized by broad interconfiguration $d^{10} \rightarrow d^9p$ transitions include $3d^{10}$ (Cu^+, Zn^{2+}, Ga^{3+}) and $4d^{10}$ (Ag^+), but none have been studied in glass using laser excitation techniques.

Intense, broadband fluorescence is observed from several post-transition-group ions in glass: $5s^2$ (Sn^{2+}, Sb^{3+}) and $6s^2$ (Tl^+, Pb^{2+}, Bi^{3+}). Of these, laser-excited studies have been reported for Bi^{3+}. Although the homogeneous linewidths are large, the site-to-site variations of the energy levels and line strengths are also large and can be investigated using tunable excitation.

Many aromatic hydrocarbons and dye molecules can be added to organic solvents and frozen to a glassy state at liquid helium temperatures. Examples of molecules and hosts and references to laser-excited spectroscopy in organic glasses are given in Table 6.2. The emission spectra generally consist of zero-phonon lines and vibronic sidebands. For good site selectivity, the zero-phonon lines should be narrow and intense. Trapped electrons generated in organic glasses by photoionization have been used for wavelength-selective laser bleaching and hole burning in absorption spectra [6.8].

6.3 Experimental Techniques

Site selection spectroscopy is possible by studying the frequency, temporal, or polarization behavior of the excitation-fluorescence process. The frequency domain is the one most commonly exploited and discussed, but the other two domains have also been employed for laser-excited studies in glass.

6.3.1 Fluorescence Line Narrowing

The absorption and emission of activator centers in glass consist of a zero-phonon line and phonon sidebands. The relative intensities of the two contributions depend upon the strength of the electron-phonon coupling. The fraction of the total intensity that is in the zero-phonon line is given by the Debye-Waller factor $\alpha(T)$ and is usually a decreasing function of temperature, T. The value of α depends on the nature of the center and ranges from near unity for many trivalent rare-earth ions to near zero for some transition-metal ions and molecular centers.

The absorption and emission spectra are described by

$$S_{abs} = \alpha Z_{abs}(\nu - \nu_i) + (1 - \alpha) V_{abs}(\nu - \nu_i) \tag{6.1}$$

and

$$S_{em} = \alpha Z_{em}(\nu_i - \nu) + (1 - \alpha) V_{em}(\nu_i - \nu), \tag{6.2}$$

where Z and V correspond to the zero-phonon and vibronic parts of the spectrum and ν_i is the center frequency of the zero-phonon line of the ith site. In many treatments the quantities Z and V are normalized to unity [6.24, 35]. As shown later, the radiative transition probabilities are also site dependent and this variation must be included in describing the line-narrowed intensity.

If $D(\nu_i)$ is the distribution of center frequencies for the different sites in the glass, then   if   the sample is irradiated by monochromatic light of frequency ν_l, a distribution of excited states $D(\nu_i)S_{abs}(\nu_l-\nu_i)$ is formed. The resulting fluorescence spectrum is given by

$$F(\nu) = \int S_{em}(\nu_i-\nu)D(\nu_i)S_{abs}(\nu_l-\nu_i)d\nu_i. \tag{6.3}$$

Substituting (6.1, 2) into (6.3) yields four terms proportional to $Z_{em}Z_{abs}$, $V_{em}Z_{abs}$, $Z_{em}V_{abs}$, and $V_{em}V_{abs}$. The first and third terms correspond to excitation into the zero-phonon line or the phonon sideband followed by emission in the zero-phonon line; the other two terms involve corresponding excitations followed by emission in the phonon sideband.

The homogeneous width of the zero-phonon line at low temperatures is generally narrow, < 1 cm^{-1}. The phonon sideband, on the other hand, is broad with any structure having homogeneous widths ranging from tens to hundreds of cm^{-1}. Therefore, significant laser-induced line narrowing occurs only for absorption into the zero-phonon line and emission into the zero-phonon line. The other three terms from (6.3) involve absorption or emission in the phonon sidebands and produce broad spectral features even with laser excitation. However, since the width of the inhomogeneous broadening in glasses is > 100 cm^{-1}, there may still be some site selection and line narrowing, but this is usually only of limited usefulness.

Using laser excitation, the homogeneous linewidth $\Delta\nu_H$ of a selected subset of ions can be measured in the presence of inhomogeneous broadening. If spectral diffusion is negligible or the observation is made in times short compared to ion-ion cross relaxation, the width of the resonant FLN signal is [6.24, 35]

$$\Delta\nu = 2\Delta\nu_H + \Delta\nu_{instr}, \tag{6.4}$$

where the instrumentation width $\Delta\nu_{instr}$ is governed by the spectral width of the excitation source and the detector resolution. When the latter are non-negligible, techniques of deconvoluting different line-shape functions given in [6.40, 41] can be used to determine $\Delta\nu_H$.

The homogeneous linewidth is related to the T_2 relaxation time by

$$T_2^{-1} = \pi\Delta\nu_H = (2T_1)^{-1} + \Gamma, \tag{6.5}$$

where T_1 is the longitudinal relaxation time of the excited-state population and Γ is the dephasing rate [6.42].

The FLN signal will be broadened by time-dependent spectral diffusion due to radiative and nonradiative transfer among ions in different sites. Radiation trapping by self-absorption of the resonance fluorescence broadens the line by $2\Delta\nu$ in a time $t = \tau'\tau/(\tau' - \tau)$, where τ' and τ are the fluorescence decay times with and without self-absorption. If W_T is the effective rate of nonradiative energy transfer, the width of the line-narrowed fluorescence will increase by $2\Delta\nu$ in a time $t \approx W_T^{-1}$ [6.24]. To avoid these distortions of the true homogeneous linewidth, low activator concentrations and small samples are used.

Line narrowing and site selection in a different form occur if one begins with an inhomogeneously broadened inverted population or gain profile instead of an absorption profile. A laser beam traversing such a medium de-excites those ions resonant with the laser frequency by stimulated emission. If the spontaneous emission spectrum from the remaining excited ions is subsequently examined, there will be a hole (or holes) in the spectrum where stimulated emission has occurred. The width of the hole is determined by the homogeneous linewidths of the subset of ions de-excited. Spectral hole burning is a well-known occurrence in inhomogeneously broadened lasing media [6.43]. The phenomena of spectral hole burning and resonant fluorescence line narrowing are complementary; the FLN signal is the antihole, but with twice the linewidth, because both absorption and emission processes are active in laser-excited FLN. Difference spectra obtained by subtracting the fluorescence before and after the passage of a giant laser pulse through an optically pumped neodymium glass amplifier [6.44] exhibit the same features observed in laser-induced FLN spectra [6.18].

6.3.2 Temporal Dependence

Ions in different sites in glass have different radiative and nonradiative relaxation rates. The existence of a distribution of transition probabilities is reflected in the time dependence of the excited state relaxation. Following pulsed excitation, the fluorescence at frequency ν is given by

$$S(\nu,t) = \sum_i n_i(t) A_i g_i(\nu), \tag{6.6}$$

where A_i is the spontaneous emission probability for the transition at the ith site, $g_i(\nu)$ is the line-shape function $[\int g(\nu)d\nu = 1]$, and $n_i(t)$ is the number of excited ions at time t. The population of each subset of sites i after excitation at $t = 0$ decays as

$$n_i(t) = n_i(0) \exp(-t/\tau_i). \tag{6.7}$$

The fluorescence lifetime τ_i of an individual site is simply

$$\tau_i^{-1} = W_i^R + W_i^{NR}, \tag{6.8}$$

where W_i^R and W_i^{NR} are the total radiative and nonradiative probabilities for site i summed over all terminal states.

Because excited ions in glass decay at different rates, the fluorescence following pulsed broadband excitation will have, from (6.6, 7), a non-single-exponential time dependence. The initial decay is composed of a superposition of fluorescence from all excited ions weighted by their absorption and emission probabilities. At later times the contributions of the longer-lived ions become increasingly prominent. If the ions are located in sites having different emission frequencies, this appears as a spectral shift. Therefore time-resolved spectra also reveal a distribution of transition probabilities and site selectivity.

The more monochromatic the excitation light, the smaller the number of spectrally different sites excited, and the less the derivation of $S(v,t)$ from a simple exponential decay. However, even for resonant excitation-fluorescence schemes, the fluorescence decay may still be nonexponential depending upon the ratio of the excitation bandwidth to the homogeneous linewidth and the possibility that some ions in different sites may still have the same excitation frequency.

For a laser amplifier there is an additional term in the exponent of (6.7) proportional to $\sigma_i(v) I (v,t)t/hv$, where $\sigma_i(v)$ is the stimulated emission cross section of the ith site at frequency v and $I(v,t)$ is the instantaneous beam intensity. The rate of energy extraction therefore varies from site to site. This causes hole burning and nonuniform energy extraction.

When the concentration of activators becomes sufficiently large, nonexponential fluorescence decays and time-dependent spectral shifts occur due to ion-ion cross relaxation [6.5]. These additional temporal effects complicate the determination of site-dependent intrinsic decay probabilities and are usually avoided by using low concentrations of activators or observing a gated fluorescence signal immediately after the excitation pulse. The investigation of the concentration-dependent temporal behavior is of considerable interest in understanding spectral and spatial energy migration in inhomogeneously broadened systems and "hole filling" in amplifying media. The use of transient decay measurements and time-resolved spectra for these studies is discussed in Sect. 6.5.4.

6.3.3 Polarization

The usefulness of polarized emission in optical spectroscopy is well recognized [6.45]. In a glass, each activator site has a set of principal axes, however, the orientation of a given geometry site is oriented randomly in space. Therefore, polarized excitation and fluorescence provides a further method for site selectivity.

Polarized fluorescence spectra have been used to investigate the Stark structure [6.46] and the dipolar nature of radiative transitions [6.47] for rare earths in glass. If the rare-earth ion has an even number of $4f$ electrons, the degeneracy of the free-ion levels is completely lifted and dipole transitions are ascribed to a linear oscillator; if the ion has an odd number of $4f$ electrons, the twofold Kramers degeneracy remains and transitions are modeled by partially anisotropic oscillators. If I is the intensity of emission polarized parallel ($\parallel$) or perpendicular ($\perp$) to the exciting light, the degree of polarization $P = (I_\parallel - I_\perp)/(I_\parallel + I_\perp)$ has a maximum value of 0.5 for a linear oscillator. This is observed for resonant transitions of Eu^{3+} ($4f^6$) [6.46]. The polarization P is small for nonresonant transitions and is affected by simultaneous excitation of centers having different orientations but the same excitation frequency. Polarization effects are also changed by ion-ion cross relaxation.

Most lasers are polarized and thus cause site selectivity in the excitation process. A depolarizer should be used between the fluorescing sample and monochromator if correct relative intensities are of interest.

6.3.4 Excitation-Fluorescence Schemes

In fluorescence line narrowing, if the same two levels are involved in both the excitation and fluorescence transition, the process is a resonant one. Figure 6.3 shows several resonant schemes. Levels 1 and 3 represent, for example, the lowest Stark levels of two rare-earth J manifolds; levels 2 and 4 then represent higher-lying Stark levels of the same or other J states. At low temperatures, the homogeneous broadening of levels 1 and 3 can be very small. If the linewidth of the FLN signal in resonant scheme I is governed by the spectral width of the source, the emission is "laser-narrow." At higher temperatures the FLN linewidth measures the natural lifetime (homogeneous) broadening of levels 1 and 3. The energies and homogeneous linewidths of levels 2 and 4 are measured using schemes II and III. Note that the latter involves excitation from a thermally populated level.

Examples of schemes where the excitation and fluorescence frequencies are not equal are shown at the right in Fig. 6.3. These nonresonant schemes complicate the studies of linewidths because the homogeneous broadening of three or

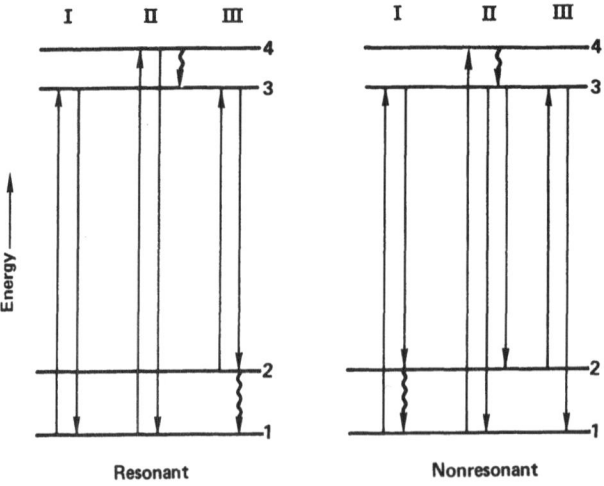

Fig. 6.3. Several possible resonant and nonresonant excitation-fluorescence schemes used in line-narrowing experiments in glass. Wavy lines indicate nonradiative transitions

more levels is involved and the relative sensitivity of the levels to inhomogeneous broadening must be considered.

In nonresonant scheme III, excitation from a thermally populated level generates upconverted anti-Stokes fluorescence [6.31]. This scheme is useful for studying transitions to the ground state in the absence of resonant excitation radiation [6.23, 31]. Other excitation-fluorescence schemes for studying energy levels of higher-lying excited states [6.48] are discussed in Sect. 6.5.1a.

6.3.5 Accidential Coincidence

Site selection spectroscopy and fluorescence line narrowing can be partially frustrated by accidential coincidence of excitation levels of ions in different sites [6.24, 49]. Figure 6.4 illustrates this phenomenon. Ions at sites A, B, C, D, and E are all excited by the same frequency but fluoresce to various terminal states at different frequencies. As a consequence, the nonresonant emission spectrum may be broad and characteristic of ions in several dissimilar environments.

Cases A and B in Fig. 6.4 illustrate the most common cause of residual inhomogeneous broadening due to spectral accidential coincidence. Excitation and fluorescence transitions may also originate or terminate on vibronic levels as shown for sites C and D. These phonon-assisted transitions generally have much larger homogeneous linewidths than zero-phonon transitions. Laser excitation into vibronic bands is more subject to accidential coincidences and less useful for site selectivity. Case E occurs only when there is a significant thermal population in an electronic or vibronic excited state.

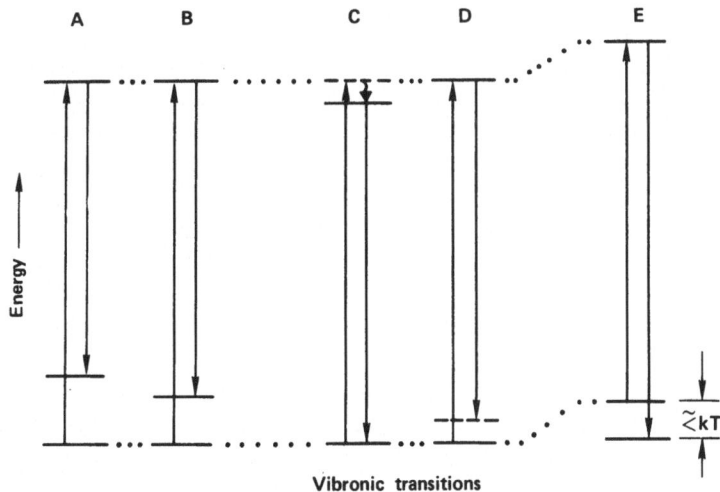

Fig. 6.4. Examples of accidental coincidences of excitation. All sites are excited by the same frequency, but fluoresce at different frequencies. Dashed levels indicate vibronic states

Accidental coincidence of two different sites may also occur in the resonant case [6.49]. This affects determinations of homogeneous linewidths because although the excitation levels may be the same, the locations of other energy levels of the two sites may be different and thereby influence the rates of phonon processes and associated lifetime broadening.

Double-frequency excitation techniques [6.50] provide further site selectivity, but residual inhomogeneous broadening may still exist because of accidental coincidences of levels from different sites. The usefulness of these techniques for glass studies has not been demonstrated.

6.4. Experimental Apparatus

The experimental apparatus for fluorescence line narrowing in glasses can be relatively simple. Because of the large inhomogeneous broadening, only modest spectral resolution is needed to achieve site selectivity, even for rare-earth ions. Whereas a broadband source and monochromator have been used as a tunable excitation source [6.51, 52], the present-day availability of tunable lasers provides the determined and resourceful experimenters with a plethora of pulsed and cw tunable sources of greatly superior resolution and usefulness [6.53]. Dye lasers with rhodamine, coumarin, and other organic dyes operate in the spectral range from the near-ultraviolet through the near-infrared. With the addition of harmonic generators and optical parameter oscillators to either dye lasers or various solid-state lasers, such as neodymium, both wavelength limits can be extended. Color-center and phonon-terminated lasers are other examples of tunable lasers [6.53].

Commercial dye lasers are available pumped by cw ion lasers (Ar, Kr) or by the following pulsed sources (pulse durations are given in parentheses) operating at repetition rates of at least 10 Hz: flashlamps ($\approx 1\ \mu s$), Nd:YAG lasers (10-100 ns), and nitrogen lasers (≈ 5 ns). The laser bandwidths are typically $1-2\ \mathrm{cm}^{-1}$ and can be narrowed by an order of magnitude or more by the addition of an etalon. Laser linewidths of 50 MHz have been obtained using an external confocal cavity and optical amplifier [6.54]. For studies in organic glasses, fixed-frequency lasers in the blue and near-ultraviolet, such as Ar (476.5, 514.5 nm) and He-Cd (325.0, 441.6 nm), have been used. With any laser source, broadband radiation from the dye or the pump lamp must be prevented from inadvertently exciting fluorescing ions in the glass.

For investigations of temperature-dependent phenomena, glass samples are mounted in variable-temperature dewars of conventional design. Because of the small thermal conductivity of most glasses, if the sample is cooled by conduction rather than immersion, it is important to monitor accurately the sample temperature and to minimize local heating by the laser.

Fluorescence spectra are scanned using a grating monochromator or, for high resolution studies, a pressure-scanned Fabry-Perot interferometer [6.54]. Photomultiplier tubes are the standard detectors. When using high resolution, low activator concentration, or excited-state absorption transitions, the signals are weak and photon counting electronics may be required. Gated integrators are used for recording time-resolved spectra.

Alternatively, FLN spectra may be recorded using a monochromator equipped with a vidicon detector. This enables one to accumulate data simultaneously over an extended wavelength interval, to scan the spectrum electronically, and to monitor continuously the improvement in signal-to-noise ratio. Gated vidicons provide an excellent means of recording time-resolved spectra [6.26].

6.5 Experimental Results

6.5.1 Energy Levels

The energy level splittings of the electronic states of ions and molecules in solids are discussed in Chap. 1. The Hamiltonian for a paramagnetic ion in a solid is

$$H = H_{el} + H_{so} + V, \tag{6.9}$$

where H_{el} is the electrostatic interaction of the electrons, H_{so} is the spin-orbit term, and V is the potential at the ion site due to its environment. The interelectron repulsion term H_{el} is described by Slater integrals or the Racah parameters E_0, E_1, E_2, E_3; the spin-orbit interaction is described by the spin-orbit constant ζ. When an ion is introduced into a solid, these parameters are reduced from their free-ion values. This is the nephelauxetic effect [6.55]. It shifts the free-ion states and causes differences in the frequency of optical transitions between SL multiplets. In a glass, the Racah and spin-orbit parameters vary from site to site because of differences in the coordination number, the ligand distances, and the degree of covalent bonding. The result is inhomogeneous broadening of optical transitions between J states and can range from a few to tens of cm^{-1}.

a) Rare-Earth Ions

For rare earths, which have thus far been the principal paramagnetic ions used for laser-excited fluorescence studies in glass, $H_{el} \gg H_{so} \gg V$. Therefore, V can be treated as a small perturbation on the free-ion energies. The local field at the ion site in a glass is described by expanding the potential V in (6.9) in terms of tensor operators $C_{-q}^{(k)}$ that transform as spherical harmonics [6.56]. Thus,

$$V = \sum_{k,q,i} B_q^k (C_{-q}^{(k)})_i, \tag{6.10}$$

where the B_q^k are parameters to be determined from experimental data and the summation involving i is over all electrons of the ion of interest. The number and type of terms in (6.10) are derived from group theory given the site symmetry.

Since the site symmetry in the glass is C_1, all terms in (6.10) are allowed. For f electrons, k is limited to values $\leqslant 6$.

The even-k terms in (6.10) remove the degeneracy of the free-ion J states of rare earths and cause Stark splitting of $\approx 10^2$ cm^{-1}. They also introduce admixing of J states; this is a small effect on energy levels but can be a large effect in relaxing selection rules for optical transitions. The principal differences in the Stark splittings of ions at various sites arise from the local field parameters B_q^k. Because the inhomogeneous broadening is comparable to the Stark splitting, the Stark structure of rare earths in glass is usually poorly resolved.

Europium. *Motegi* and *Shionoya* [6.5], in their investigation of Eu^{3+} in a calcium metaphosphate glass, $Ca(PO_3)_2$, used tunable laser excitation to pump the $^7F_0 \rightarrow {}^5D_0$ transition and observed that the $^5D_0 \rightarrow {}^5F_1$ emission lines were narrowed and the positions of the peaks varied with excitation wavelength. This was a striking demonstration of site-dependent Stark splitting.

Brecher and *Riseberg* [6.22] subsequently made a more extensive study of the $^5D_0 \rightarrow {}^7F_1$ and $^5D_0 \rightarrow {}^7F_2$ emission bands of Eu^{3+} in a silicate glass following pulsed tunable excitation into the $^5D_0$ level. The spectra, measured at low tem-

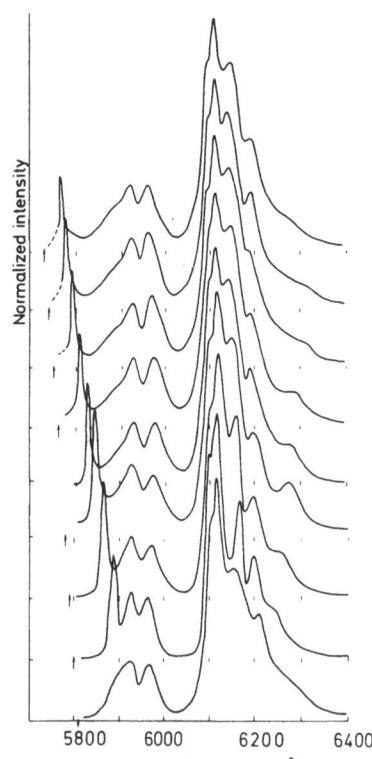

Fig. 6.5. Fluorescence spectra of Eu^{3+} in a Na-Ba-Zn silicate glass at 20 K as a function of $^7F_0 \rightarrow {}^5D_0$ excitation wavelength (denoted by arrows). Intensities are normalized to the most intense peak at each excitation wavelength [6.22]

peratures (20 K) and in times (100 μs) short compared to cross-relaxation rates, are shown in Fig. 6.5. Transitions from $^5D_0$ to the three Stark levels of $^7F_1$ appear in the region 575–600 nm and show a dramatic shift in the frequency of the highest-energy component. Fluorescence peaks corresponding to transitions from $^5D_0$ to the five levels of $^7F_2$ appear in the 605–630 nm region but show only small variations in position and intensity with excitation wavelength. Variations in the $^5D_0 \rightarrow {}^7F_4$ band are even less pronounced and have not been useful for crystal-field analysis of Stark structure [6.57]. Note that the spectra in Fig. 6.5 are recorded under nonresonant conditions and therefore exhibit considerable residual inhomogeneous broadening due to accidental coincidences.

Following the method of *Motegi* and *Shionoya* [6.5], the energies of the fluorescence peaks in Fig. 6.5 measured with respect to the $^5D_0$ level at constant energy are plotted as a function of excitation energy at the left of Fig. 6.6. As the laser frequency is tuned through the inhomogeneously broadened $^7F_0 \rightarrow {}^5D_0$ band, the splittings of the three Stark components of the $^7F_1$ manifold vary systematically from approximately 150 cm^{-1} to 600 cm^{-1}. Thus, in a single oxide glass, ions reside in low-symmetry sites having crystal-field strengths ranging from among the smallest to the largest values reported for any oxide crystals.

To analyze the crystalline Stark splitting, *Brecher* and *Riseberg* considered C_{2v} site symmetry, the highest symmetry in which complete splitting of the $^7F_1$

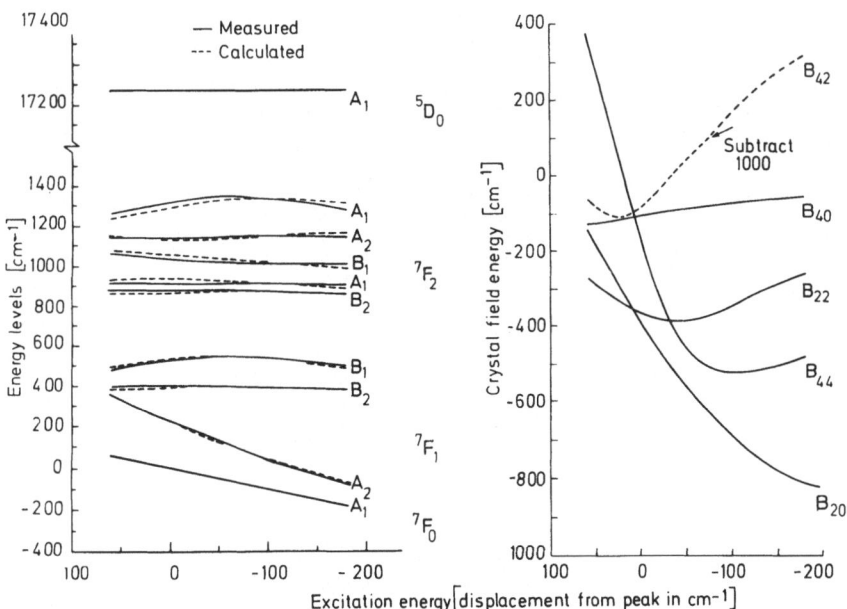

Fig. 6.6. Variations of the crystalline Stark levels (*left*) and crystal-field parameters (*right*) for Eu^{3+} in a silicate glass as a function of excitation energy [6.57]. Level assignments are based on C_{2v} site symmetry

and $^7F_2$ states is allowed and the lowest symmetry for which crystal-field calculations are routinely performed. The assignments of levels to C_{2v} symmetry states
were based upon selection rules and observed relative intensities. Using these
assignments, a best fit of measured and calculated $^7F_1$ and $^7F_2$ energy levels
yielded a set of B_{kq} parameters for each excitation wavelength. These are plotted
at the right of Fig. 6.6. The large variation of the axial symmetry parameter B_{20}
accounts for the large change of the $^7F_1(A_2)$ energy level. The agreement between
measured and calculated energy levels, indicated by the solid and dashed lines in
Fig. 6.6, is good even though J-state mixing was not included. The variation of the
$^7F_0$ energy, which was not treated, is a consequence of J-state mixing with $^7F_2$.

The FLN spectra of Eu^{3+} have now been investigated in several different
glasses. For silicate and phosphate glass of comparable compositions [mol.%: 60
(SiO_2 or P_2O_5), 27.5 Li_2O, 10 CaO, 2.5 Al_2O_3], the spectra indicated remarkably
little variation of the crystal fields [6.25]. The $^7F_1$ energy levels of Eu^{3+} in an

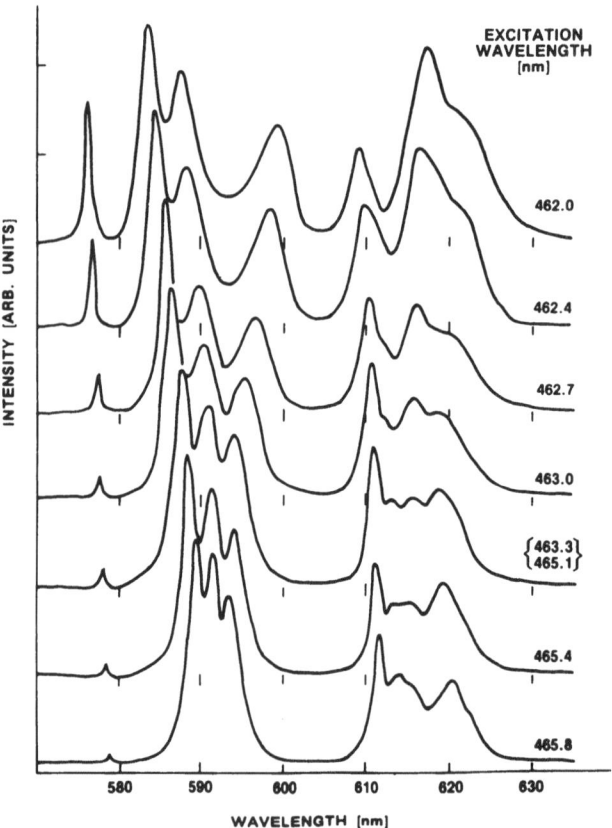

Fig. 6.7. Fluorescence spectra of Eu^{3+} in a K-Ca-Al fluoroberyllate glass at 80 K as a function
of $^7F_0 \rightarrow {}^5D_2$ excitation wavelength. Intensities are normalized to the most intense $^5D_0 \rightarrow {}^7F_1$
peak [6.28]

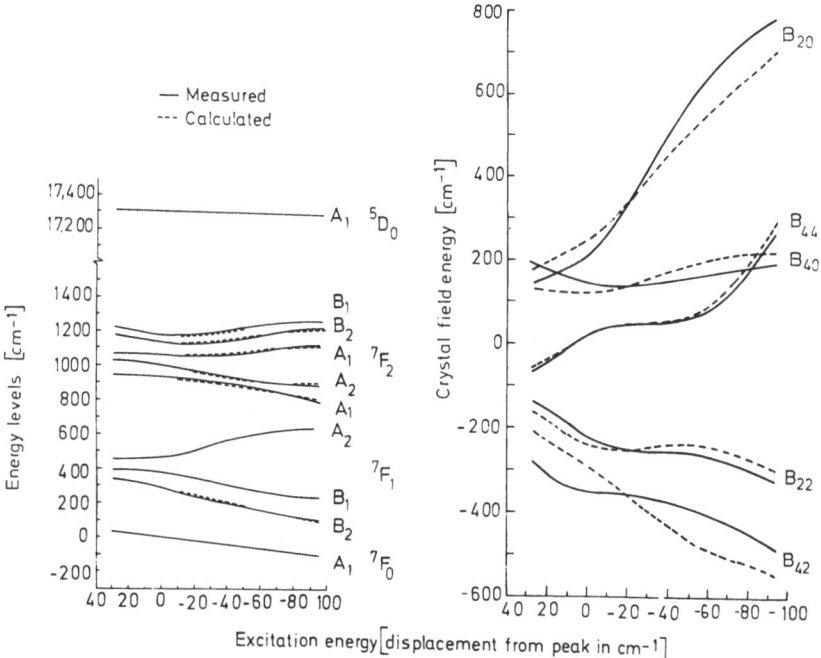

Fig. 6.8. Variations of the crystalline Stark levels (*left*) and crystal-fold parameters (*right*) of Eu^{3+} in a fluoroberyllate glass as a function of excitation energy [6.57]. Level assignments are based on C_{2v} site symmetry

alkali borate glass (40 Li_2O, 60 B_2O_3) exhibited a splitting pattern similar to that of the silicate glass in Fig. 6.6 but with a slightly larger $B_1 - B_2$ separation [6.13]. *Alimov* et al. [6.23] measured the $^7F_1$ energy levels in a borosilicate glass and observed larger splittings than in silicate glass. In all of these oxide glasses, the local field at the Eu^{3+} is determined predominantly by the oxygen ligands; compositional differences in the network former cations or modifier cations, which share the second coordination sphere cause smaller variations in spectroscopic properties.

Large differences in the FLN spectra appear when the network forming anion is changed. This can be seen from the spectrum for a pure fluoride glass in Fig. 6.7. Here, because the $^7F_0 \rightarrow {}^5D_0$ transition is very weak, excitation was via the low- and high-energy sides of the $^7F_0 \rightarrow {}^5D_2$ transition [6.28]. Ions excited into the $^5D_2$ state rapidly decay to $^5D_0$ from which fluorescence is again observed to levels of the ^7F multiplet. Although there are more possibilities of accidental coincidences when using $^5D_2$ excitation, the spectra still show well-defined peaks that vary with excitation wavelength. Comparing Figs. 6.5 and 6.7 reveals a marked difference in the splitting patterns. Using the same assignment procedures as for the silicate glass, *Brecher* and *Riseberg* [6.28] derived the level assignments and crystal-field parameters shown in Fig. 6.8. Note the differences in the signs of several of the parameters compared to those found for an oxide glass (Fig. 6.6).

Recently *Videau* et al. [6.29] reported FLN spectra for Eu^{3+} in two Na-Al-fluorophosphate glasses. The oxygen-to-fluorine ratios were approximately 1 and 2. Examination of the $^5D_0 \rightarrow {}^7F_1$ and $^5D_0 \rightarrow {}^7F_2$ fluorescence bands suggested the existence of distinctly different Eu^{3+} sites. The spectra were interpreted in terms of two types of sites: one, the more ionic, was attributed to Eu^{3+} surrounded by fluorine ions with approximate O_h symmetry; the other, more covalent site was assumed to be a mixed oxygen-fluorine environment of C_{2v} symmetry.

Lebedev et al. [6.46] have shown that variations in Stark splittings can also be detected using polarized excitation and fluorescence spectra. Their spectra for Eu^{3+} in phosphate glass are shown in Fig. 6.9. Differences in the Stark components and intensities are attributed to polarization site selectivity. (A more detailed discussion of the experiments is given in Sect. 6.5.2a.)

Laser-induced FLN techniques have been extended to measure site-dependent variations in the spectroscopic properties of excited states. *Alimov* et al. [6.23] excited the $^7F_1 \rightarrow {}^5D_1$ transition of Eu^{3+} and observed a distinct three-line $^5D_0 \rightarrow {}^7F_1(1)$ fluorescence spectrum. This arose from the accidental coincidence of excitation transitions of three groups of sites with different nonresonant fluorescence lines. The process was analyzed in greater detail by *Hegarty* et al. [6.48] using $^7F_0 \rightarrow {}^5D_1$ excitation and observing nonresonant $^5D_0 \rightarrow {}^7F_{0,1,2}$ emission. The FLN spectra for various excitation energies are shown in Fig. 6.10. In addition to the characteristic wavelength shift of the $^5D_0 \rightarrow {}^7F_0$ line and the splitting of the

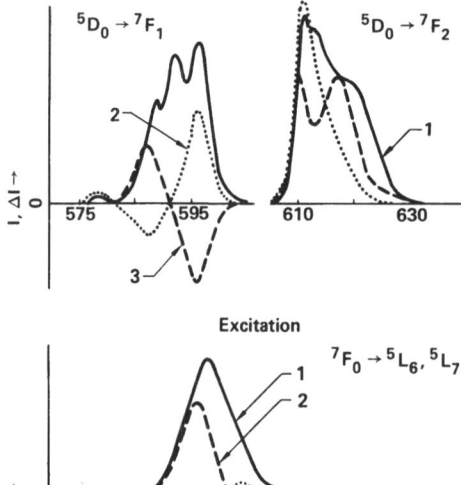

Fig. 6.9. Polarized fluorescence and excitation spectra of Eu^{3+} in phosphate glass. *Top*: Curve 1 - $(I_\parallel + I_\perp)$; Curve 2 - $(I_\parallel - I_\perp)$, 464-nm excitation; Curve 3 - $(I_\parallel - I_\perp)$, 394-nm excitation. *Bottom*: Curve 1 - $(I_\parallel + I_\perp)$; Curve 2 - $(I_\parallel - I_\perp)$, 587.5-nm fluorescence; Curve 3 - $(I_\parallel - I_\perp)$, 569-nm fluorescence [6.46]

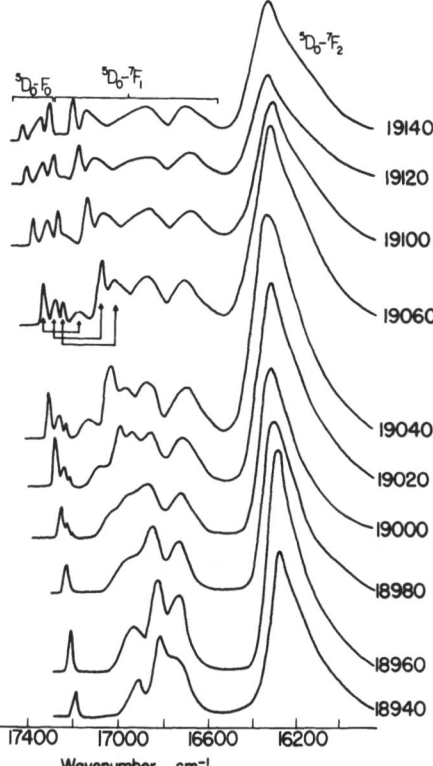

Fig. 6.10. $^5D_0 \rightarrow {}^7F_{0,1,2}$ fluorescence spectra of Eu^{3+} in 60 $B_2O_3 \cdot 40$ Li_2O glass at 1.7 K as a function of $^7F_0 \rightarrow {}^5D_1$ excitation energy (cm^{-1}). The correspondence between the $^5D_0 \rightarrow {}^7F_0$ and $^5D_0 \rightarrow {}^7F_1$ (1) lines is shown for the 19,060 cm^{-1} spectrum [6.48]

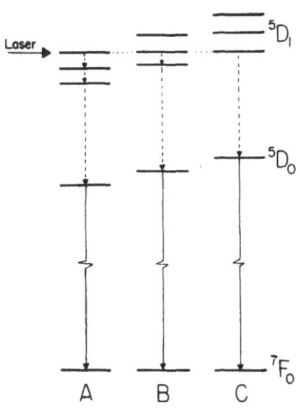

Fig. 6.11. Partial energy-level scheme of Eu^{3+} in a glass. Ions *A*, *B*, and *C* represent three subsets of ions which are in resonance with the laser via different $^5D_1$ Stark components. Solid and dashed lines denote radiative and nonradiative transitions, respectively. Only the $^7F_0$ level within the ^7F_J group of terminal levels is shown for simplicity [6.48]

$^7F_1$ manifold, the $^5D_0 \rightarrow {}^7F_0$ line splits first into two and then into three lines. The origin of these additional lines can be seen from the Eu^{3+} energy-level diagram for three different sites in Fig. 6.11. A careful comparison of the $^5D_1$ splitting showed that it was identical to the $^7F_1$ splitting pattern but reduced in magnitude by a factor of about 0.2 across the entire spectrum of sites. This factor is a reasonable expectation for the relative splitting of the $^5D_1$ and $^7F_1$ manifolds based on the ratio of the matrix elements for the two states.

Neodymium. Laser-excited fluorescence of Nd^{3+} has been studied extensively in several glasses (silicate, phosphate, borate, fluoroberyllate, and fluorophosphate) by *Brecher* et al. [6.18]. Excitation was via the $^2P_{1/2}$ state. The $^4I_{9/2} \rightarrow {}^2P_{1/2}$ absorption spectra of the different glasses at liquid helium temperature are shown in

Fig. 6.12. For the more covalently bonded oxide glasses, the spectrum is shifted to lower energy as a result of the nephelauxetic effect. Ions excited into the $^2P_{1/2}$ state decay nonradiatively to the $^4F_{3/2}$ state from which fluorescence is observed to the ^4I_J states ($J = 9/2, 11/2, 13/2, 15/2$). Examples of the $^4F_{3/2} \rightarrow {}^4I_{9/2}$ and $^4I_{11/2}$ spectra from Nd^{3+} in a phosphate glass for several $^4I_{9/2} \rightarrow {}^2P_{1/2}$ excitation wavelengths are shown in Fig. 6.13. This is again a nonresonant scheme. The variations in the five peaks in the $^4I_{9/2}$ spectra are of comparable clarity to the five peaks in the $^7F_2$ spectra of Eu^{3+}. The individual Stark components of Nd^{3+} are most clearly resolved in phosphate and fluoroberyllate glasses; in the other glasses the expected number of peaks was not observed. The $^4F_{3/2}$ fluorescence of Nd^{3+} terminates on manifolds having multiplicities of ≥ 5. In addition, since $J \geq 9/2$, all crystal-field terms through sixth order in (6.10) operate on these states. Attempts to derive crystal-field parameters for Nd^{3+}, as was done for Eu^{3+}, have thus far not been successful because of insufficient data [6.57].

The fluorophosphate glass is an example of a mixed anion glass in which Nd^{3+} ions may have combinations of oxygen and fluorine ligands. The wide range of physically different local environments yields absorption sites that straddle the wavelength regions of both pure fluoride (fluoroberyllate) and pure oxide (phosphate) glasses. This is seen from the spectra in Fig. 6.12. The fluorescence spectra exhibits a similar behavior [6.58].

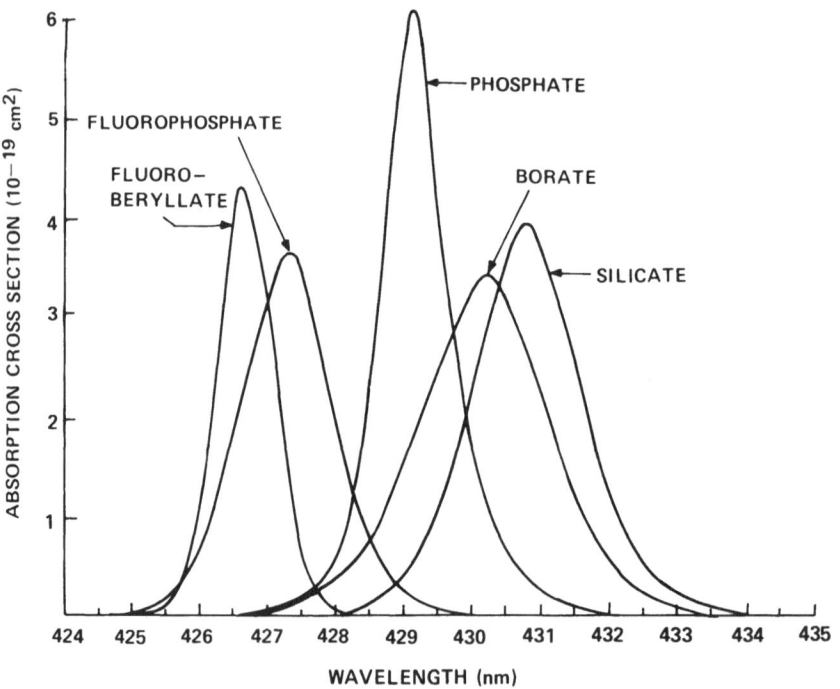

Fig. 6.12. Absorption cross sections of the inhomogeneously broadened $^4I_{9/2} \rightarrow {}^2P_{1/2}$ transition of Nd^{3+} in different glasses at liquid helium temperature [6.18]

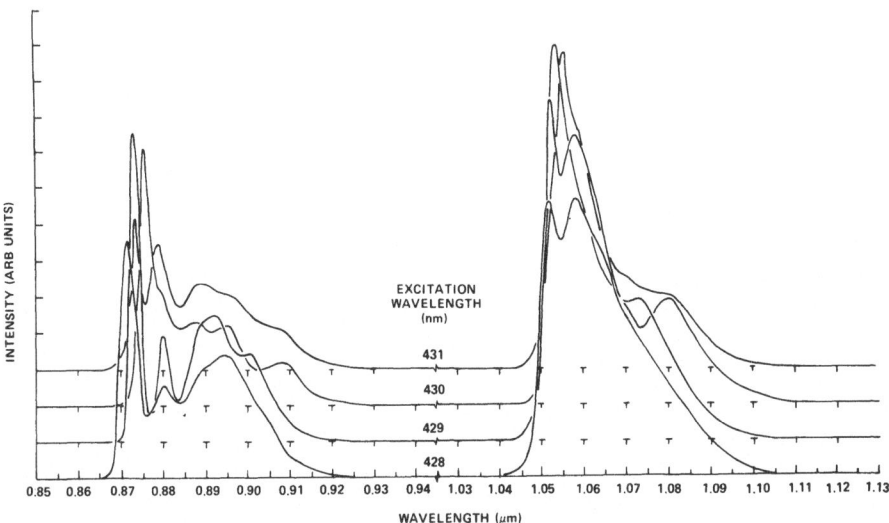

Fig. 6.13. Line-narrowed fluorescence spectra of Nd^{3+} in a phosphate glass as a function of excitation wavelength. The intensities (at liquid-helium temperature) are given in terms of photons emitted for equal numbers of excitation photons absorbed and are normalized to the highest emission peak as unity. The fluorescence was measured 100 μs after the excitation pulse; the spectra are not normalized for differences in decay times [6.18]

Alimov et al. [6.16] examined the FLN spectra of Nd^{3+} in a commercial silicate laser glass (LGS-28) using rhodamine 6G dye laser excitation of the $^4G_{5/2}$, $^2G_{7/2}$ band. The greatest line narrowing was obtained for excitation in the long- and short-wavelength wings of the absorption band; little or no narrowing was observed for excitation at the band center where various Stark levels of different sites overlap. Not all Stark levels of the $^4I_{9/2}$ and $^4I_{11/2}$ states were resolved, however a plot of three levels of $^4I_{9/2}$ and one of $^4F_{3/2}$ as a function of excitation energy is given.

Recently resonant FLN of Nd^{3+} in a silicate glass has been obtained via $^4I_{9/2} \rightarrow {}^4F_{3/2}$ transitions [6.19, 20]. The resonant line was line narrowed but the remaining nonresonant $^4F_{3/2} \rightarrow {}^4I_{9/2}$ and $^4F_{3/2} \rightarrow {}^4I_{11/2}$ transitions continued to show residual inhomogeneous broadening. The resolution of nonresonant peaks for $^4F_{3/2}$ excitation was not noticeably different than that observed for $^2P_{1/2}$ excitation [6.18].

In an experiment analogous to fluorescence line narrowing, *Nikitin* et al. [6.44, 59] examined the $^4F_{3/2} \rightarrow {}^4I_{11/2}$ fluorescence spectrum from a Nd:glass laser amplifier immediately before and after the passage of an intense laser pulse. Since stimulated emission was most probable from those ions in sites resonant with the laser wavelength, selective site de-excitation occurred and a hole was observed in the fluorescence profile. The difference spectrum, obtained by subtracting the spectra before and after the pulse, showed a narrow peak resonant with the laser wavelength and a broad, slightly structured background spectrum. This is the counterpart of the usual FLN spectrum obtained by absorption. The width of the resonant peak is equal to the averaged homogeneous linewidths of the ions de-excited; the background corresponds to the remaining nonresonant $^4F_{3/2} \rightarrow ^4I_{11/2}$ transitions.

Samarium. Selective laser excitation of Sm^{3+} in a lanthanum aluminosilicate glass was reported by *Basiev* et al. [6.21]. Both resonant and nonresonant excitation-fluorescence schemes were used. They observed substantial line narrowing, the appearance of additional structure, and a continuous variation of energy levels as the wavelength of the $^6H_{5/2} \rightarrow ^4G_{5/2}$ excitation light was varied. A diagram of the overall splitting of the Stark levels of the $^4G_{5/2}$, $^6H_{5/2}$, and $^6H_{7/2}$ states for different Sm^{3+} sites was included. Data of the homogeneous linewidth behavior are given in Sect. 6.5.3.

Ytterbium. Site-dependent Stark energy levels of the $^2F_{7/2}$ and $^2F_{5/2}$ states of Yb^{3+} have been measured in a silicate glass [6.32]. Tunable radiation in the 0.9 − 1.0 μm spectral region suitable for exciting Yb^{3+} was obtained from a pulsed Nd: YAG-laser-pumped optical parametric oscillator. The overall splittings of both $^2F_{7/2}$ and $^2F_{5/2}$ exhibited only small variations ($< 15\%$) as a function of the excitation energy of the transition between the lowest Stark levels of the two manifolds. Because of its relative simple energy level scheme, Yb^{3+} has been used for time-resolved energy transfer studies (see Sect. 6.5.4).

b) Other Ions

Chromium. Small variations in the $^4T_2 \rightarrow ^4A_2$ fluorescence spectra of Cr^{3+} in a silicate glass were noted by *Brawer* and *White* [6.12] when they used different laser wavelengths for excitation. Since a tunable laser was not used, no systematic study of these variations was attempted. However, the experiment demonstrated that even in systems with large homogeneous linewidths, by using monochromatic laser excitation subsets of sites were excited which exhibited different fluorescence behavior.

Molybdenum. As shown in Fig. 6.2, the first excited state of Mo^{3+} in oxide glasses is ^2E. The width of the $^4A_2 - ^2E$ transition in absorption and emission is several hundred cm^{-1} at 77 K and increases with increasing temperature [6.60]. This width is larger than expected from site-to-site variations in the splitting of ^2E in a noncubic field. In the strong-field limit, the $^4A_2 - ^2E$ separation is independent of the cubic field paramter Dq and is given by $9B + 3C$, where B and C are the Racah parameters for d^3. The nephelauxetic effect is more pronounced for $3d$ and $4d$ electrons of Cr^{3+} and Mo^{3+} than for $4f$ electrons of rare-earth ions. For example, whereas the free-ion value of B for Cr^{3+} is $920\ cm^{-1}$, the value for Cr^{3+} in solids is $600–700\ cm^{-1}$. Thus, part of the observed $^2E - ^4A_2$ linewidth originates in site-dependent values in the Racah parameters. In addition, the absorption and emission spectra of Mo^{3+} in phosphate glass are composed of both zero-phonon and vibronic transitions. This was confirmed by attempts to observe fluorescence line narrowing following tunable laser excitation in the $^4A_2 \rightarrow {}^2E$ and $^4A_2 \rightarrow {}^2T_1$ bands [6.14]. Only a small amount of line narrowing was observed and the Stokes-shifted emission peak changed monotonically with excitation wavelength. Both the site-to-site variation of the $^4A_2 - {}^2E$ energy separation and the extent of the vibronic spectra of the phosphate glass were found to be in the 500–1000 cm^{-1} range. Thus, the inhomogeneous broadening is masked by overlapping vibronic transitions and little line narrowing is observed. The inhomogeneous nature of the Mo^{3+} sites is better revealed by the site-dependent transition probabilities discussed in Sect. 6.5.2.

Bismuth. The absorption and emission spectra of Bi^{3+} in germanate glass are the broadest studied thus far using laser-excited fluorescence techniques. In a series of experiments, *Boulon* et al. [6.33] investigated both the fluorescence spectra and decay properties. Excitation was via the $^1S_0 \rightarrow {}^3P_1$ transition; emission was from the metastable $^3P_0$ state to vibrational levels of $^1S_0$. The absorption and emission bands are shown in Fig. 6.14. The Stokes shift is very large. At low temperatures (4 K) no zero-phonon line is observed because of the large Huang-Rhys parameter.

The Bi^{3+} fluorescence spectrum depends on the excitation wavelength and bandwidth; this is illustrated for three cases in Fig. 6.14. The peak emission wavelength increased with increasing excitation wavelength and varied by 4200 cm^{-1}, more than one order of magnitude larger than for rare-earth ion spectra. In addi-

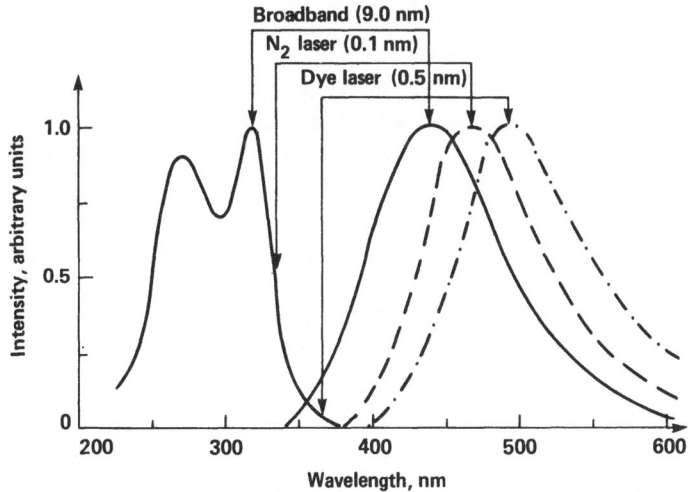

Fig. 6.14. $^3P_0 \rightarrow {}^1S_0$ fluorescence spectra of Bi^{3+} in germanate glass at 4 K for selective excitation into the $^3P_1$ absorption band. The excitation wavelengths and resulting emission peaks are indicated by arrows; the excitation bandwidth is given in parenthesis [6.33]

tion, the emission bandwidth is larger for broadband pumping (5350 cm^{-1}) than for N_2 laser (4500 cm^{-1}) or dye laser (3900 cm^{-1}) pumping. Therefore, site selectivity in glass occurs even for broadband spectra.

The large variation of Bi^{3+} energy levels in glass arises from the site-dependent nephelauxetic effect. The stronger the covalency of the $Bi^{3+}-O^{2-}$ bond, the smaller the $6s^2-6s6p$ separation and the longer the wavelength of the $^3P_0 \rightarrow {}^1S_0$ emission. The $^3P_1 - {}^3P_0$ separation is also site dependent.

6.5.2 Transition Probabilities

Transition probabilities are proportional to the square of an interaction and therefore are potentially more sensitive to site-dependent variations in the interaction strength than are energy levels. This comment applies to electric-dipole transitions between states of the same electronic configuration that become allowed by admixing of states of opposite parity by the crystal field, and to nonradiative transitions arising from electron-phonon coupling.

a) Radiative Transitions

Radiative probabilities for intraconfigurational transitions of paramagnetic ions in glass are site dependent. Admixing of states of opposite parity is required to make normally Laporte-forbidden electric-dipole transitions allowed; this occurs via the odd-k terms in the crystal field expansion in (6.10). The $B^{k\text{-odd}}$ parameters, interconfigurational radial integrals, and the energy separation of the states of opposite parity all enter into the electric-dipole transition probability. (For f-f transitions of rare earths, these quantities are combined in the Judd-Ofelt parameters [6.56].) Since these three quantities are dependent on the local environment, paramagnetic ions in glass generally exhibit a distribution of radiative relaxation rates. With broadband excitation and observation, the fluorescence from different sites is summed according to (6.6). If the distribution of transition probabilities is large, the decay will be highly nonexponential. On the other hand, with laser excitation, selected subsets of ions are excited and the fluorescence decays are more exponential.

Of the three site-dependent factors in the electric-dipole transition probability mentioned above, the largest site-to-site variation is probably due to the crystal field, although this has not been demonstrated by a careful comparison of parity-allowed and parity-forbidden transitions in the same glass. Phonon-assisted vibronic transitions which involve the dynamic crystal field are expected to be strongly site dependent. Variations in the radiative rates of interconfigurational transitions may also be due to site-dependent electron-phonon coupling.

Studies of fluorescence decay of rare-earth ions show that the degree of nonexponential character differs with glass type. The decay in glasses such as phosphates [6.52] and tellurites [6.61] more nearly approximates a simple exponential, whereas in borates, silicates, and germanates the observed decay rate may vary by $50-100\%$ or more during the relaxation [6.17, 62].

For ions such as Nd^{3+}, Eu^{3+}, and Yb^{3+} which relax predominantly by radiative processes, tunable laser-excited fluorescence studies reveal large variations in decay rates. For example, the $^2F_{5/2}$ lifetime of Yb^{3+} in silicate glass ranged from 1.1 to 2.6 ms for different excitation wavelengths [6.49]. As part of their investigation of laser-excited Eu:silicate glass, *Brecher* and *Riseberg* measured lifetime variations of a factor of 5 by exciting into the extreme high-energy wing of the $^7F_0 \rightarrow {}^5D_0$ absorption. The fastest decay rates occurred for those sites having the largest $^7F_1$ splitting. Thus there is a correlation between the strength of the even- and odd-order crystal-field parameters in (6.10). The crystal-field model used by *Brecher* and *Riseberg* to account for the energy-level variations also showed that the B_{31}, B_{32}, and B_{50} parameters varied by more than a factor of 4. Detailed correlation of radiative transition probabilities and branching ratios with a site-dependent crystal-field model has not been done, however.

As shown in Fig. 6.9, the spectra of rare earths in glass are polarized. Polarization properties for various transitions of Pr^{3+}, Eu^{3+}, Tb^{3+}, Er^{3+}, and Tm^{3+} in phosphate glass are given in [6.46]. The experiments were done using a mercury lamp and monochromator for excitation (resolution: 2 nm). The polarization of the fluorescence emitted in the direction of the exciting light was analyzed (excitation and fluorescence radiation were separated by a two-disk chopper). The degree of polarization is dependent on the Stark components involved in the absorption and emission. The results in Fig. 6.9 are for a nonresonant scheme, but both resonant and nonresonant transitions were examined. The degree of polarization P for the resonant $^7F_0 \rightarrow {}^5D_0$ transition of Eu^{3+} was 0.5; the value for nonresonant transitions was smaller, $|P| \gtrsim 0.1$.

Although most radiative transitions are crystal-field or vibrationally-induced electric-dipole transitions, occasional transitions such as the $^5D_0 \rightarrow {}^7F_1$ transitions of Eu^{3+} or intramultiplet transitions are of magnetic-dipole nature. *Kushida* et al. [6.47] have shown that the dipolar nature of optical transitions in glasses can be established from polarized excitation and emission studies.

The fluorescence decay of the $^4F_{3/2}$ state of Nd^{3+} has been the subject of study in several glasses. *Przhevuskii* et al. [6.52], using a pulsed mercury lamp and monochromator, found that the decays in a series of silicate and phosphate laser glasses depended on the glass composition, the excitation wavelength, and the spectral width of the excitation and fluorescence. This agrees with the behavior expected in a glass. (In one glass the lifetime was very long, 2.8 ms, which suggests a highly symmetric site and forbidden transitions.)

Detailed laser-excited decay measurements of Nd^{3+} were conducted in silicate, phosphate, borate, fluoroberyllate, and fluorophosphate glasses by *Brecher* et al. [6.18]. The samples were cooled to 77 K and excited to the $^2P_{1/2}$ state using a tunable pulsed dye laser. To determine W^R, the fluorescence intensity was measured under conditions of constant absorbed excitation into $^2P_{1/2}$. The radiative quantum efficiency, $\eta_i = W^R \tau_i$, was then normalized to unity at one wavelength and the resulting variations of W^R and W^{NR} plotted as shown in Fig. 6.15 for a silicate and a phosphate glass. Whereas in the silicate glass the transition pro-

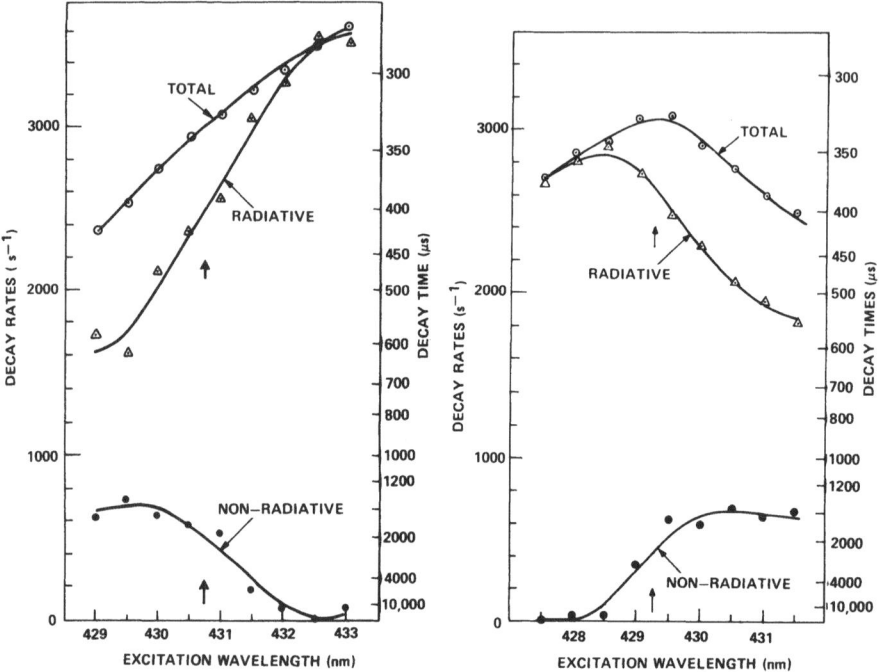

Fig. 6.15. Excitation wavelength dependence of the radiative and nonradiative decay rates of the $^4F_{3/2}$ state for Nd^{3+} in a silicate glass (*left*) and a phosphate glass (*right*). The wavelength of the excitation absorption peak in Fig. 6.12 is indicated by an arrow [6.18]

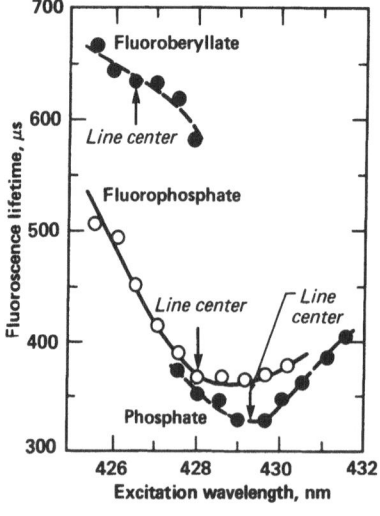

Fig. 6.16. Variations of the $^4F_{3/2}$ fluorescence decay time of Nd^{3+} in phosphate, fluoride, and fluorophosphate glasses at 4 K following pulsed laser excitation into the $^2P_{1/2}$ band [6.58]

babilities vary monotonically, in the phosphate glass the shortest lifetimes occur near the high-population-density center of the excitation band. This accounts in part for the more nearly single exponential decay found in phosphate glasses.

In the Nd fluorophosphate glass, which contained large numbers of both fluorine and oxygen anions, comparison of the laser-excited fluorescence spectra of Nd^{3+} with the corresponding results from pure oxide and pure fluoride glasses demonstrated the presence of Nd^{3+} sites having both fluorine and oxygen nearest-neighbor coordination. Thus, in fluorophosphate glasses, some sites are expected to have lifetimes and quantum efficiencies similar to those in phosphate glasses, whereas other sites have lifetimes and quantum efficiencies similar to those in pure fluoride glasses. This was studied   by pumping throughout the $^4I_{9/2} \rightarrow {}^2P_{1/2}$ band shown in Fig. 6.12; the results are presented in Fig. 6.16. The lifetimes for the short- and long-wavelength extremes of the fluorophosphate glass absorption do not equal those for the pure phosphate and fluoride glasses for two reasons: 1) the network modifier ions in the various glasses were not identical in number density or type, and 2) the refractive indices n of the glasses, which enter into the spontaneous emission probability as $n(n^2 + 2)^2$, were different. Using the index correction for the local fields, the lifetime of 500 μs in a fluorophosphate glass increased to 620 μs in a fluoroberyllate glass; similarly, the lifetime of 360 μs in a fluorophosphate glass decreased to 320 μs in a phosphate glass. The latter values in both cases are in good agreement with experimental values [6.58].

The fluorescence of Bi^{3+} involves an interconfigurational transition and is parity allowed. The probability of $^3P_0 \rightarrow {}^1S_0$ emission, however, is dependent on the admixing with other ^3P and ^1P states and on the vibronic nature of the transition, both of which are site dependent. The fluorescence decay under broadband excitation is indeed nonexponential. Using laser excitation, *Boulon* et al. [6.33] observed that the decay became more exponential with decreasing excitation bandwidth and changed with emission wavelength. With increasing wavelength of the $^3P_0 \rightarrow {}^1S_0$ emission, the fluorescence lifetime increased from 0.5 to 1.0 ms at 4 K. This variable lifetime was also evident in the time evolution of the spectra which shifted to longer wavelengths.

b) Nonradiative Transitions

Guest ions and molecules interact with the local vibrations in glass via the electron-phonon coupling. Nonradiative transitions between energy levels separated by less than the maximum phonon energy occur by direct and Raman relaxation processes. These one- and two-phonon mechanisms cause rapid transitions and lifetime broadening. The electron-phonon coupling and the resultant homogeneous broadening at individual ion sites varies because of differences in the separation of energy levels, strength of the local fields, and character of the vibrational modes. As described in the next section, fluorescence line narrowing provides a method for studying homogeneous linewidths in glass and site-dependent variations in the electron-phonon coupling.

When the energy difference between an excited state and the next-lower level exceeds the maximum single-phonon energy, multiphonon processes are required to conserve energy in a purely nonradiative transition. The larger the energy gap, the more phonons required and the less probable the process. Studies of nonradiative relaxation of excited rare earths in crystals and glasses have shown that for a given rare earth and host, the rates of these high-order processes are determined principally by the size of the energy gap rather than the particular electronic states or phonon modes involved [6.63]. The dependence of multiphonon relaxation rates on the phonon spectrum and temperature has been studied for a series of oxide glasses (borate, phosphate, silicate, germanate, tellurite) with progressively smaller phonon energies [6.64] and for one fluoride glass [6.65]. Since the electron-phonon coupling is site dependent, there is also a range of multiphonon emission rates. For example, the relaxation of the $^4S_{3/2}$ state of Er^{3+} in most glasses is predominantly by multiphonon emission to $^4F_{9/2}$. For nonselective excitation, this decay in a fluoroberyllate glass is nonexponential with rates varying by a factor of 2.5 [6.13].

Nonradiative processes are generally only a small fraction of the total decay rate for the $^4F_{3/2}$ state of Nd^{3+} in silicate, phosphate, and fluoride glasses. In the laser-excited decay studies of *Brecher* et al. [6.18], this contribution was found to vary with excitation wavelength and was usually larger for the high-energy or high-field sites. In borate glasses, the vibrational frequencies are higher and nonradiative processes are the dominant source of the $^4F_{3/2}$ relaxation with a lifetime of ≈ 50 μs. The variation in this rate with wavelength was small, however. There does not appear to be a universal trend of multiphonon emission rates and excitation wavelengths for the various oxide and fluoride glasses investigated thus far.

A more dramatic variation of multiphonon decay rates was observed for the ^2E lifetime of Mo^{3+} in phosphate glass [6.14]. The energy gap to the $^4A_2$ ground state is large, $\approx 10,000$ cm^{-1}, but the electron-phonon coupling for the outer $4d$ electrons of Mo^{3+} is greater than for the shielded $4f$ electrons of rare earths, and hence multiphonon processes are competitive with radiative processes. From the $^4A_2 \rightarrow {}^2E$ absorption intensity, the average radiative lifetime of the ^2E state is estimated to be ≈ 5 ms [6.66]. For broadband excitation into the ^4T bands in Fig. 6.2, the $^2E \rightarrow {}^4A_2$ fluorescence decay is very nonexponential with initial and final decay rates differing by about one order of magnitude at both 300 and 77 K. Using a monochromator to observe the decay at selected wavelengths, *De Groot* measured the lifetimes shown in Fig. 6.17. The difference between the fastest to the slowest e-folding times was ≈ 25. Since all decay rates were temperature dependent and at low temperatures are less than the predicted radiative rate, nonradiative phonon processes contribute to the decay.

The large variation of fluorescence lifetime of Mo^{3+} in glass with wavelength and temperature is due to site-dependent electron-phonon coupling. The presence of intense vibronic transitions, absence of pronounced FLN, and temperature-dependent spectral and decay properties [6.14] confirm that the coupling of the $4d^3$ electrons of Mo^{3+} via the time-dependent crystalline field is strong.

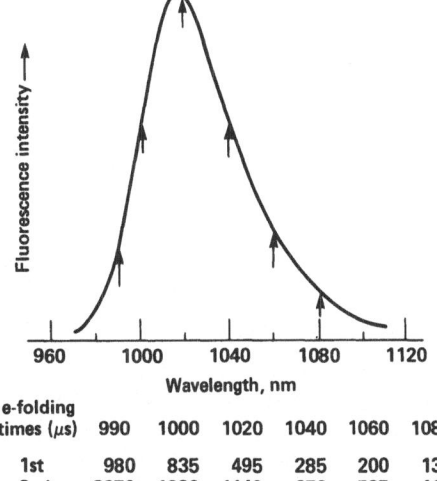

Fig. 6.17. Broadband-excited $^2E \rightarrow {}^4A_2$ fluorescence of Mo^{3+} in phosphate glass at 77 K. The first three e-foldings times of the fluorescence decay are listed for various wavelengths [6.13]

e-folding times (μs)	990	1000	1020	1040	1060	1080
1st	980	835	495	285	200	135
2nd	2270	1830	1140	670	525	410
3rd	3450	2725	1720	1200	930	900

Both phonon-assisted radiative transitions and multiphonon emission contribute to the temperature dependence of the lifetime. Thermally populating the $^2T_1$ state is not expected to contribute to a large difference in radiative decay rates because the $^2E \rightarrow {}^4A_2$ and $^2T_1 \rightarrow {}^4A_2$ transition probabilities are of comparable magnitude.

Figure 6.17 shows that the lower-energy Mo^{3+} sites have the faster decay rates. (This is contrary to the behavior expected if cross relaxation was active.) The ≈ 800 cm^{-1} range of energies corresponds to a change of one phonon in a multiphonon decay process. The ratio of transition probabilities for an n- and an $(n$-1$)$-phonon process could account for much of the observed site-dependent difference in decay rates, however, the theory [6.67] and parameters for multiphonon relaxation of transition metal ions are insufficient to establish this conclusively. Since the sites with smaller ^2E–$^4A_2$ splittings are those for which the Racah parameters are smaller, the bonding is more covalent. This may also be reflected in stronger electron-phonon coupling. Thus, for sites with smaller energy gaps, fewer phonons are required to conserve energy and the electron-phonon coupling contributing to phonon-assisted transitions may be greater, both of which increase the decay rate, as observed experimentally.

6.5.3 Homogeneous Linewidth

Using laser-excited fluorescence techniques, the homogeneous linewidth of transitions involving selected subsets of ions or molecules in a glass can be investigated in the presence of inhomogeneous broadening. To minimize accidental coincidences, zero-phonon transitions and resonant excitation-fluorescence schemes (Fig. 6.3) should be used. In the case where the laser linewidth is much less than the homogeneous linewidth of the transition, (6.3) reduces to a Lorentzian line shape with a width of $2\Delta\nu_H$ [6.24, 35].

Personov et al. [6.4, 68] observed line narrowing of the fluorescence from perylene in glassy ethanol using laser excitation. The zero-phonon linewidth was reduced from tens of cm^{-1} to ≈ 0.4 cm^{-1} [6.7]. Vibronic structure normally obscured by inhomogeneous broadening was also observed with laser excitation [6.6, 7]. From the vibronic bandwidth, the lifetime of excited vibrational states can be determined [6.69].

The linewidths of photochemical and nonphotochemical holes burned in the spectra of organic glasses have been the subjects of several studies [6.36-39]. The magnitude and behavior of the hole widths are attributed to a two-impurity-site model characteristic of glassy solvents. Measurements of the temperature dependence of the hole width of tetracene in alcoholic glasses were related to the rate of site interconversion [6.38]. The ion configurations are described by two (or more) levels in a local double-well potential having a distribution of energy minima, barrier heights, and asymmetries. Physically, these two-level systems (TLS) may involve rearrangements of ions and bonding electrons. Various properties of amorphous materials measured at low temperatures have been interpreted using this phenomenological model originally proposed by *Anderson* et al. [6.70] and *Phillips* [6.71].

Concurrent with the investigations of the linewidth behavior of molecules in organic glasses were similar investigations of line broadening of rare-earth ions in inorganic glasses. *Kushida* and *Takushi* [6.24] used a cw dye laser and a double chopper to study resonant fluorescence at various delays after the cutoff of excitation. Applying (6.3), they showed that the FLN line shape of the $^5D_0-{}^7F_0$ transition of Eu^{3+} in a calcium metaphosphate glass was in good agreement with the Lorentzian form expected for a lifetime broadening mechanism. They also observed resonant $^5D_0-{}^7F_1$, and $^5D_0-{}^7F_2$ transitions at room temperature (see Fig. 6.3, resonant scheme III). Only transitions to the lowest Stark levels of each $\bar{J}$ manifold were narrowed and had widths of 44 cm^{-1} at 587.5 nm and 25 cm^{-1} at 609 nm. The homogeneous width of the $^5D_0-{}^7F_0$ transition, in comparison, was only 3 cm^{-1} due to the weaker electron-phonon coupling for these states.

When a nonresonant excitation-fluorescence scheme is used, only partial line narrowing occurs because of accidental coincidences. (This is particularly prevalent when broad vibronic transitions are involved [6.6].) In Fig. 6.5, for example, the highest-frequency component of the $^5D_0 \rightarrow {}^7F_1$ transition is narrowed by nonresonant $^7F_0 \rightarrow {}^5D_0$ laser excitation, but the width is still larger than the true homogeneous linewidth. The linewidth of the Eu^{3+} $^5D_0 \rightarrow {}^7F_1(1)$ transition in a silicate glass has been measured at 300 K using both resonant and nonresonant excitation [6.49]. The linewidth in the latter case was approximately twice as large, thereby indicating significant residual inhomogeneous broadening. The linewidths were also measured at 20 K where, because of the strong temperature dependence of the homogeneous broadening, the difference between the resonant and nonresonant excitation linewidths was larger (a factor of ≈ 9). In addition, the width of the nonresonant line changed with wavelength from a value of 55 cm^{-1} for the low-field sites to 12 cm^{-1} for the high-field sites. This correlates with a reduction in the degree of accidential coincidences.

Studies of the homogeneous linewidths of several rare earths and glasses show that the width varies with excitation wavelength. This arises from site-to-site differences in the electron-phonon coupling. Factor-of-two linewidth variations of the $^5D_0 - ^7F_0$ transition of Eu^{3+} were measured in a silicate glass at 35 K by *Selzer* et al. [6.72]. Similar results were obtained in a phosphate glass at 373 K by *Avouris* et al. [6.42] and are shown in Fig. 6.18. The largest linewidths are observed for ions in sites having the largest $^7F_1$ splittings. The behavior, however, could not be explained in terms of the $^7F_0 - ^7F_1$ energy separation and harmonic phonons; other possible processes are discussed in [6.26].

The narrowest homogeneous linewidth observed in a glass is the ≈ 20 MHz width of the $^5D_0 \rightarrow ^7F_0$ line of Eu^{3+} in a silicate glass measured at 1.7 K [6.54, 72]. The width was limited by the 50 MHz laser width. The radiative lifetimes of excited electronic states of trivalent rare earths are relatively long, typically $\approx 10^{-4} - 10^{-2}$ s. This is the T_1 time in (6.5). For the optical linewidth to approach this limit, any dephasing due to hyperfine coupling between the rare earth and neighboring atoms, given by the rate Γ in (6.5), must be reduced or eliminated. Spin decoupling and additional line narrowing were observed recently by *Rand* et al. [6.73] for Pr^{3+} in a LaF_3 crystal with the use of an optical free-induction decay. By irradiating the ^{19}F nuclei with an appropriate rf field, they undergo forced precession about an effective field at the magic angle in the rotating frame. The fluctuating *F-F* dipolar interaction was thereby quenched and the optical linewidth reduced to ≈ 2 kHz. Comparable linewidths have also been measured recently by *Macfarlane*

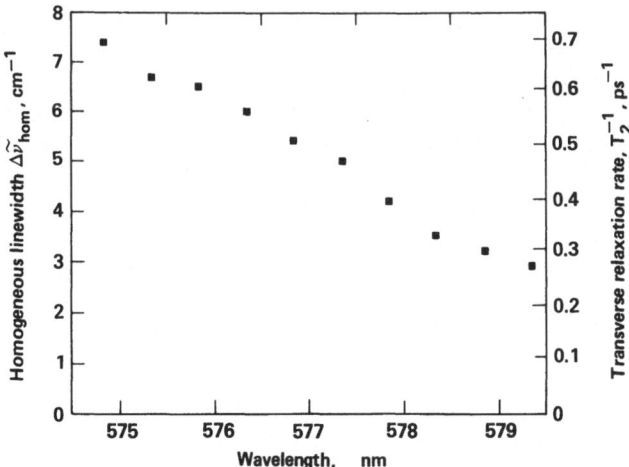

Fig. 6.18. Variation of the homogeneous linewidth of the $^5D_0 \rightarrow ^7F_0$ resonance fluorescence of Eu^{3+} in silicate glass at 393 K with laser excitation wavelength [6.42]

et al. [6.74] for Pr^{3+} in a $YAlO_3$ crystal using a delayed heterodyne photon echo technique. Relationships between optical linewidths measured in the frequency domain and dephasing rates of coherent optical transients measured in the time domain have been discussed by *Nettel* and *Lempicki* [6.75].

The homogeneous linewidth of the $^4F_{3/2} \rightarrow {}^4I_{11/2}$ laser transition of Nd^{3+} in glass is of importance for energy extraction under large-signal or saturated gain operation (see Sect. 6.6.3). *Brawer* [6.19] has measured this width resonantly at several wavelengths using the small fractional ($\approx 10^{-5}$) thermal population in the $^4I_{11/2}$ state at 300 K. A line-narrowed peak with a width ≈ 25 cm^{-1} was observed in ED-2 silicate laser glass. Measured linewidths for resonant $^4F_{3/2} - {}^4I_{9/2}$ transitions in the same glass had comparable values. For both ^4I manifolds there was little variation of linewidth with wavelength. This is reasonable since the Stark level splittings are $< kT$ and in the high-temperature limit all levels have approximately equal lifetimes. The homogeneous linewidth of the Nd^{3+} laser transition can also be determined from the width of the hole burned in the gain profile. *Nikitin* et al. [6.58] observed values of approximately 25 cm^{-1} at 300 K; measurements of the temperature dependence of the hole width in a phosphate glass were also reported.

a) Temperature Dependence

The temperature dependence of the homogeneous linewidth in glass has been the subject of several investigations. *Selzer* et al. [6.54, 72] found that the homogeneous linewidths of the same ion and transition were more than an order of magnitude larger in a glass than in a crystal and had a different temperature dependence. In crystals, the temperature-dependent linewidths are explained by one- and two-phonon relaxation processes. At low temperatures the $^5D_0 \rightarrow {}^7F_0$ linewidth of Eu^{3+} exhibits an exponential temperature dependence due to the thermal activation of an Orbach process involving the $^7F_1$ state. In contrast, the same linewidth in glass exhibits a $T^{1.85 \pm 0.2}$ dependence from liquid helium temperatures to ≈ 100 K. The homogeneous linewidth of the $^4G_{5/2} \rightarrow {}^6H_{5/2}$ transition of Sm^{3+} in an aluminum silicate glass, determined using FLN techniques, decreased from 19 cm^{-1} to $\leqslant 9$ cm^{-1} when the temperature was lowered from 300 to 194 K [6.21], again an approximate T^2 dependence. More recent measurements by *Hegarty* and *Yen* [6.15] of the homogeneous linewidth of the $^3P_0 \rightarrow {}^3H_4(1)$ resonance fluorescence of Pr^{3+} in fluoroberyllate and germanate glasses were made in the range from liquid helium up to 300 K. As shown in Fig. 6.19, a T^2 behavior was observed over the entire temperature range.

Two-level systems have been proposed as a possible source of the additional homogeneous broadening observed in glass at low temperatures. However, a predicted linear dependence of the linewidth on temperature T, does not agree with experiment [6.54]. *Lyo* and *Orbach* [6.76] have recently shown that a T^2 behavior can arise from phonon modulation of the interaction. *Reinecke* [6.77] has also treated the fluorescence linewidths in terms of thermal transitions between states of the TLS in glass. This shifts the electronic energy levels via the strain field coup-

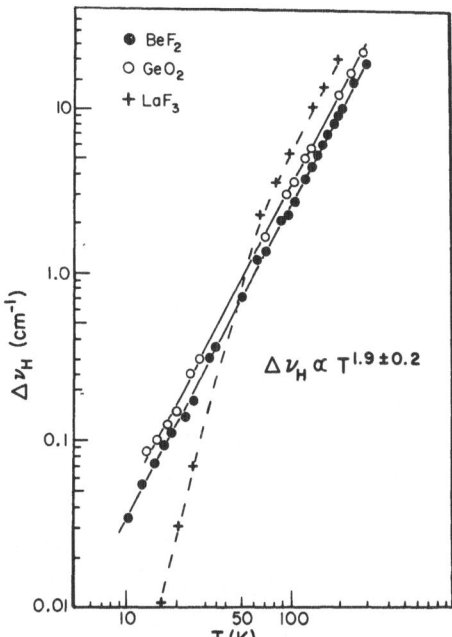

Fig. 6.19. Temperature dependence of the homogeneous linewidth of the Pr^{3+} $^3P_0 \rightarrow$ $^3H_4(1)$ transition in fluoroberyllate and germanate glasses and a lanthanum trifluoride crystal [6.15]

ling; the resulting fluctuations increase the effective optical linewidth. Reasonable agreement is obtained for the magnitude and temperature dependence of the Eu^{3+}: silicate glass linewidth data.

6.5.4 Energy Transfer

When the activator ion concentration in glass becomes sufficiently high, ions interact by multipolar or exchange coupling and ion-ion energy transfer occurs. Because of the disordered nature of glass, ions in nearby sites may be in physically different local environments with greatly varying spectroscopic properties. Therefore the transfer, in addition to causing a spatial migration of energy, may also produce spectral diffusion within the inhomogeneously broadened spectral profile.

In this section we review experimental studies of ion-ion energy transfer in glass which illustrate the essential features of the process. The theory and treatment of energy transfer in ordered and disordered systems have already been reviewed in this volume. Holstein, Lyo and Orbach discuss various possible transfer processes in Chap. 2 and Huber treats the rate equations governing the time evolution of spectral properties in Chap. 3.

The use of pulsed monochromatic radiation enables one to selectively excite subsets of ions within an inhomogeneous distribution and thereby monitor the time evolution of their fluorescence. This possibility was noted in the first

FLN experiments in glass by *Denisov* and *Kizel* [6.1] using incoherent light. Later, *Riseberg* [6.3], using laser light, observed a concentration dependence of the line narrowing; the line-narrowed fraction of the emission decreased monotonically for rare-earth oxide concentrations greater than ≈ 1 wt.%. This was attributed to the onset of ion-ion energy transfer.

Motegi and *Shionoya* [6.5] examined energy migration among Eu^{3+} in a $Ca(PO_3)_2$ glass using time-resolved spectroscopy. A pulsed tunable dye laser was used for the first time to probe throughout the $^7F_0 \rightarrow {}^5D_0$ absorption profile; the nonresonant $^5D_0 \rightarrow {}^7F_1$ fluorescence was observed. The latter spectra changed with excitation wavelength and with time following the excitation pulse. The behavior was analyzed in terms of a phonon-assisted dipole-dipole process; all ions were assumed to have the same line strengths. In glass, energy transfer is characterized by different probabilities for different spectral components. If the process involves phonon participation, it has the additional features that the transfer rates are temperature dependent and the broadband emission exhibits a red shift with time following a short excitation pulse [6.23].

The influence of energy transfer on the position of luminescence bands of rare-earth ions in glass has been studied using nonlaser excitation [6.79]. A shift to longer wavelengths and a deformation of the inhomogeneously broadened fluorescence were observed for Yb^{3+} and Nd^{3+} ions by increasing the rare-earth content. Hosts included both oxide (silicate, phosphate, germanate), and fluoride (fluoroberyllate) glasses. At lower temperatures, the fluorescence bands showed a small shift to longer wavelengths as Yb^{3+} content increased from 0.5 to 5.0 wt.%. Energy therefore migrated to sites having the lowest energies. Conversely, the observed shift to higher energies with time of the Mo^{3+} $^2E \rightarrow {}^4A_2$ emission [6.14] and the Yb^{3+} $^2F_{5/2} \rightarrow {}^2F_{7/2}$ emission at low concentrations [6.80] indicate that it is due to site-dependent transition probabilities and not energy transfer.

The time evolution of laser-induced *resonant* line-narrowed spectra was studied in Eu^{3+} by *Yen* et al. [6.30, 31] and in Yb^{3+} by *Paisner* et al. [6.32]. As an example, the time-resolved spectra of 0.5% Yb in a silicate glass [6.49] are shown in Fig. 6.20. The emission corresponds to the transition between the lowest Stark levels of the $^2F_{5/2}$ and $^2F_{7/2}$ manifolds. Following the laser pulse, the initial line-narrowed fluorescence decays due to a combination of radiative decay and nonradiative transfer to other Yb^{3+} ions without any pronounced line broadening. Subsequent fluorescence from the acceptor ions replicates the inhomogeneously broadened equilibrium emission profile, thus demonstrating that transfer occurred to spatially nearby but spectrally dissimilar sites. In this case the transfer is not to resonant sites but to the full range of sites within the inhomogeneous profile. The transfer is attributed to multipolar processes with an effective range ≈ 1 nm. The effective transfer rate (decrease of the narrowed line and growth of the inhomogeneous profile) of Yb^{3+} varied with concentration and the excitation wavelength.

Because of its relatively simple energy scheme, energy transfer of Eu^{3+} ions has been the subject of many investigations. The behavior of the $^5D_0 \rightarrow {}^7F_0$ emission is similar to that of Yb^{3+} and has been studied by *Yen* et al. [6.31],

Avouris et al. [6.81] and *Alimov* et al. [6.23]. The time evolution of the line-narrowed spectra has, in several instances, been compared to that predicted for electric multipole processes [6.82]. Studies by *Avouris* attribute the transfer in a phosphate glass to a dipole-dipole process because of observed $\exp(-at^{1/2})$ time dependence. The dependence on temperature and energy mismatch was also used to define the dominant transfer mechanism. They concluded that a two-phonon-assisted transfer, with one phonon active at each of two coupled sites, was operative. *Alimov* et al. [6.23], on the other hand, in a study of Eu^{3+} in a borosilicate glass, concluded that transfer occurred by a quadrupole-quadrupole process. However, normal quadrupolar transfer rates are reduced because $0 \leftrightarrow 0$ and $0 \leftrightarrow 1$ transitions are forbidden by ΔJ selection rules and become allowed only by J-state mixing. *Takushi* and *Kushida* [6.83] considered, in addition, the possible importance of pairs of $^5D_0 \rightarrow ^7F_1 : ^7F_1 \rightarrow ^5D_0$ transitions involving transitions from thermally populated $^7F_1$ levels. Although the population is small at low temperatures, the transition probabilities to $^5D_0$ from $^7F_1$ are greater than from $^7F_0$ which enhances the transfer [6.78].

Neodymium energy transfer, because of its relevance to the efficiency of Nd laser action, has also been the subject of several investigations. It has been studied extensively in random disordered crystal systems such as the yttrofluorides CaF_2-YF_3. These systems form mixed solid solutions and the Nd^{3+} spectra exhibit inhomogeneous broadening comparable to that in some glasses. *Basiev* et al. [6.84] examined the kinetics of spectral migration by using a rhodamine 6G dye laser for

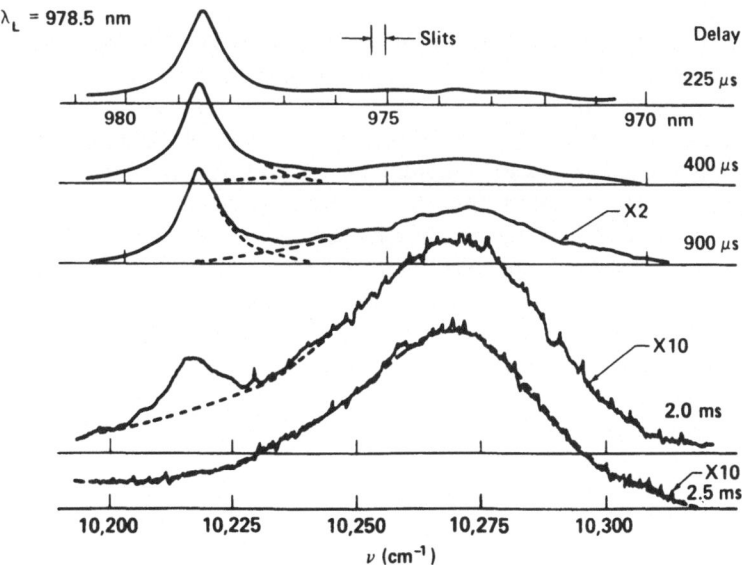

Fig. 6.20. Time evolution of the resonant line-narrowed $^2F_{5/2} \rightarrow ^2F_{7/2}$ fluorescence of 0.5% Yb^{3+} in silicate glass at ≈ 100 K [6.49]

excitation, and described the transfer satisfactorily based on a dipole-dipole mechanism. Additional studies by *Dianov* et al. [6.85] using a frequency-doubled pulsed Nd:YAG laser found that the transfer was completed within ≈ 1 ms at 300 K at concentrations of 1 wt.% NdF_3.

Comparable energy-transfer rates were measured by *Brawer* [6.19] for Nd^{3+} in a silicate glass. In these glass studies, Nd^{3+} ions were excited directly into the $^4F_{3/2}$ emitting state and the resonant $^4F_{3/2} \rightarrow {}^4I_{9/2}$ transition was observed. Measurements were made using concentrations ranging from 1 to 6 wt.% Nd_2O_3; the transfer rate increased approximately as the square of the Nd concentration. Cross relaxation between Nd^{3+} ions in spectrally different sites reestablishes an equilibrium distribution and leads to hole filling and partial gain recovery following an intense saturating pulse in a Nd glass amplifier. The results in [6.19] demonstrate that the rate of hole filling is too slow to be of any importance for pulsed lasers operating in the nanosecond time domain.

As we have seen, in glass there are large site-to-site variations in the energy level structure, probabilities for radiative and nonradiative transitions, and homogeneous linewidths. These variations are generally much larger than those encountered in crystalline hosts. Thus the transitions of the two ions may be either resonant, to within their homogeneous line shapes, or phonon assisted. When broadband radiation is used, all ions are excited in varying degrees. Energy transfer is then usually treated by ascribing average rates to the system, both in the case of direct, one-phonon processes and multistep, diffusion-limited relaxation [6.78]. However, the kinetics of the relaxation, migration probability, and time evolution of the inhomogeneous line shape are not fully describable by a single parameter.

Quantitative comparison of transfer rates and the time development of the fluorescence with theory is very complicated for amorphous materials where the inhomogeneities are large. It requires, as a prerequisite, a mapping of the site variations of the energy levels because in multilevel systems many pairs of resonant and phonon-assisted transitions are possible. In addition, site-to-site variations in the line strengths and homogeneous linewidths of the two transitions must be taken into account. The number density of the site distribution in specific glasses is generally unknown. Different time dependences of the spectral profiles are predicted, for example, for Gaussian and Lorentzian densities of state. A random spatial distribution of dopant ions may not always apply in some glasses. These complexities combine to render quantitative descriptions of energy transfer in glass very difficult. Theoretical treatment of transfer processes including the full range of possible spectroscopic variations remains to be done.

6.6 Applications of Fluorescence Line Narrowing

6.6.1 Glass Structure

Our understanding of the spin resonance and optical properties of paramagnetic ions in glass has been plagued in the past by inhomogeneous broadening of the absorption and emission spectra. This broadening has been the bane of the solid-state spectroscopist and theorist seeking information about specific sites and how the local structure varies throughout the glass and with changes of composition [6.86]. The location of spectral peaks and linewidths are of little value for structural determinations because they merely reflect the wide variation of environments.

Two approaches have been used to relate inhomogeneously broadened spectra in glass to a structural model [6.87]. One is to compare the transition in the glass with the same transition in a crystal and, from the known crystal symmetry, to ascribe an "average" site to the glass. Alternatively, the crystal spectrum is convoluted with some function characteristic of the inhomogeneous broadening to reproduce the broadness of the spectrum observed from the glass.

With the advent of laser-excited FLN, we have a unique microscopic probe of the environment at an impurity site with which to test any proposed structural model. However, as a tool to investigate the structure of glass, spectroscopy of dilute paramagnetic ions must contend with the fact that the ion usually constitutes a defect. The presence of the activator ion perturbs the local field from that of the original perfect glass. While this may temper one's enthusiasm for the use of impurities as probes of ideal glass structure [6.88], it is precisely the field at the activator site that governs the luminescence properties. An understanding of this field and the ability to alter luminescence properties by changing the glass composition are the keys to tailoring glasses for applications. In this light, FLN studies are of paramount importance.

a) Geometric Models

In the past, several attempts have been made to analyze inhomogeneously broadened glass spectra in terms of a small number of sites having some predominant symmetry. The actual spectrum is attributed to perturbations of the average environment(s). The interpretations of optical spectra are numerous and varied. For example, *Kurkjian* et al. [6.89], suggested that Eu^{3+} is sevenfold coordinated in alkali silicate glass. *DeShazer* and co-workers [6.90, 91] deconvolved observed absorption and fluorescence spectra of Eu^{3+} and Nd^{3+} into groups of overlapping lines corresponding to transitions between individual Stark levels. In addition, by analogy with the spectra of the rare-earth sesquioxides, they concluded that the site symmetries in borosilicate and phosphate glasses were predominantly trigonal C_{3v} with a small distortion of triclinic C_1. From the low-temperature optical spectra of Yb^{3+} in phosphate, silicate, and germanate glasses, *Robinson* and *Four-*

nier [6.92] found evidence of near-octahedral site symmetry. For phosphate glasses, both sixfold and eightfold nearest-neighbor oxygen coordination arising from distortions of 3 or 4 nearby PO_4 tetrahedra has been proposed for rare-earth sites [6.93, 94]. Evidence of sixfold coordination of Nd^{3+} was derived from a study of a Ba-Rb-silicate glass [6.95]. In a study of Er^{3+} in a multicomponent alkali silicate glass, *Robinson* [6.96] discussed sixfold near-octahedral sites, but in simple alkali disilicate glasses three more sites were required to describe the observed spectra [6.97].

The actual number of experimentally distinguishable sites is dependent on the ratio of the homogeneous to the inhomogeneous linewidth. To date the narrowest linewidth reported is that of Eu^{3+} in a silicate glass where, at liquid helium temperatures, a resolution-limited $^5D_0 \rightarrow {}^7F_0$ linewidth of 20 MHz was measured [6.54]. Since the inhomogeneous linewidth was $\approx 3 \times 10^{12}$ Hz, this corresponds to the possible observation of $\sim 10^5$ distinct sites. Assuming a tunable laser is available with a spectral width smaller than the natural linewidth, are there laser excitation frequencies which are not resonant with any site? This would appear as a hole in the absorption or excitation spectra. Scanning with a spectral resolution of 1/2000 of the inhomogeneous linewidth, *Selzer* et al. [6.72] found no discontinuities in the FLN signal. This is not surprising. The spectroscopic properties of activator ions are usually influenced by ions beyond the first coordination sphere. Therefore, based on a random network model of a glass, a quasi continuum of local field strengths within the extremes of the inhomogeneous broadening is expected. Thus an approach beginning with a continuous distribution of local fields and interaction strengths rather than a small number of discrete sites is more realistic and consistent with existing FLN data.

Armed with FLN measurements of site-to-site variations in energy levels and transition probabilities, can one invert this information to create a geometric model of the glass structure, and one less ambiguous than those in the reports cited above? This was attempted by *Brecher* and *Riseberg* using FLN data of Eu^{3+} first in a silicate glass [6.22] and later in a fluoroberyllate glass [6.28]. Inspection of Figs. 6.5 and 6.7 quickly reveals that the environments of Eu^{3+} in the two glasses must be different. Additional differences are apparent when glasses of mixed glass formers, such as fluorophosphates, are examined. To rationalize the magnitudes and variations of the crystal-field parameters and their structural implications, *Brecher* and *Riseberg* considered a simple point-charge model of the immediate environment surrounding the emitting ion. They concentrated on structures involving eightfold or ninefold coordination for three major reasons: 1) appropriate ionic size and available bonding orbitals, 2) ubiquitousness of eightfold and ninefold coordination in crystalline rare-earth compounds, and 3) absence of chemical or structural constraints that might prevent the rare earth from achieving its maximum coordination. Since this is no less true in the fluoride case than in the oxide, the same procedure was followed for both glasses.

The structural model for the coordination in oxide glasses involved eight coordinators initially equidistant from the Eu^{3+} ion, arranged in a distorted Archimedes antiprism with C_{2v} symmetry, and a ninth coordinator introduced along the C_2 axis at a variable distance from the central ion. The model is shown at the left in Fig. 6.21. By adjusting the ion positions, reasonable agreement was obtained between the calculated crystal-field parameters and those derived from the measured Stark splitting of the $^7F_1$ and $^7F_2$ states, assuming no J-state mixing.

For the fluoroberyllate glass, the same geometrically derived parameters have the opposite sign from what is observed. While the signs are essentially arbitrary for the B_{22} and B_{42} terms (depending only on the naming of the x and y axes), the other three terms (B_{20}, B_{40}, B_{44}) are uniquely determined by the spectral assignments and must be satisfied. It is feasible to satisfy the requirements of sign and magnitude for any of the crystal-field parameters by suitable variations in the geometric parameters of the structural model. What proved impossible to do, however, was to satisfy the signs and magnitudes of all of them — along with their relative changes as functions of distortion (caused by the ninth coordinator) — and still maintain the packing density and C_{2v} symmetry of the eight equidistant primary coordinators. Moreover, structures with seven or fewer primary coordinators gave even less satisfactory results, while larger numbers of primary coordinators required an intolerably large radial distance to the central coordinating ion.

Seeking a suitable structural model for the fluoride glass, *Brecher* and *Riseberg* considered various kinds of structural modifications possible within the context of close-packed eightfold or ninefold coordination having C_{2v} symmetry. This analysis revealed that with only a small alteration in the starting assumptions, a structural model can be generated that exhibits appropriate behavior in both sign and magnitude for all five crystal-field parameters. This change involves the segregation of the primary coordinators into two different classes based on radial distance. The resultant new model is a ninefold coordinated structure, with three of the coordinators lying 5% closer to the central ion than do the other six. The structure resembles a capped antiprism (the oxide model) elongated in the y direction (perpendicular to the C_2 axis) and is shown at the right in Fig. 6.21. The parametric variation corresponding to the range of excitation energies is reflected in the difference between the average distances of the two coordinator classes, with the higher crystal field associated with the larger radial difference and hence the closer approach to the peri-coordinators. The crystal-field parameters calculated with this model generally agree with the spectroscopic values to within 20%.

As for the oxide glasses, no independent support for this hypothetical model is available. The differences can be attributed to the chemical differences between oxygen and fluorine themselves. The degree of covalency in the metal-oxygen bond should be considerably greater than that in the metal-fluorine bond. Furthermore, to the extent that covalency enters the picture, the greater electronegativity of the fluorine ion and its univalent bonding should tend toward more bridging asymmetry than in the oxide case. Thus, the model can be viewed as an EuF_3 entity that has become solvated by BeF_2 groups through the formation of

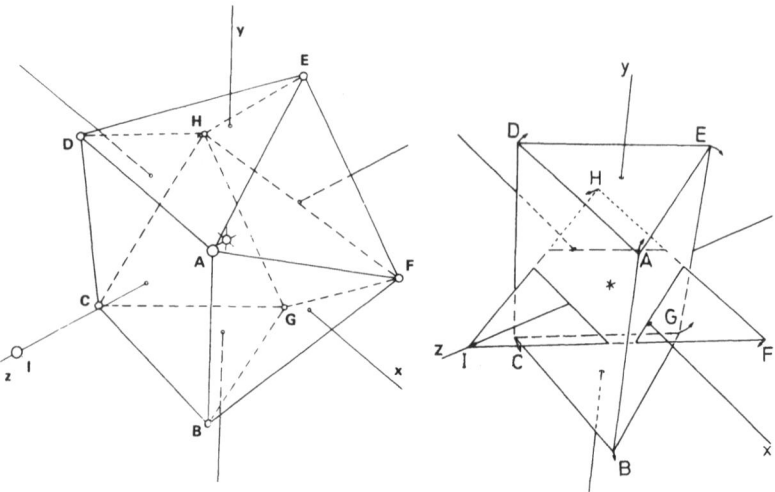

Fig. 6.21. Geometric models of the nearest neighbor environment of the Eu^{3+} ion in oxide and fluoride glasses. The structure at the left for a silicate glass is derived from the square Archimedean antiprism with the Eu^{3+} ion at the origin and eight equidistant oxygens at the vertices. A 9th coordinating oxygen (*I*) is introduced along the *z* direction. As it moves inward, the structure distorts from the low-field configuration shown to the high-field model in which *ABCD* becomes larger and more rectangular and the points *E* and *G* move to larger negative values of *z* [6.22]. The structure at the right for a fluoroberyllate glass is different. Nine fluorine ions are initially located at vertices of equilateral triangles. The arrows at the vertices show the distortions required to reach a structure representing the long wavelength, low crystal-field spectra. The faces *ADE* and *BCG* become elongated in the *z* direction and inclined so that the prism edge *EG* is shorter than the edges *AB* and *CD*; triangle *FHI* becomes deformed with isosceles by compression in the *z* direction [6.28]

nine asymmetric fluorine bridges. The model is chemically consistent and is the least complicated arrangement that can give reasonable agreement to the spectroscopically derived parameters.

A further test of the correctness of any model of the local crystal field would include consideration of the odd-order terms in the expansion in (6.10). These terms enter into predictions of relative fluorescence intensities and polarization.

The above modeling was done by replacing the glass structure by a single effective coordination shell of simple point charges. But the second coordination shell, either in the form of network former or network modifier cations, is known to affect energy levels and transition probabilities of rare earths [6.98]. To model the effects of the entire glass, local site structures can be derived from computer simulations. Molecular dynamics has been used to generate simple glasses [6.99], such as BeF_2, SiO_2, and $ZnCl_2$, and more complex binary glasses [6.100]. Monte Carlo methods of statistical mechanics have also been used to generate glass structures [6.101].

Monte Carlo computer simulations of BeF_2 glass [6.102] have recently been extended to include a rare-earth ion [6.103]. The result of each simulation is a different geometric arrangement of ions about the rare earth. This process is repeated many times. The collection of site configurations represents the glass. The results show very vividly that there is no single geometry or coordination number; instead, there is a wide and continuous variation of local structures. Furthermore, it can be shown that details of the glass structure at rare-earth ion sites cannot be deduced unambiguously from optical spectra alone [6.104].

Using the simulated glass structure and a point charge model of the local field, large site-to-site differences in the energy level splitting of the $^7F_1$ manifold of Eu^{3+} are predicted. The smallest and largest splittings are calculated to differ by at least a factor of 4; this is in good agreement with the splittings observed from FLN spectra of Eu^{3+}:BeF_2 [6.103]. These preliminary results are encouraging. A molecular orbital, self-consistent-field calculation of the energy levels and transition probabilities, combined with the computer glass simulations, should replicate both inhomogeneously broadened and FLN spectra. Comparison with observed spectra will provide a critical test of the accuracy of structural and local field models of simple and multicomponent glasses.

6.6.2 Structural Changes

Structural changes in the microscopic environments of glass due, for example, to phase transitions or hydrostatic pressure, are in principle detectable from the behavior of FLN spectra. Two structural phenomena occur in alkali borate glasses that modify local sites [6.105]. First, in simple glassy B_2O_3, boron has three coordinating oxygen ions, and a glass is formed from a random network of connected BO_3 triangles or boroxyl groups. If an alkali oxide is added as a network modifier, boron also exists with four coordinating oxygens in a tetrahedral complex. Second, subliquidus immiscibility is well established in binary alkali borates. Separation into B_2O_3-rich and alkali-rich phases occurs in regions on a scale of 5-500 nm. Therefore, if a paramagnetic ion is introduced into these glasses, the position of lines in the laser-excited fluorescence spectra should reflect differences in the number of nearby BO_3 triangles, BO_4 tetrahedra, and alkali network-modifier cations.

Evidence of the above effects has been observed [6.27]. Two Eu^{3+}-doped lithium-borate glasses were studied: one with 40 mol.% Li_2O, a concentration at which the fractional number of borons with four coordinating oxygens is a maximum ($\approx 45\%$), and a second with 10 mol.% Li_2O, which is at the maximum of the coexistence curve. Laser-excited fluorescence for the $10\ Li_2O \cdot 90\ B_2O_3$ glass is shown in Fig. 6.22. The $^5D_0 \rightarrow {}^7F_0$ line appears in the region 17,200–17,400 cm^{-1}, the three $^5D_0 \rightarrow {}^7F_1$ transitions in the region 16,700–17,150 cm^{-1}, and five $^5D_0 \rightarrow {}^7F_2$ transitions in the region 15,800–16,600 cm^{-1}. At low excitation energies, the spectra are nearly equal to those observed in the 40% Li_2O borate glass; at high-excitation energies, however, the splittings are different and additional lines appear. There are two groups of lines characterized by different fluorescence lifetimes. The first group consists of lines 1, 2, 5, 6, and 8 and has a shorter lifetime than the second group consisting of lines 3, 4, and 7.

Spectral shifts arise from the nephelauxetic effect. When the europium-oxygen interactions is large, the Racah parameters and the $^7F-^5D$ separation are reduced. Therefore, the low-excitation-energy spectra arise from ions in sites where the Eu-O interaction is strong. This occurs when there are many lithium cations or four-coordinated borons in the second coordination shell. Conversely, if three-coordinated borons predominate, the effective Eu–0 interaction is weaker and the $^7F_0 \rightarrow {}^5D_0$ transition occurs at higher energies. This suggests that the new lines appearing at higher excitation energies in the 10% Li_2O glass arise when BO_3 triangles are more numerous. Spectra similar to those in Fig. 6.22 were also found in a 3% Li_2O borate glass [6.106].

The above results and their interpretation are still incomplete but are suggestive of the types of structural studies that can be investigated using FLN techniques. Additional studies are needed in which the alkali ion is changed and the alkali ion concentration varied between the minimum and maximum values possible in the glass. The effects of phase separation should also be investigated as a function of thermal history. The resulting spectroscopic parameters would provide a test of geometric models for the local coordination at the probe ion site.

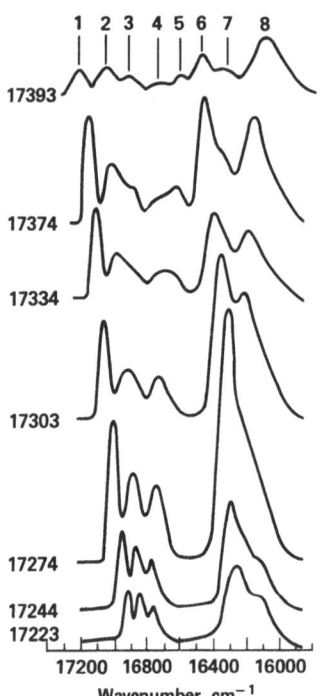

Fig. 6.22. Laser-excited $^5D_0 \rightarrow {}^7F_{1,2}$ fluorescence spectra of Eu^{3+} in a 10 $Li_2O \cdot 90$ B_2O_3 glass at 2 K as a function of $^7F_0 \rightarrow {}^5D_0$ excitation energy (cm^{-1}) [6.27]

6.6.3 Laser Glass

All glass lasers to date have used trivalent rare earths as the active ion. The inhomogeneous nature of the rare-earth spectra in glass is beneficial for optical pumping with broadband sources such as xenon flashlamps because more of the spectrum is absorbed. In addition, the effective cross section for amplified spontaneous emission, which limits the excited state population, is less. Both of these features combine to provide greater energy storage in glass than in crystalline materials. For energy extraction, however, the inhomogeneous spectroscopic properties can be detrimental. This is not evident under small-signal gain conditions, but under large-signal or saturated gain conditions, hole burning and reduced output energy occur in inhomogeneous laser systems. This arises because of the existence of a distribution of stimulated emission cross sections σ and extraction rates $\sigma I/h\nu$ for different sites in the glass, where I is the intensity. Whereas for homogeneous systems the small-signal gain coefficient is given by $g = \sigma \Delta N$, where ΔN is the population inversion of the initial and final lasing states, for an inhomogeneous system, $g = \Sigma\ \sigma_i\ \Delta N_i$. The gain is no longer simply proportional to the stored energy.

Spectral hole burning in the gain profile of Nd laser glass is well established. In the 1960's *Snitzer* and co-workers [6.62] used streak camera recordings of time-resolved lasing spectra to show the time evolution of holes. More recently *Nikitin* et al. [6.44] presented a particularly clear demonstration of spectral hole burning in a Nd:silicate laser glass at 300 K. They examined the $^4F_{3/2} \to {}^4I_{11/2}$ fluorescence from a rod amplifier immediately before and after the passage of a giant laser pulse. A hole resonant with the laser wavelength developed. In addition, the difference spectrum revealed the presence of several additional weaker holes corresponding to other transitions between the $^4F_{3/2}$ and $^4I_{11/2}$ manifolds. In further experiments, *Nikitin* et al. [6.58] examined both silicate and phosphate glasses and the effects of temperature on the hole width.

FLN experiments provide a conceptually simple demonstration of the inhomogeneous nature of a laser transition and a prediction of spectral hole burning, because the observation of resonant fluorescence line narrowing is a necessary and sufficient condition for the occurrence of spectral hole burning. This has been confirmed recently by resonant FLN measurements of the $^4F_{3/2} \to {}^4I_{11/2}$ transition of Nd^{3+} in a silicate glass [6.19]. Similar results are expected for other glasses.

For a Nd:glass amplifier operating at room temperature, spectral hole burning should be characterized by twelve holes in the gain profile corresponding to all possible transitions between the two Kramers-degenerate levels of $^4F_{3/2}$ and the six levels of $^4I_{11/2}$. The resonant hole will be most pronounced because the remaining holes will still exhibit residual inhomogeneous broadening due to accidental coincidences of de-excitation transitions at different sites. Observation of a difference in the saturation behavior of the Nd^{3+} laser gain for two different wavelengths is a sufficient but not a necessary condition for the existence of spectral hole burning.

The reduction in the energy extracted from an inhomogeneous system compared to that from a homogeneous system of the same small-signal gain is a function of the ratio of the homogeneous and inhomogeneous linewidths $(\Delta\nu_H/\Delta\nu_{IH})$. If this ratio is $\gtrsim 1$, ions in different regions of the inhomogeneous distribution act independently and severe spectral hole burning will result. *Hall* et al. [6.107] has quantified this effect for the case of a Gaussian frequency distribution of Lorentzian lines. For the small-signal case, where the input fluence Φ_{in} is very much less than the saturation fluence Φ_s, the deviation of the output energy from that for a homogeneous system is small, even for large inhomogeneities, because the depth of the hole is small. For $\Phi_{in} \approx \Phi_s$, however, a deep hole is burned when $(\Delta\nu_H/\Delta\nu_{IH}) < 1$ and the gain coefficient is reduced. If $(\Delta\nu_H/\Delta\nu_{IH}) \ll 1$, the energy extracted under large-signal conditions is greatly reduced. This would occur, for example, for glass at low temperatures where the homogeneous linewidths become very small. To minimize these effects, the dependence of both the homogeneous and inhomogeneous linewidths of rare earths on glass composition are being studied.

Similar reductions in extracted energy occur whenever the glass has a distribution of stimulated emission cross sections at the laser wavelength. This can arise not only from site-to-site variations in the center frequency, but also from site-to-site variations in the homogeneous linewidth or in the transition probability for stimulated emission at a specific wavelength and polarization. Transition probabilities can vary simply from different orientations of a given site. If linearly polarized radiation is used, preferential site excitation or de-excitation will occur if the parallel or perpendicular polarization cross sections for a site differ. Thus in a glass laser amplifier, a hole can be burned in the site distribution without any evidence of *spectral* hole burning [6.108]. The energy extracted from such a system will again be less than that expected assuming a homogeneous system of the same small-signal gain. In laser glass, the effective cross section for stimulated emission can also change in time due to differences in site-dependent decay rates [6.109].

6.7 Concluding Remarks

Laser-excited spectroscopy in glass combines the sensitivity of fluorescence techniques with the selectivity possible with monochromatic laser radiation. As a spectroscopic tool, laser-excited fluorescence provides a unique method for interrogating subsets of ions in disordered media. Using fluorescence line narrowing and site selection spectroscopy techniques developed during the past decade, a variety of spectroscopic features in glasses have been measured. The existence of large site-to-site differences in spectroscopic properties have been demonstrated with exceptional clarity from these measurements. Analysis of the results provides an interpretation of the local structure and physical processes active at paramagnetic ion and molecular sites in glasses.

The increasing availability of different types of tunable lasers expands the spectral range and number of probe ions and molecules that can be used for these experiments. To realize the full potential for frequency selectivity of sites, ions having narrow zero-phonon lines and samples at low temperatures are required. For energy transfer studies, a simple energy-level scheme is desirable. Recently, fluorescence line narrowing of Gd^{3+} has been observed in several glasses [6.110]. The S-state ground state of this ion has a very small splitting. Hence at low temperatures, where only the lowest Stark level of the $^6P_{7/2}$ manifold is populated, the $^6P_{7/2} \rightarrow {}^8S_{7/2}$ transition approximates a two-level system. In addition, the fluorescence was accompanied by vibronic sidebands free of any overlapping zero-phonon transitions to other Stark levels. Therefore, this rare-earth ion should be useful for investigating site-dependent electron-phonon coupling.

Accidental coincidences of excitation transitions of ions and molecules in different environments continue to plague the degree of site selectivity achievable. Multiple resonance techniques such as double resonance or two-photon absorption spectroscopy, using equal or unequal photon energies, may increase selectivity significantly, although accidental coincidences may still occur. Application of these techniques to glasses should be explored.

Other studies are needed to determine the number density of ions in physically different sites and to relate the results to models or computer simulations of the site-to-site inhomogeneities in glass. Possible clustering of ions, a phenomenon known to occur in rare-earth-doped crystals, may also occur in glasses and should be evident from careful laser-excited fluorescence studies. The presence of phase separation and microcrystallinity in glass also warrant further investigation using the arsenal of techniques now available. Interpretation of these results in terms of local environments should provide new insights into the microscopic structures of glasses.

Recent studies have shown that the homogeneous linewidths of Eu^{3+} [6.111] and Nd^{3+} [6.112] vary widely for different glass hosts. The observed widths are strongly correlated with the velocity of sound in the material. However, the dependence of the linewidth on velocity does not agree with predictions based on mechanisms for line broadening of rare-earth ions in crystal. Additional work is needed to understand both the temperature and the sound velocity (or host) dependencies of the line-broadening processes in glass. However, even without such knowledge, the general correlation of the homogeneous linewidth with sound velocities and the experimentally determined temperature dependence already provide valuable criteria for selecting glass hosts to maximize energy extraction from laser ions.

This review has concentrated on laser-excited fluorescence studies in inorganic glasses. As noted in the introduction, site selection spectroscopy has also been applied to study organic glasses [6.113]. The emphasis of the investigations of organic materials has been on photochemical reactions, molecular configurations, and vibrational structure. Research on organic and inorganic materials developed independently and, as evident by the lack of cross referencing [6.114], with little mutual awareness of others' activities. Further progress in studies of luminescence properties of both organic and inorganic glasses using laser-excited techniques would benefit from more interaction between those working on these two classes of materials.

Acknowledgements. Preparation of this chapter was performed under the auspices of the Materials Science Program of the U.S. Department of Energy Office of Basic Energy Sciences and the Lawrence Livermore National Laboratory under Contract No. W-7405-Eng-48.

References

6.1   Yu V. Denisov, V.A. Kizel: Opt. Spectrosc. **23**, 251 (1967)
6.2   A. Szabo: Phys. Rev. Lett. **25**, 924 (1970); **27**, 323 (1971)
6.3   L.A. Riseberg: Phys. Rev. Lett. **28**, 789 (1972); Solid State Commun. **11**, 469 (1971); Phys. Rev. A 7, 671 (1973)
6.4   R.I. Personov, E.I. Al'Shits, L.A. Bykovskaya: Opt. Commun. **6**, 169 (1972)
6.5   N. Motega, S. Shionoya: J. Lumin. **8**, 1 (1973)
6.6   J.H. Eberly, W.C. McColgin, K. Kawaoka, A.P. Marchetti: Nature (London) **251**, 215 (1974)
6.7   B.M. Kharlamov, R.I. Personov, L.A. Bykovskaya: Opt. Commun. **12**, 191 (1974); Opt. Spectrosc. **39**, 240 (1975)
6.8   S.L. Hager, J.E. Willard: J. Chem. Phys. **61**, 3244 (1974)
6.9   W.H. Zachariasen: J. Am. Chem. Soc. **54**, 3841 (1932)
6.10   H. Rawson: *Inorganic Glass-Forming Systems* (Academic Press, New York, 1967)
6.11   D.R. Uhlmann, A.G. Kolbeck: Phys. Chem. Glasses **17**, 146 (1976)
6.12   S.A. Brawer, W.B. White: J. Chem. Phys. **67**, 2043 (1977)
6.13   M.J. Weber: *Proc. 7th Intern. Conf. Amorphous and Liquid Semiconductors* ed. by W.E. Spear (Edinburgh 1977) p. 645
6.14   M.J. Weber, S.A. Brawer, A.J. De Groot: Phys. Rev. B **23**, II (1981)
6.15   J. Hegarty, W.M. Yen: Phys. Rev. Lett. **43**, 1126 (1979)
6.16   O.K. Alimov, T.T. Basiev, Yu K. Voron'ko, Yu V. Gribkov, A. Ya. Karasik, V.V. Osiko, A.M. Prokhorov, I.A. Shcherbakov: Sov. Phys. JETP **47**, 29 (1978)
6.17   Y. Kalisky, R. Reisfeld, T. Haas: Chem. Phys. Lett. **61**, 19 (1979)
6.18   C. Brecher, L.A. Riseberg, M.J. Weber: Phys. Rev. B **18**, 5799 (1978)
6.19   S.A. Brawer, M.J. Weber: Appl. Phys. Lett. **35**, 31 (1979)
6.20   M.V. Glushkov, Yu. V. Kosichkin, V.V. Osiko, Zh. A. Pukhlii, I.A. Shcherbakov: Sov. J. Quantum Electron. **9**, 1296 (1979)
6.21   T.T. Basiev, M.A. Borik, Yu. K. Voron'ko, V.V. Osiko, V.S. Federov: Opt. Spectrosc. (USSR) **46**, 510 (1979)
6.22   C. Brecher, L.A. Riseberg: Phys. Rev. B **13**, 81 (1976)
6.23   O.K. Alimov, T.T. Basiev, Yu. K. Voron'ko, L.S. Gaigerova, A.V. Dmitryuk: Sov. Phys. JETP **45**, 690 (1977)
6.24   T. Kushida, E. Takushi: Phys. Rev. B **12**, 824 (1975)
6.25   C. Brecher, L.A. Riseberg, M.J. Weber: Proc. 12th Rare Earth Research Conf. (1976) p. 351
6.26   P. Avouris, A. Campion, M.A. El-Sayed: Proc. Soc. Photo-Opt. Instrum. Eng. **113**, 57 (1977)

6.27   M.J. Weber, J. Hegarty, D.H. Blackburn: In *Borate Glasses,* ed. by L.D. Pye, V.D. Frechette, N.J. Kreidl (Plenum Press, New York 1978) p. 215

6.28   C. Brecher, L.A. Riseberg: Phys. Rev. **B21,** 2607 (1980)

6.29   J.J. Videau, J. Portier, B. Blanzat, C. Burthov: Mat. Res. Bull. **14,** 1225 (1979)

6.30   S.S. Sussman, J.A. Paisner, W.M. Yen, M.J. Weber: Bull Am. Phys. Soc. **20,** 44 (1975)

6.31   W.M. Yen, S.S. Sussman, J.A. Paisner, M.J. Weber: Lawrence Livermore Laboratory Rpt. UCRL-76481 (1975)

6.32   J.A. Paisner, S.S. Sussman, W.M. Yen, M.J. Weber: Bull Amer. Phys. Soc. **20,** 447 (1975); Laser-Fusion Annual Report - 1974, Lawrence Livermore Laboratory Rpt. UCRL-50021-74, pp. 273-277

6.33   G. Boulon, B. Moine, J.C. Bourcet: Phys. Rev. B **22,** 1163 (1980)

6.34   J. P. Hessler, W. T. Carnall: *Lanthanide and Actinide Chemistry and Spectroscopy,* ed. N.M. Edelstein, (American Chemical Society, Washington, D. C. 1980), p. 349

6.35   I. Abram, R.A. Auerbach, R.R. Birge, B.E. Kohler, J.M. Stevenson: J. Chem. Phys. **63,** 2473 (1975)

6.36   U. Bogner: Phys. Rev. Lett. **37,** 909 (1976)

6.37   J.M. Hayes, G.J. Small: Chem. Phys. **27,** 151 (1978)

6.38   J.M. Hayes, G.J. Small: Chem. Phys. Lett. **54,** 435 (1978)

6.39   F. Graf, H.K. Hong, A. Nazzal, D. Haarer: Chem. Phys. Lett. **59,** 217 (1978)

6.40   G.K. Wertheim, M.A. Butler, K.W. West, D.N.E. Buchanan: Rev. Sci. Instrum. **45,** 1369 (1974)

6.41   J. Hegarty, R. Brundage, W.M. Yen: Appl. Opt. **19,** 1889 (1980)

6.42   P. Avouris, A. Campion, M.A. El-Sayed: J. Chem. Phys. **67,** 3397 (1977)

6.43   See, for example, A. Yariv: *Introduction to Optical Electronics,* 2nd ed. (Holt, Rinehart, and Winston, New York 1976)

6.44   V.I. Nikitin, M.S. Soskin, A.I. Khizhnyak: Sov. Tech. Phys. Lett. **2,** 64 (1976)

6.45   P.P. Feofilov: *The Physical Basis of Polarized Emission* (Consultants Bureau, New York 1961)

6.46   V.P. Lebedev, A.K. Przhevuskii: Sov. Phys. Solid State **19,** 1389 (1977)

6.47   T. Kushida, E. Takushi, Y. Oka: J. Lumin. **12/13,** 723 (1976)

6.48   J. Hegarty, W.M. Yen, M.J. Weber: Phys. Rev. B **18,** 5816 (1978)

6.49   M.J. Weber, J.A. Paisner, S.S. Sussman, W.M. Yen, L.A. Riseberg, C. Brecher: J. Lumin. **12/13,** 729 (1976)

6.50   J. Hegarty, E. Strauss, W.J. Miniscalco, W.M. Yen: Bull. Amer. Phys. Soc. **24,** 894 (1979)

6.51   Yu. V. Denison, B.F. Dzhurinskii, V.A. Kizel: Izv. Akad. Nank SSSR, Neorg. Materialy **2,** 693 (1966)

6.52   A.K. Przhevuskii, V.A. Savostyanov, M.N. Tolstoi: Sov. J. Quantum Electron. **8,** 54 (1978)

6.53   See *Methods of Experimental Physics,* Vol. 15, Part A, ed. by C.L. Tang (Academic Press, New York 1979) for a review of properties of various lasers.

6.54   P.M. Selzer, D.L. Huber, D.S. Hamilton, W.M. Yen, M.J. Weber: Phys. Rev. Lett. **36,** 813 (1976)

6.55   C.K. Jorgensen: *Modern Aspects of Ligand Field Theory* (North-Holland, Amsterdam 1971)

6.56   B.G. Wybourne: *Spectroscopic Properties of Rare Earths* (Wiley Interscience, New York 1965)

6.57   C. Brecher, L.A. Riseberg: J. Non-Cryst. Solids **40,** 469 (1980)

6.58   C. Brecher, L.A. Riseberg, M.J. Weber: J. Lumin. **18/19,** 651 (1979)

6.59   V.I. Nikitin, M.S. Soskin, A.I. Khizhnyak: Sov. Tech. Phys. Lett. **3,** 5 (1977)

6.60   S. Parke, S. Gomolka, J.N. Sandee: J. Non-Cryst. Solids **20,** 1 (1976)

6.61   S. Singh, L.G. Van Uitert, W.H. Grodkiewicz: Opt. Commun. **17,** 315 (1976)

6.62   E. Snitzer, C.G. Young: *Lasers,* Vol. 2, ed. by A.K. Levine (Dekker, New York 1966) p. 191

6.63   L.A. Riseberg, M.J. Weber: In *Progress in Optics,* Vol. XIV, ed. by E. Wolf (North-Holland, Amsterdam 1976) p. 89

6.64   C.B. Layne, W.H. Lowdermilk, M.J. Weber: Phys. Rev. B **16**, 10 (1977)

6.65   C.B. Layne, M.J. Weber: Phys. Rev. B **16**, 3259 (1977)

6.66   S. Parke, A.I. Watson: Phys. Chem. Glasses **10**, 37 (1969)

6.67   M.D. Sturge: Phys. Rev. B **8**, 6 (1973)

6.68   R.I. Personov, B.M. Kharlamov: Opt. Commun. **7**, 417 (1973)

6.69   A.P. Marchetti, W.C. McColgin, J.H. Eberly: Phys. Rev. Lett. **35**, 387 (1975)

6.70   P.W. Anderson, B.I. Halperin, C.M. Varma: Philos. Mag. **25**, 1 (1972)

6.71   W.A. Phillips: J. Low Temp. Phys. **7**, 351 (1972)

6.72   P.M. Selzer, D.L. Huber, D.S. Hamilton, W.M. Yen, M.J. Weber: In *Structure and Excitations in Amorphous Solids,* AIP Conference Proc. No. 31 (1976) p. 328

6.73   S.C. Rand, A. Wokaun, R.G. DeVoe, R.G. Brewer: Phys. Rev. Lett. **43**, 1868 (1979)

6.74   R.M. Macfarlane, R.M. Shelby, R.L. Shoemaker: Phys. Rev. Lett. **43**, 1726 (1979)

6.75   S.J. Nettel, A. Lempicki: Am. J. Phys. **47**, 987 (1979)

6.76   S.K. Lyo, R. Orbach: Phys. Rev. B **22**, 4223 (1980)

6.77   T.L. Reinecke: Solid State Commun. **32**, 1103 (1979)

6.78   See, for example, M.J. Weber: Phys. Rev. B **4**, 2932 (1971) and references cited therein

6.79   L.E. Ageava, A.K. Przhevuskii, M.N. Tolstoi, V.N. Shepavalov: Sov. Phys. Solid State **16**, 1082 (1974)

6.80   V.A. Savostyanov, V.A. Malyshev, A.K. Przhevskii, A.S. Troshin: Opt. Spectrosc. **47**, 295 (1979)

6.81   P. Avouris, A. Campion, M.A. El-Sayed: Chem. Phys. Lett. **50**, 9 (1977)

6.82   M. Inokuti, F. Hirayama: J. Chem. Phys. **43**, 1978 (1965)

6.83   E. Takushi, T. Kushida: J. Lumin. **18/19**, 661 (1979)

6.84   T.T. Basiev, Yu. K. Voron'ko, A. Ya. Karasik, V.V. Osiko, I. A. Shcherbakov: Sov. Phys. JETP **48**, 32 (1978)

6.85   E.M. Dianov, A. Ya. Karasik, I.A. Shcherbakov: Sov. J. Quantum Electron. **7**, 588 (1977)

6.86   J. Wong, C.A. Angell: *Glass Structure by Spectroscopy* (Dekker, New York 1976)

6.87   G.E. Peterson, C.R. Kurkjian, A. Carnevale: Phys. Chem. Glass **15**, 52 (1974)

6.88   D.L. Grisson: In *Borate Glasses,* ed. by L.D. Pye, V.D. Frechette, N.J. Kreidl (Plenum Press, New York 1978) p. 11

6.89   C.R. Kurkjian, P.K. Gallagher, W.R. Sinclair, E.A. Sigety: Phys. Chem. Glasses **4**, 239 (1963)

6.90   D.K. Rice, L.G. DeShazer: Phys. Rev. **186**, 387 (1969)

6.91   M.M. Mann, L.G. DeShazer: J. Appl. Phys. **41**, 2951 (1970)

6.92   C.C. Robinson, J.T. Fournier: Chem. Phys. Lett. **3**, 517 (1969); J. Phys. Chem. Solids **31**, 895 (1970)

6.93   J.T. Fournier, R.H. Bartram: J. Phys. Chem. Solids **31**, 2615 (1970)

6.94   R. Reisfeld: Structure and Bonding **13**, 53 (1973)

6.95   C.C. Robinson: J. Chem. Phys. **54**, 3572 (1971)

6.96   C.C. Robinson: J. Non-Cryst. Solids **15**, 1 (1974)

6.97   C.C. Robinson: J. Non-Cryst. Solids **15**, 11 (1974)

6.98   R.R. Jacobs, M.J. Weber: IEEE J. **QE-12**, 102 (1976)

6.99   L.V. Woodcock, C.A. Angell, P. Cheeseman: J. Chem. Phys. **65**, 1565 (1976)

6.100   T.F. Soules: J. Chem. Phys. **71**, 4570 (1979), **73**, 4032 (1980)

6.101   K. Binder, D. Stauffer: In *Monte Carlo Methods in Statistical Physics,* ed. by K. Binder, Topics in Current Physics, Vol. 7 (Springer, Berlin, Heidelberg, New York 1979) p. 301

6.102   S.A. Brawer: J. Chem. Phys. *72*, 4264 (1980)
6.103   S.A. Brawer, M.J. Weber: Phys. Rev. Lett. **45**, 460 (1980)
6.104   S.A. Brawer, M.J. Weber: J. Non-Cryst. Solids **38/39**, 9 (1980)
6.105   See *Borate Glasses,* ed. by L.D. Pye, V.D. Frechette, M.J. Kreidl (Plenum Press, New York 1978)
6.106   J. Hegarty, W.M. Yen, M.J. Weber, D.H. Blackburn: J. Lumin. **18/19**, 657 (1979)
6.107   D.W. Hall, R. Haas, W.F. Krupke, M.J. Weber, IEEE J. Quantum Electron. **QE-19**, 1704 (1983)
6.108   M.J. Weber: J. Non-Cryst. Solids **47**, 117 (1982)
6.109   A.R. Kangro, Ya E. Kariss, A.K. Przheuskii, V.A. Savostyanov, and M.N. Tolstoi: Sov. Tech. Phys. Lett. **2**, 255 (1976)
6.110   D.W. Hall, S.A. Brawer, M.J. Weber: Phys. Rev. B **25**, 2828 (1982)
6.111   J. Morgan, E.P. Chock, W. Hopewell, M.A. El-Sayed, R.L. Orbach: J. Phys. Chem. **85**, 747 (1981)
6.112   J.M. Pellegrino, W.M. Yen, M.J. Weber: J. Appl. Phys. **51**, 6332 (1980)
6.113   B.E. Kohler: In *Chemical and Biochemical Application of Lasers,* Vol. IV, ed. by C.B. Moore (Academic Press, New York 1979) p. 31
6.114   See, for example, W. DeW. Horrocks, Jr., D.R. Sudnick, Science **206**, 1194 (1979)

Additional References

Section 6.2.2. J. Friedrich, J. D. Swalen, D. Haarer: Electron-phonon coupling in amorphous organic host materials as investigated by photochemical hole burning. J. Chem. Phys. **73**, 703 (1980)
A. R. Guitierrez: Optical hole burning and matrix effects in a phthalocyanine complex of ruthenium (II). Chem. Phys. Lett. **74**, 293 (1980)
Section 6.3.3. V. P. Lebedev, A. K. Przhevuskii: Determination of multipole orders of optical transitions in the spectra of glasses activated with Eu^{3+} ions by the method of polarized luminescence. Opt. Spectrosc. (USSR) **48**, 513 (1980)
Section 6.5.1. T. T. Basiev, Yu. K. Voron'ko, S. B. Mirov, A. M. Prokhorov: Frequency selection of Nd^{3+} ions in glass excited by monochromatic laser radiation at the resonant transition $^4I_{9/2} \rightarrow ^4F_{3/2}$. JETP Lett. **29**, 639 (1979)
Section 6.5.4. B. Moine, J. C. Bourcet, G. Boulon, R. Reisfeld, Y. Kalisky: Interaction mechanisms in the $Bi^{3+} -Eu^{3+}$ energy transfer in germanium glass at low temperature. J. de Phys. (in press)
Section 6.6.3. V. I. Nikitin, M. S. Soskin, and A. I. Khizhnjak: Deformation of the 1.06 μ Nd^{3+} band profile of glasses under free-oscillation conditions. Sov. J. Quantum Electron. **9**, 1314 (1979)

7. Excitation Dynamics in Molecular Solids

A.H. Francis and R. Kopelman

With 24 Figures

Excitation dynamics in molecular crystals have much in common with excitation dynamics of ionic crystals and glasses. While the optical spectroscopy of the electronic centers in solids (Chap. 1) was treated for both ionic and molecular solids in a unified way, the review of the excitation dynamics in this chapter has to be compared with somewhat parallel reviews dealing predominantly with inorganic solids (Chaps. 2-6). This is done partly for pragmatic reasons and partly in view of the unique opportunities afforded by the study of molecular crystals. In isotopic substituted molecular crystals one has essentially "perfect randomness" in the substitutional composition, as well as minimal effects due to local heterogeneities. Thus, such studies may be the best proving grounds for the testing of theoretical concepts. Also, the ease of *excitation fusion* in such crystals provides clear-cut criteria for the testing of long-range excitation transport in undoped crystals.

We first expand on the brief discussion (Chap. 1) of spectroscopic line shapes. These are of prime importance for the interpretation of excitation dynamics experiments because they are related to both the *exciton-phonon interactions* (see Chap. 2) and to lattice *heterogeneities*. Both the above concepts are basic to current theoretical formalisms on the transport of excitations and other elementary particles in solids. The empirical methods for the study of excitation migration in molecular crystals are described for both ordered and disordered systems. Special emphasis is placed on the phenomena of critical excitation transport, scaling and universality, as these have not been reviewed before. The concepts of Anderson-Mott transition and dynamic percolation receive special attention, in view of their potential relevance to the excitation dynamics of inorganic solids (Chaps. 5, 6). Finally, some new approaches to the problem of excitation kinetics in disordered crystals (see Chap. 3) are presented via their application to molecular solids.

7.1 Lineshape Studies of Excitation Dynamics

The origin of linewidth and the details of lineshape reflect a range of interactions of varying magnitude whose individual importance depends upon the particular molecular solid studied. Thus, inhomogeneities caused by chemical and isotopic

---

Supported by NSF Grant DMR 800679 and NIH Grant 2 R01 NS 08116-10A1

impurities or mechanical strain, intermolecular vibrational and electronic coupling, as well as coupling between electronic excited states and lattice vibrational
modes, all may play an important role in the determination of linewidth and lineshape. In principle, therefore, a considerable amount of insight into the details
of the interactions between electronically excited molecules and their environments may be derived from a study of linewidths and shapes in the optical spectra
of molecular solids. Moreover, in addition to their experimental simplicity relative to other methods of studying excited dynamical processes, lineshape studies
are particularly well suited to the investigation of processes which are too fast to
measure in real time.

The following expression may be obtained from first-order time-dependent
perturbation theory [7.1] for the lineshape associated with a transitioh between
two states (i) and (j) with populations $N_i(t)$ and $N_j(t)$ and separated in energy by
$\Delta E_{ij} = E_i - E_j = \hbar \omega_{ij}$

$$I(\omega) = \frac{4\pi(N_i - N_j)\omega_{ij}}{ch} |M_{ij}|^2 \left[\frac{1/T_2}{(\omega - \omega_{ij})^2 + (1/T_2)^2 + (T_1/T_2)\left(\frac{E_0 \cdot M_{ij}}{h}\right)^2} \right]$$

(7.1)

Here $|M_{ik}|$ is the matrix element of the transition dipole moment between states
(i) and (j) and T_1 and T_2 are first-order relaxation times associated with relaxation
of the populations and the dielectric polarization of the sample, respectively.
E_0 is the amplitude of the electric field vector.

The "longitudinal" relaxation time, T_1, is related to energy changing processes and, in electronic spectroscopy, is frequently associated with the decay of
the excited-state population, $N_j(t)$. T_2, the "transverse" relaxation time, is related
to processes which change the phase of the individual dipole oscillators. Since
first-order processes which contribute to T_1 may also contribute to T_2, it is useful
to define T_2^*, the transverse relaxation time excluding T_1 processes, as

$$1/T_2^* = 1/T_2 - 1/2T_1.$$

(7.2)

At optical power levels sufficiently low that the power saturation term (in E_0^2) is
much smaller than T_2^{-2}, the lineshape function given in (7.1) is that of a homogeneously broadened Lorentzian centered about $\omega = \omega_{ij}$, with a linewidth (FWHM)
given by

$$\Delta \omega = 2/T_2.$$

(7.3)

Through (7.3), the linewidth of homogeneously broadened lines in the spectra of
molecular crystals can be related to the dephasing time T_2.

The relaxation times T_1 and T_2 which appear in (7.1) are added to the
time-dependent perturbation theory phenomenologically. Therefore, if one wishes

to relate the observed lineshape to dynamical processes, it is necessary that the line be homogeneous (i.e., all molecules have the same T_1 and T_2) and that a microscopic theory for the intermolecular interactions be developed to relate the experimentally determined T_1 and T_2 to intermolecular potential functions. In this regard, the interpretation of lineshape studies is less straightforward than the interpretation of time-domain measurements.

7.1.1 Pure Crystal Lineshape Theory

Detailed studies of lineshapes in pure crystals reveal that they depart markedly, in many cases, from the Lorentzian shape. The most commonly observed absorption lineshape associated with a purely electronic excitation is probably the asymmetrical Lorentzian, which is accurately Lorentzian on only one side and exhibits *Urbach* rule behavior [7.2], or an exponential lineshape, on the other. Additionally, lineshape is strongly temperature dependent and may exhibit a transition from Lorentzian to Gaussian as a function of temperature [7.3].

It is useful, therefore, before proceeding to the detailed theories of lineshape, to examine the physical reason for the lineshape variants in terms of the autocorrelation function for the microscopic transition electric dipole moment M_{ij}. We employ an illustrative example due to *Kubo* [7.4]. Consider a randomly modulated electric dipole oscillator with the equation of motion

$$\dot{M}(t) = i\omega(t) M(t), \tag{7.4}$$

where the instantaneous frequency of oscillation $\omega(t)$ varies about its mean value ω_0 according to

$$\omega(t) = \omega_0 + \omega_1(t). \tag{7.5}$$

The fluctuations of $\omega(t)$ about its mean are assumed to be random and may be due to lattice potential fluctuations arising from exciton motion in a disordered lattice, lattice phonons, etc. The autocorrelation function for the transition dipole moment is

$$<M(t) M^*(0)> = |M(0)|^2 \exp(i\omega_0 t)\, \phi(t), \tag{7.6}$$

where $M(0)$ is the dipole moment at $t=0$ and $\phi(t)$ is the relaxation function of the dipole, given by

$$\phi(t) = <\exp\left[i \int_0^t \omega_1(t')dt' \right]>. \tag{7.7}$$

The absorption lineshape for the oscillator is obtained from the Fourier transform of the autocorrelation function. If $\omega_1(t)$ has a Gaussian probability distribution about ω_0, as is frequently assumed, then the mean squared frequency displacement is $<\omega_1^2> = \Delta^2$. The correlation time of the potential fluctuation is given by

$$\tau_c = 1/\Delta^2 \int_0^\infty <\omega_1(t)\,\omega_1(t+t')> dt' \tag{7.8}$$

and is a measure of the dwell time for a frequency displacement. In the limit that $\tau_c\Delta \gg 1$ the frequency of oscillation is much greater than the frequency of fluctuation and the absorption lineshape reflects directly the distribution of the fluctuations. This is the strong coupling limit, which, for the present example, yields a response function

$$\phi(t) = \phi(0) \, \exp[-(t/\tau_c)^2] \tag{7.9}$$

corresponding to a Gaussian lineshape with linewidth (FWHM) of about 2Δ

$$I(\omega) = (1/\sqrt{2\pi}\,\Delta) \exp[-(\omega-\omega_0)^2/2\Delta^2]. \tag{7.10}$$

In the weak coupling limit $\tau_c\Delta \ll 1$, any frequency fluctuation persists for only about ω_1^{-1}, so that the fluctuations are smoothed out and the lineshape narrows. In the limit that $\tau_c\Delta \to 0$ the response function becomes

$$\phi(t) = \phi(0) \exp(-t/\tau_c) \tag{7.11}$$

and the lineshape approaches a Lorentzian with linewidth (FWHM) $\Delta^2\tau_c$;

$$I(\omega) = \frac{\Delta^2\tau_c}{(\omega-\omega_0)^2 + (\Delta^2\tau_c)^2}. \tag{7.12}$$

Deviations from symmetrical Lorentzian or Gaussian lineshapes will occur when the probability distribution function for the frequency fluctuations becomes asymmetrical.

a) Phonon Damping and Random Lattice Models

In molecular crystals above $T=0$, excitons move in a fluctuating potential due to thermally excited lattice vibrations which, in addition to broadening the (rigid lattice) exciton states, produce localized states below the exciton band that are responsible for the Urbach rule behavior. In molecular crystals, within the adiabatic and elastic scattering approximations, lattice coordinates are taken to be fixed during the exciton-phonon encounter, so that the exciton is regarded as moving in a static random lattice.

The degree of exciton localization may be parameterized in terms of the relative magnitudes of the free exciton bandwidth $(2B)$ and the mean amplitude of the lattice potential fluctuation (D). *Toyozawa* [7.3] has treated the problem of exciton-phonon coupling using damping theory in the weak coupling limit for which

$$D^2/B^2 = g \ll 1. \tag{7.13}$$

The lineshape of the $k=0$ exciton band is essentially Lorentzian, but with an asymmetric tail due to indirect transitions on the high or low frequency edge depending on the position of the $k=0$ exciton state at the bottom or top of the exciton band, respectively. The results are appropriate for $k=0$ at the extreme of the band only, and assume that $k=0$ is nondegenerate [i.e., $E(k)$ is a smooth function about $k=0$].

In the strong coupling limit $(g \gg 1)$, the exciton becomes localized by the fluctuating lattice potential and a Gaussian lineshape was predicted for the $k=0$ transition, with halfwidth varying as $T^{1/2}$.

More recently, *Sumi* [7.5] has treated the exciton-phonon coupling problem using the coherent potential approximation, previously employed to describe the motion of an electron in a random lattice. In Sumi's treatment, the exciton is coupled to phonons at single sites in the crystal. The interaction with phonons appears as a random lattice problem in which random potential fluctuations take on a continuous distribution. The lattice potential, V_n, at each lattice site n is assumed to have a Gaussian distribution, given by

$$P(V_n) = (1 / \sqrt{2\pi}\, D)\, \exp(-V_n^2/2D^2) \tag{7.14}$$

with root-mean-square amplitude D. The energy D represents the average amplitude of the potential fluctuation due to lattice vibrations and is given by

$$D^2 = \frac{1}{2} \sum_q |V(q)|^2 \frac{\coth(\hbar\omega_q/2kT)}{\hbar\omega_q}, \tag{7.15}$$

where V_q is the linear exciton-phonon coupling function for a phonon of frequency ω_q. For $2kT \gg \hbar\omega_q$,

$$D^2 \simeq \sum_q |V_q|^2 \frac{kT}{(\hbar\omega_q)^2}. \tag{7.16}$$

In the weak coupling limit $(g \ll 1)$ the $k=0$ adsorption band has a sharp Lorentzian peak in its main part. Since the lineshape is somewhat asymmetric, it is useful to define two half-widths at half-maximum, $\Delta\omega_1$ and $\Delta\omega_2$, given by

$$(1/2)\, I(\omega_p) = I(\omega_p + \Delta\omega_1) = I(\omega_p - \Delta\omega_2) \tag{7.17}$$

and illustrated in Fig. 7.1. ω_p is the frequency of the intensity peak. *Sumi* has obtained the following approximations for $\Delta\omega_1$ and $\Delta\omega_2$ in the limit $g \ll 1$

$$\Delta\omega_1 \simeq 3.18\, g^2 B, \quad \Delta\omega_2 \simeq 0.61\, g^2 B. \tag{7.18}$$

Both $\Delta\omega_1$ and $\Delta\omega_2$ vary as T^2 at high temperatures. The frequency shift of the peak intensity from its position when $g=0$ has also been given by *Sumi*

$$\Delta\omega_p = -2B\sqrt{g}, \tag{7.19}$$

$\Delta\omega_p$ varies as T' at high temperatures.

In the weak coupling limit, the long wavelength absorption edge of the $k=0$ band falls off exponentially below the absorption peak in agreement with Urbach rule behavior. Moreover, the Lorentzian peak merges smoothly into the Urbach tail to describe the change in character of the exciton from extended to localized.

In the limit of strong exciton-phonon coupling $(g \gg 1)$, the $k=0$ absorption band has a Gaussian shape over its entire frequency range given by

$$I(\omega) = (1/\sqrt{2\pi D})\exp(-\hbar^2\omega^2/2D^2). \tag{7.20}$$

The linewidth (FWHM) is given by

$$\Delta\omega = (2D/\hbar)\sqrt{2\ln(2)} \tag{7.21}$$

and varies as $T^{1/2}$ at temperatures for which (7.16) is a good approximation. This is the same result obtained by *Toyozawa* from a damping model [7.3]. The

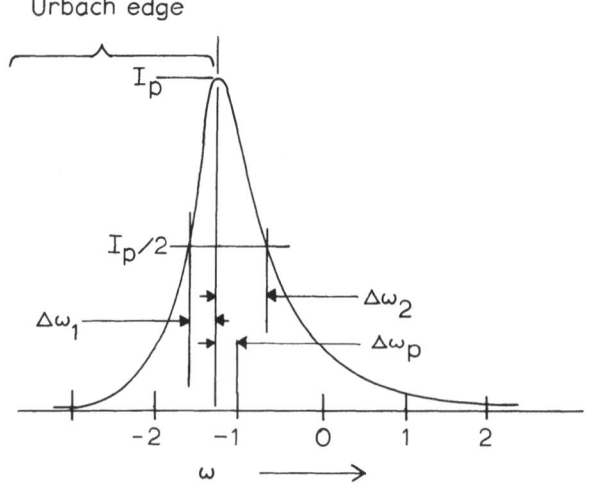

Fig. 7.1. Schematic lineshape for an exciton absorption line when $k = 0$ lies at the bottom of the band

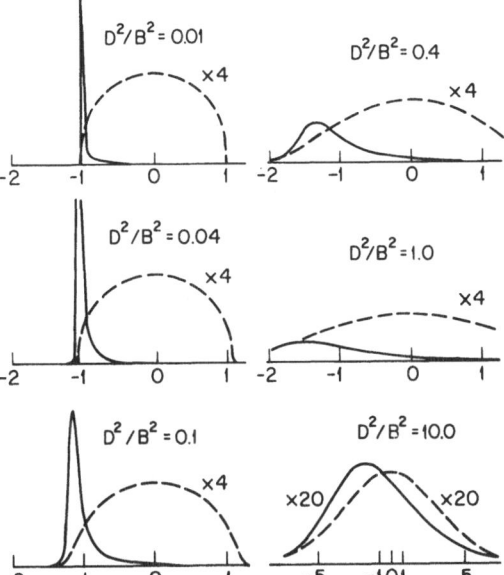

Fig. 7.2. Exciton absorption band (—) and exciton density of states (- - -) calculated for g = 0.01, 0.04, 0.4, 1, 10. The energy is in units of B on the abscissa [7.5]

k=0 exciton absorption lineshape has been computed by *Sumi* for several values of the exciton-phonon coupling strength parameter and the results, appropriate for k=0 at the bottom of the exciton band, are shown in Fig. 7.2.

b) Polaron Models

The theoretical models employed by *Toyozawa* and by *Sumi* deal with the form of the k=0 zero-phonon band and cannot be used to describe the phonon sideband structure which originates in the dynamical character of exciton-phonon coupling.

The details of lineshape due to exciton-phonon coupling in pure crystals are determined by the behavior of the lineshape function

$$I(k,\omega) = \frac{\gamma(k,\omega)}{[\hbar\omega - E(k) - \Delta(k,\omega)]^2 + [\gamma(k,\omega)]^2}, \qquad (7.22)$$

where $\gamma(k,\omega)$ and $\Delta(k,\omega)$ are the real and imaginary parts of $\Sigma(k,\omega)$, the exciton self-energy, which is, in general, both frequency and wave-vector dependent. Comparison of (7.22) with (7.1) reveals that the real part of the self-energy corresponds to the frequency shift term, while the imaginary part determines the linewidth. If $\Sigma(k,\omega)$ is a slowly varying function of frequency, then the lineshape will be approximately Lorentzian, with linewidth (FWHM) of $2\gamma(k,\omega)$ and peak intensity at $\omega=[E(k)+\Delta]/\hbar$. However, the frequency and wave-vector dependence of $\Sigma(k,\omega)$ must be determined in order to ascertain the extent to which deviations from

Lorentzian lineshape will occur. If the exciton-phonon interaction consists of single phonon absorption and emission events only, then the real and imaginary parts of the exciton self-energy are given by the following equations [7.63]:

$$\Delta(k,\omega) = \sum_{s,q} |F_s(kq) + X_s(q)|^2 \left\{ \frac{<n_q>}{\hbar\omega - E(k+q) - \hbar\omega_q} \right.$$
$$\left. + \frac{<n_q> + 1}{\hbar\omega - E(k+q) + \hbar\omega_q} \right\} \tag{7.23}$$

$$\gamma(k,\omega) = \frac{2\pi}{N} \sum_{s,q} |F_s(kq) + X_s(q)|^2 \left\{ <n_q> \delta\,[\hbar\omega - E(k+q) - \hbar\omega_q] \right.$$
$$\left. + (<n_q> + 1)\,\delta\,[\hbar\omega - E(k+q) + \hbar\omega_q] \right\}. \tag{7.24}$$

$E(k+q)$ is the energy of the exciton after interaction with the sth branch phonon of frequency ω_q and $<n_q>$ is the Bose-Einstein phonon distribution function. $F_s(k,q)$ and $X_s(q)$ are the resonance and dispersion exciton-phonon coupling functions, respectively. The exciton-phonon coupling is seen to consist of both a modification of the resonance coupling for energy transfer (resonance coupling) and a lattice deformation about the excited molecule (dispersion coupling).

When the exciton bandwidth is greater than any lattice frequency, excitation causes no local lattice distortion. Changes in lattice structure are distributed over a region of the crystal within which the exciton is delocalized. Under these circumstances, the only exciton-phonon coupling is through the dependence of the resonance interaction on lattice displacement. This is the case of weak coupling, characterized by the absence of phonon sidebands accompanying the $k=0$ transition.

In the strong coupling limit, the exciton bandwidth is comparable to the acoustic phonon dispersion and dispersion coupling causes lattice relaxation about an electronically excited molecule. When dispersion coupling is significant, instead of a single peak at the energy of the $k=0$ transition, there appears a progression corresponding to two-particle exciton-phonon excitations. The lattice deformation accompanying electronic excitation leads to Franck-Condon factors in the transition probability which determine the profile of the exciton-phonon absorption spectrum, those phonon modes becoming spectrally active whose origins are most shifted by the electronic excitation.

Theoretical studies of the exciton-phonon spectrum using Green's function methods [7.9] have been conducted by *Fisher* and *Rice* [7.6], *Grover* and *Silbey* [7.7] and, most recently, by *Craig* and *Dissado* [7.8], the collective exciton-phonon excitation is treated as a polaron-type quasiparticle and both resonance and dispersive exciton-phonon coupling are included explicitly.

In the strong dispersive coupling limit at high temperatures, individual phonon structure is not resolved and the absorption spectrum consists of a Gaussian
envelope. *Craig* and *Dissado* have observed that a Gaussian absorption envelope
is obtained in the high-temperature limit only when the lifetime imposed by phonon damping is less than or comparable to $1/\omega_q$, for the spectrally active phonons. In general, they regard the occurrence of a Gaussian envelope as the result
of an unusual combination of circumstances. The results of *Craig* and *Dissado*
may be summarized in terms of the parameters, $F(k,q)$, the resonance exciton-
phonon coupling function, $X(q)$, the dispersive coupling function, ω_q, the frequency of the spectrally active phonons and $\gamma(k)$ the phonon damping constant.

When $\gamma > (\omega_q/2)$ the exciton-phonon spectrum shows no phonon structure; when $F/X < 0.1$, the spectrum has a near-Gaussian shape; when $F/X > 1$,
the band profile has a Lorentzian shape for values of $\gamma > (\omega_q/2)$ with linewidth
equal to 2γ (FWHM); when $F/X \sim 0.2$ the structureless band profile is intermediate in character and asymmetric.

c) Lineshape Moments

Morris and *Sceats* [7.10] have applied the method of moments to the extraction
of the parameters $F(k,q)$ and $X(q)$ for exciton-phonon coupling in molecular crystals. The moments of the absorption lineshape function are given by the relation

$$\mu^m = (1/2\pi) \int_{-\infty}^{+\infty} \omega^m I(\omega) d\omega. \tag{7.25}$$

Thus, the zeroth moment is the integrated intensity of the band; the first moment is the mean frequency of the absorption band, the second moment is the
mean squared width of the absorption band and the third moment is a measure
of the asymmetry of the absorption band. Experimentally, it is difficult to determine lineshapes with sufficient accuracy to determine moments greater than the
third. *Morris* and *Sceats* have obtained expressions for the first five lineshape
moments using three different models for exciton-phonon coupling: A weak
coupling model, a strong coupling model and a polariton model. For the weak
coupling model,

$$\mu^{(1)} = E(k) = \omega_{AV} \tag{7.26}$$

$$\mu^{(2)} = \sum_{s,q} |F_s(k,q)|^2 \coth[\hbar\omega_s(q)/2kT] \tag{7.27}$$

where $F_s(q,k)$ is the exciton-phonon coupling function for an sth branch phonon
of wave vector q and frequency $\omega_s(q)$. The zeroth and first moments of the spectral lineshape $I(\omega)$ are independent of temperature in the weak coupling model.
At $T=0$, the second moment is

$$\mu^{(2)} = \sum_{s,q} |F_s(kq)|^2.$$ (7.28)

When $kT \gg \hbar\omega_s$, the second moment is approximately given by

$$\mu^{(2)} \simeq \sum_{s,q} |F(kq)|^2 [2kT/\hbar\omega_s(q)]$$ (7.29)

so that a plot of $\mu^{(2)}$ versus T should be linear at high T with a slope of

$$\sum_{s,q} |F(kq)|^2 [2k/\hbar\omega_s(q)].$$ (7.30)

d) Disorder Scattering

Klafter and *Jortner* [7.9] have investigated the effects of structural disorder on the absorption lineshape of the $k=0$ transition at $T=0$. The effect of structural disorder is simulated by a Gaussian distribution of site excitation energies about the average crystal excitation energy $<E>$. Thus,

$$E_n = <E> + V_n$$ (7.31)

where V_n is given by (7.14) with D the width of the distribution. To low energies of the direct $k=0$ transition edge, assumed to be at the bottom of the exciton band, the lineshape is determined by the density of states of the ideal crystal and therefore depends critically upon the dimensionality of the exciton band structure. In the weak coupling approximation ($g \ll 1$) a quasi-Urbach behavior is found in two- and three-dimensional exciton band structures over a limited energy range. In the one dimensional case the tail exhibits a nearly Gaussian energy dependence

$$I(\omega) \propto \exp[-(\hbar^2\omega^2-1)/2D^2]$$ (7.32)

e) Phonon Exchange Model

Harris [7.11] has developed a theory for exciton absorption lineshapes based on a phonon exchange which is similar to the exchange theory of *Kubo* [7.12] and *Anderson* [7.13], employed in magnetic resonance. The exchange process involves the resonant absorption and emission of a single, dispersionless optic branch phonon by the $k=0$ exciton. The temperature-dependent line shift and linewidth arise from the inclusion of quadratic exciton-phonon coupling which gives rise to a frequency shift between the ground and excited state phonon frequencies and leads to dephasing of the $k=0$ exciton. This mechanism does not lead to a loss of coherence of the $k=0$ exciton and *Harris* has concluded that optical linewidths associated with the $k=0$ state need not, therefore, measure features of the exciton coherence at low temperatures. The following expressions for the temperature dependence of the linewidth and frequency shift are obtained from the exchange model:

$$\Delta\omega = \frac{(\delta\omega)^2 \tau^2 W^+}{1+(\delta\omega)^2\tau^2} \tag{7.33}$$

$$\Delta\omega_p = \frac{\delta\omega\tau W^+}{1+(\delta\omega\tau)^2} \tag{7.34}$$

where $\delta\omega$ is the phonon frequency shift due to quadratic exciton-phonon coupling, τ is the lifetime of the excited state phonon of frequency ω_i and W^+ is the phonon absorption rate, given by

$$W^+\tau \simeq [\exp(\hbar\omega_i/kT{-}1]^{-1}. \tag{7.35}$$

7.1.2 Lineshape Studies in Pure Crystals

The number of experimental lineshape investigations in pure molecular crystals, although not excessively large, is too great to attempt to cover in detail in this chapter. We give, therefore, some results representative of the body of the experimental work in this area.

a) Naphthalene

In naphthalene, the lowest singlet state exciton bandwidth is approximately 160 cm^{-1} ($-2B$) with the $k{=}0$ level at the bottom of the exciton band. *Robinette* et al. [7.14] have conducted a careful investigation of the width and frequency shifts of several absorption bands in the $S_1 \leftarrow S_0$ absorption spectrum of naphthalene single crystals. Of particular interest are the results obtained for the low-frequency a-polarized Davydov component of the electronic origin. Since there is little or no phonon sideband structure associated with this transition, it is classified in the weak coupling limit. At low temperatures and in strain-free crystals, the linewidth is on the order of 0.4 cm^{-1} (FWHM) and exhibits a marked asymmetry. The high frequency edge is close to Lorentzian at all temperatures, however, the low frequency edge falls more *slowly* than Lorentzian, as illustrated in Fig. 7.3. Thus, the lineshape theories of *Toyozawa* and *Sumi* , while correctly predicting the appearance of asymmetry in the a-band, incorrectly predict the behavior of the low frequency edge.

Based on the available data, the a-band appears to be homogeneously broadened and therefore the linewidth is a direct measure of the optical dephasing time for the $k{=}0$ exciton. The linewidth (FWHM) is observed to increase linearly

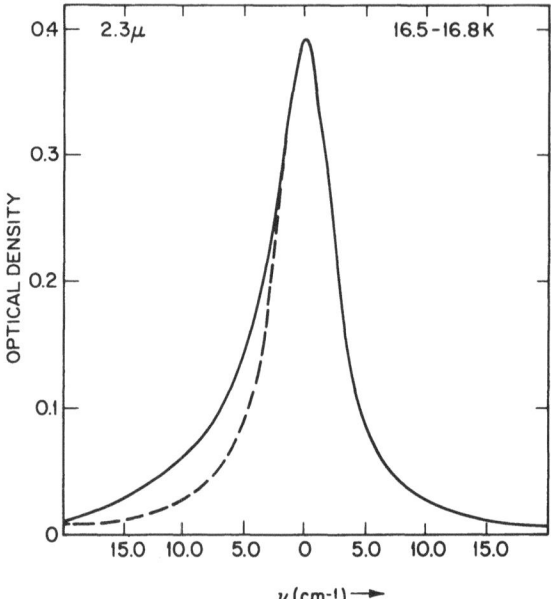

Fig. 7.3. Naphthalene *a*-band optical density profile at 16.6 K [7.14]

with temperature above 15 K (Fig. 7.4), consistent with the upward scattering of the $k=0$ exciton via one-phonon absorption processes. The 13 ps optical dephasing time of the $k=0$ exciton at 2.2 K is inferred from the linewidth of 0.4 cm^{-1} (FW-HM).

The integrated intensity of the *a*-band in crystals which are free from strain and of high chemical purity, exhibits a temperature dependence in agreement with that predicted for polariton excitation. For a crystal assumed to have infinite dimensions, each exciton mode is coupled to only one radiation mode and, according to time-dependent perturbation theory, no transition can occur in the combined system [7.15]. Excitation depends upon a final density of crystal states created in a two-step process by scattering of the crystal polariton. The scattering mechanism may include impurity scattering, phonon scattering or energy transfer. However, in a sufficiently pure crystal, with $k=0$ at the bottom of the exciton band, and for very low temperatures, single-phonon upward scattering may dominate. In this event, absorption is expected to increase with increasing temperature, in marked contrast to the temperature dependence of the zero-phonon line behavior predicted for chemically mixed crystals from a linear phonon coupling model (see Sect. 7.1.3a). This is precisely the behavior observed [7.16] for the *a*-band of naphthalene. The linear dependence of the linewidth upon temperature indicates that the principal exciton-phonon scattering mechanism is single phonon absorption in the temperature range 1.6-50 K. Over this temperature range, the integrated intensity of the *a*-band is observed to increase tenfold. It is noteworthy that this behavior is not observed in naphthalene crystals doped with beta-methyl-

naphthalene chemical impurity traps, where the absorption intensity is independent of temperature. Evidently energy transfer is sufficiently rapid to damp the
initially created polariton.

b) 1,2,4,5-Tetrachlorobenzene (TCB)

TCB crystallizes in a monoclinic space group with two molecules per unit cell.
At 188 K a phase transition occurs to a triclinic modification, closely related in
unit cell dimensions and molecular orientation to the high temperature monoclinic structure. In both structures, translationally equivalent molecules stack
very nearly plane-to-plane along the crystal a-axis with the molecular out-of-plane
axis almost parallel to a. The structure, with its short interplanar spacing of 3.76
Å between translationally equivalent molecules along the a axis, exhibits one-dimensional exciton behavior. It has been determined that triplet exciton behavior in
TCB is essentially one-dimensional with the direction of propagation parallel to
a [7.17]. The one-dimensional exciton bandwidth is estimated to be 1.4 cm^{-1}
with the $k=0$ level at the *top* of the band. *Burland* et al. [7.18] have examined
the temperature dependence of the absorption line-width of the electronic origin

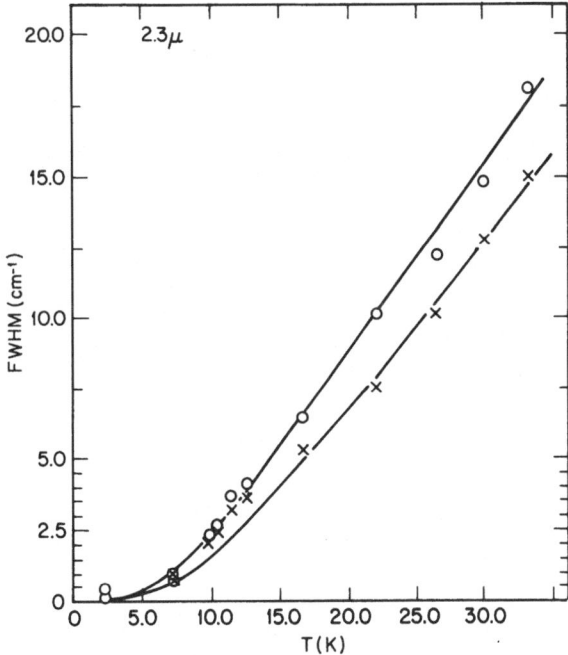

Fig. 7.4. Thermal broadening of the naphthalene a-band. Circles are experimental linewidth
(FWHM) and crosses are the linewidth (FWHM) inferred from the Lorentzian high energy
absorption edge. [7.14]

band for the $T_1 \leftarrow S_0$ transition. The lineshape below about 10 K (Fig. 7.5) is described as Lorentzian on the low frequency edge and Gaussian on the high frequency edge with a linewidth of 1.3 cm^{-1} (FWHM). The reversal of the Lorentzian edge is consistent with the position of the $k=0$ level at the top of the exciton band. The Gaussian high frequency edge, however, suggests that the lineshape is inhomogeneously broadened, at least on the high frequency edge. The authors estimate the low temperature homogeneous linewidth to be approximately 0.6 cm^{-1}, corresponding to an optical T_2 on the order of 20 ps. As the temperature is increased, the origin lineshape becomes more symmetrical until, at 14 K, it is accurately Lorentzian with a linewidth of 1.3 cm^{-1} (FWHM) corresponding to an optical dephasing time of T_2 = 8 ps for the $k=0$ exciton.

At still higher temperatures (> 20 K) the lineshape becomes Gaussian. *Burland* et al. suggested that the Gaussian lineshape obtained at 20 K is the result of a transition from weak to strong exciton-phonon coupling. In the weak coupling limit $D^2/B^2 = g \ll 1$, where $2B$ is the free exciton bandwidth and D is the mean amplitude of the lattice potential fluctuation given by (7.14). D will increase rapidly with temperature and for B = 0.7 cm^{-1}, it is likely that the condition for strong exciton-phonon interaction ($D^2/B^2 = g > 1$) will be fulfilled at moderately low temperatures at which point the Gaussian lineshape given by (7.20) should be observed.

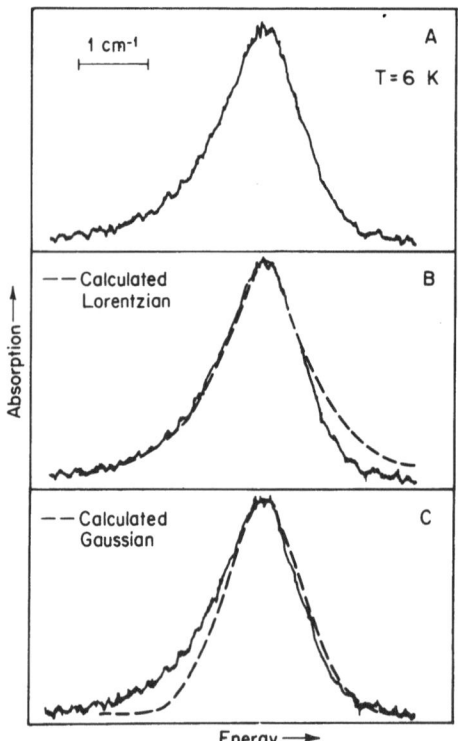

Fig. 7.5. Triplet exciton absorption lineshape for the electronic origin of *TCB* at 6 K [7.18]

Burland et al. have determined the temperature dependence of both the peak absorption frequency and linewidth for the electronic origin. Interpretation of the data is complicated by the fact that the magnitude of the Davydov splitting is unknown. It is therefore possible that the exciton bands of the two Davydov components are nearly overlapping in the region of the $k=0$ absorption band so that inter-branch exciton scattering is possible. The observed temperature dependence of the shift and linewidth are not readily interpreted within existing theoretical models.

c) 1,4 Dibromonaphthalene

There has been a considerable amount of experimental information gathered concerning the behavior of triplet excitons in 1,4 dibromonaphthalene (DBN). It has been established [7.19] that the exciton behavior is essentially one-dimensional and propagation is parallel to the crystallographic c axis. The bandwidth of the 0–0 transition has been determined to be 29.6 cm^{-1} with $k=0$ at the bottom of the exciton band.

The lineshape of the $k=0$ transition in DBN is asymmetric at low temperatures but quite accurately Lorentzian on its high energy edge (Fig. 7.6a). At higher temperatures (25 K) the lineshape is accurately Lorentzian (Fig. 7.6b). The linewidth of the 0–0 transition band at 2 K is 0.3 cm^{-1} FWHM. At low temperatures optical branch phonon scattering of the $k=0$ excitons is not possible in DBN. Since the triplet exciton dispersion in DBN is smaller than the acoustic phonon dispersion, scattering by single acoustic phonons with simultaneous conservation of crystal momentum and energy is also not possible.

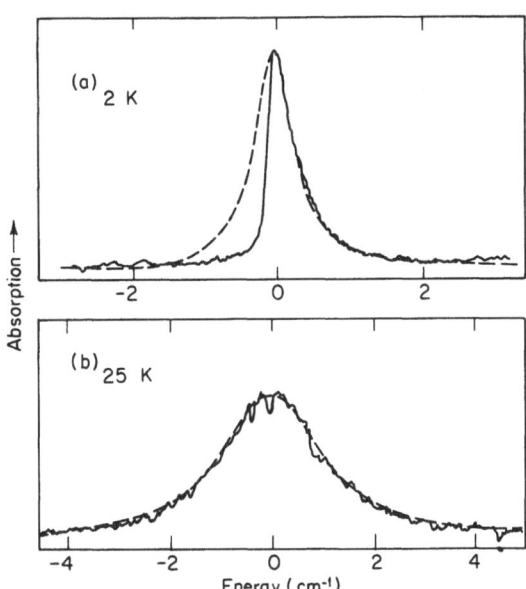

Fig. 7.6. Triplet exciton absorption lineshapes for the electronic origin of DBN. The dashed line in both spectra is a symmetric Lorentzian. [7.20]

Burland et al. [7.20] have investigated the scattering of $k=0$ triplet excitons in DBN single crystals through the temperature dependence of the absorption lineshape and concluded that at temperatures below 10 K scattering is due to lattice defects to which C^{13} isotopic impurity makes a significant contribution. Above 10 K, $k=0$ excitons are scattered by optical branch phonons. The correlation time for the exciton-impurity process is about 60 ps at 2 K. From the band dispersion the group velocity of the triplet exciton at 2 K is calculated to be $2.7 \cdot 10^4$ cm/s corresponding to a mean free path of 160 Å. *Harris* [7.21] has analyzed the DBN $T_1 \leftarrow S_0$ absorption lineshape function in terms of exchange theory (see Sect. 7.1.1d). The application of (7.33, 34) to the data suggests that very little dephasing of the optical transition occurs below about 15 K. Above 40 K, the temperature dependence of the frequency shift and linewidth can be analyzed in terms of optical dephasing by absorption and emission of a single optical branch phonon of about 38 cm^{-1}. *Harris'* theoretical fit to the data obtained by *Burland* et al. is shown in Fig. 7.7. A good fit to the experimental data was also obtained by *Burland* et al. using (7.23, 24) and assuming a scattering by absorption of a single dispersionless optical branch phonon of about 17 cm^{-1}.

d) Anthracene

Morris and *Sceats* [7.10] have examined exciton-phonon interactions in the 0–0 band of the 4000 Å, $S_1 \leftarrow S_0$ transition of anthracene. The bare exciton bandwidth in anthracene is about 1250 cm^{-1} [7.22]. There are two molecules per unit cell resulting in a Davydov splitting of the origin into a low frequency b-polarized component with $k=0$ near the bottom of the band and an a-polarized component to higher frequency with $k=0$ within the band. *Morris* and *Sceats* have conducted a moments analysis of the spectral lineshape function obtained from measurement of the reflectance spectrum of the a-band and b-band.

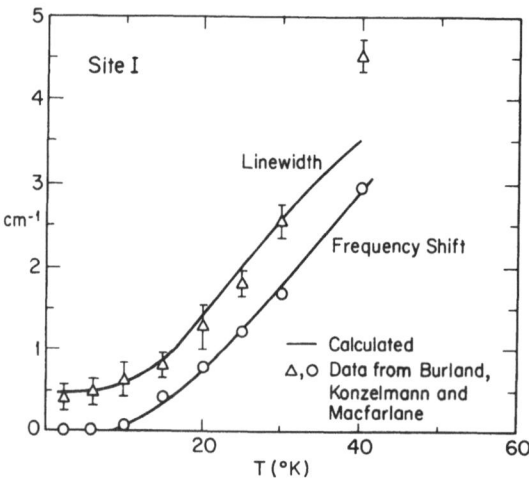

Fig. 7.7. Experimental and exchange model theoretical fit of the temperature-dependent linewidth and frequency shift of DBN. [7.21]

From the temperature dependence of the second moment, $\mu^{(2)}$, (7.27), they concluded that the exciton-phonon interaction is in the weak coupling limit. The exciton-phonon coupling strength, $F_s(k,q)$, obtained from extrapolation of $\mu^{(2)}$ to $T=0$, (7.28), is 140 cm^{-1} and is consistent with the weak coupling model employed. Assuming wave-vector independent scattering by a single dispersionless phonon, a frequency of 90 cm^{-1} is obtained for the "active" phonon from the slope of a plot of $\mu^{(2)}$ versus T in the high-temperature limit (7.29). *Morris* and *Sceats* [7.23] have also determined the correlation time for the $k=0$ dipole oscillator (7.8) from the response function (7.7) for the a-band and b-band $k=0$ excitons. The response functions are obtained from the Fourier transformation of the experimental lineshape functions (see Sect. 7.1.1).

The exciton correlation time determined by phonon scattering is given by

$$\tau_c^{-1} = \gamma(k,q) \tag{7.36}$$

where $\gamma(k,q)$ is defined in (7.24). Exciton-phonon interaction for the b-band $k=0$ excitons in anthracene must involve principally single-phonon absorption. Therefore, when the high-temperature approximation for $< n_q >$ is appropriate, $(\tau_c)^{-1}$ varies linearly with temperature, provided $F_s(k,q)$ is temperature independent. At $T=0$, $1/\tau_c=0$ for b-band $k=0$ excitons and the exciton lifetime due to phonon scattering is infinite. For the a-band, however, $k=0$ does not lie at the bottom of the band and both phonon absorption and emission processes contribute to scattering of the $k=0$ exciton. Therefore, even at $T=0$ a finite τ_c is observed due to the possibility of phonon emission.

An approximately linear variation of $1/\tau_c$ with temperature is observed for both the a-band and b-band $k=0$ excitons. The b-band correlation time extrapolates to infinity at $T=0$. *Morris* and *Sceats* have reported correlation times for the a-band and b-band $k=0$ excitons of 0.1 ps and 0.4 ps, respectively.

7.1.3 Chemically Mixed Crystal Lineshape Theory

In chemically mixed crystals, the details of lineshape of the zero-phonon band and phonon sideband spectrum may be developed from the lineshape function

$$I(\omega) = B < \sum_m |<in |M_{if} |fm>|^2 \, \delta \, (\omega_{fm} - \omega_{in} - \omega)>_n \tag{7.37}$$

where $|in>$ and $|fm>$ are the lattice vibrational wave functions for the initial and final electronic states of the guest molecule, respectively, and $< >_n$ denotes a thermal average over initial lattice vibrational levels. B is a weakly frequency dependent factor, which in the narrow-band approximation is taken to be a constant and M_{if} is the transition electric dipole moment operator. The ground-state lattice vibrational wave functions are of the form

$$|in> = \prod_s |in_s> \tag{7.38}$$

where the functions $| \, in_s >$ are harmonic oscillator wave functions for the sth independent oscillator normal mode in the nth vibrational state and ith electronic state.

In order to evaluate the matrix elements of the dipole transition moment operator appearing in (7.37), it is necessary to obtain the excited-state adiabatic vibrational wave functions $| \, fm >$. Since the excited-state normal coordinates are not in general known, it is usual to expand the excited-state adiabatic potential in the ground-state normal coordinates about the ground-state equilibrium lattice position. Thus, we obtain for the excited-state adiabatic potential

$$E_f(Q) = E_{if}^0 + \sum_s A_s Q_s + \frac{1}{2} \sum_{sr} B_{sr} Q_s Q_r \qquad (7.39)$$

where A_s and B_{sr} are the linear and quadratic electron-phonon coupling parameters. If all B_{sr} with $s \neq r$ were zero, the vibrational wave functions for the excited state would be a product of harmonic oscillator functions differing from the ground-state functions in their force constants, due to the quadratic terms B_{ss} with $s=r$, and in their equilibrium positions, due to the terms linear in Q_s. The cross terms in Q_s and Q_r arise from a mixing of normal coordinates in the electronic excited state. In certain cases of high site symmetry they may be shown to vanish identically; however, in many cases of interest they will not.

a) Linear Electron-Phonon Coupling Model

The linear term in the expansion of the adiabatic potential leads directly to a temperature dependence of the intensity of the zero-phonon line and the multiphonon sidebands associated with the zero-phonon band. The quadratic term produces a temperature-dependent width and position of the zero-phonon line. The simplest possible model for the electron-phonon interaction assumes only a linear electron-phonon coupling and was first considered by *Muto* [7.24], *Huang* [7.25], and *Pekar* [7.26]. These calculations were later extended to include both linear and diagonal quadratic phonon coupling by *Lax* [7.27], *O'Rourke* [7.28], and *Keil* [7.29]. Much of the above work is summarized in a review by *Dexter* [7.30]. Electron-phonon coupling in the linear coupling approximation has been treated extensively by *Rebane* [7.31]. The quadratic phonon coupling has been investigated in molecular crystals by *Small* [7.32].

The dependence of the transition-dipole moment upon lattice coordinates may be included by expanding the matrix element of the transition-dipole moment in the ground-state lattice coordinates and retaining only the first two terms

$$M_{if}(Q) = M_{if}^0 + \sum_s \left(\frac{\partial M}{\partial Q_s} \right)_0 Q_s. \qquad (7.40)$$

The second term in (7.40) will be important only for those cases where the first term vanishes due to electric-dipole selection rules. For an electric-dipole-allowed transition, only the first term is retained and leads to the following expression for the intensity and position of the band corresponding to the transition from the state $|in>$ to the state $|fm>$:

$$I_{nm}(\omega) = B \mid M^0 \mid^2 Z^{-1} \prod_s \exp(-\hbar\omega_s n_s/kT) \mid <in_s \mid fm_s>\mid^2 \delta\,(\omega_{fm}-\omega_{in}-\omega) \tag{7.41}$$

where Z is the partition function for the lattice vibrational states of the initial electronic state. Thus, the line corresponding to an electronic transition from an initial state characterized by a set of quantum numbers n for the independent lattice oscillators, to a final state characterized by a set of quantum numbers m, consists of a single delta function at frequency $\omega = \omega_{fm} - \omega_{in}$. The intensity of the zero-phonon line is obtained from (7.41) by setting $n=m$ and summing over all possible n to obtain

$$I_{Z.P.} = B \mid M^0 \mid^2 \exp\left[-\sum_s \frac{P_s}{\hbar\omega_s}\,\coth(\hbar\omega_s/2kT)\right] \tag{7.42}$$

where P_s is the Stokes loss for the sth lattice oscillator, defined as

$$P_s = 1/2 \;[\omega_s^2 m_s (A_s/B_{ss})^2]. \tag{7.43}$$

The intensity of the zero-phonon line is, therefore, expected to decrease rapidly with increasing temperature.

The ratio of the integrated intensity of the zero-phonon band to the total integrated intensity of the entire band system (i.e., zero-phonon and phonon sidebands) depends only upon temperature and the total Stokes loss at the present level of approximation. Moreover, the total integrated intensity of the system is independent of temperature; intensity which is lost from the zero-phonon line with increasing temperature is distributed to the phonon sidebands.

b) Electric-Dipole-Forbidden Transitions

When the electronic transition is forbidden by dipole selection rules, the first term of (7.40) vanishes and the transition dipole strength is determined by the dependence of the transition dipole moment operator upon the lattice coordinates contained in the second term. In the linear phonon coupling approximation, *Rebane* has shown that the integrated intensity of the entire forbidden band will vary as

$$I = \sum_s \left(\frac{Bh}{2m_s\omega_s}\right)\left(\frac{\partial M_{if}}{\partial Q_s}\right)_0^2 \coth(\hbar\omega_s/2kT). \tag{7.44}$$

Therefore, the integrated intensity for a forbidden electronic transition is a strongly increasing function of temperature.

c) McCumber-Sturge Model

The lineshape theory outlined above is useful in predicting the temperature dependence of the zero-phonon line intensity and the intensity distribution of the multiphonon sideband; however, it does not deal with the lineshape or the temperature dependence of the frequency shift of the zero-phonon line. The individual features are delta functions which, in reality, would be broadened by phonon interactions. The frequency shift and linewidth of the zero-phonon line originate in the diagonal (B_{ss}) and off-diagonal $(B_{rs}, r{\neq}s)$ electron-phonon coupling terms in (7.39). More specifically, the diagonal electron-phonon coupling leads to a temperature-dependent frequency shift when $B_{ss}-k_s = \Delta_s{\neq}0$, where k_s is the ground-state harmonic force constant for the sth lattice oscillator [7.38]. The lineshape and frequency shift of the zero-phonon line due to quadratic electron-phonon coupling has been considered by *McCumber* and *Sturge* [7.33]. If the phonon interaction is restricted to Raman processes associated with the scattering of phonons by the impurity molecule, a Lorentzian lineshape is obtained. *McCumber* and *Sturge* obtained the following expression for the temperature dependence of the Lorentzian halfwidth $(\Delta\omega)$ (FWHM) and the peak frequency shift $(\Delta\omega_p)$ of the zero-phonon line:

$$\Delta\omega \cong \int_0^{\omega D} \frac{\Delta(\omega)g(\omega)}{\omega[\exp(\hbar\omega/kT)-1]} \, d\omega \tag{7.45}$$

$$\Delta\omega_p \cong \int_0^{\omega D} \left(\frac{B(\omega)\exp(\hbar\omega/2kT)g(\omega)}{\omega[\exp(\hbar\omega/kT)-1]} \right)^2 \, d\omega \tag{7.46}$$

where ω_D is the effective Debye frequency of the lattice. The diagonal $[\Delta(\omega)]$ and off-diagonal $[B(\omega)]$ electron-phonon coupling have been treated as continuous functions of frequency and $g(\omega)$ is the phonon density of states. If $g(\omega)$ is approximated by a Debye density of states and both $\Delta(\omega)$ and $B(\omega)$ assumed to vary quadratically with frequency, then $\Delta\omega \propto T^7$ and $\Delta\omega_p \propto T^4$ in the limit $\hbar\omega \gg kT$. *Small* [7.38] has pointed out that the McCumber-Sturge theory for optical linewidths in chemically mixed crystals leads directly to the prediction of an exponential dependence of linewidth upon temperature if the off-diagonal quadratic electron-phonon coupling becomes deltoid in its frequency spectrum. If we let $B(\omega) = \delta(\omega-\omega_0)$, *Small* has shown that

$$\Delta\omega \simeq \exp(-\hbar\omega_0/kT) \quad \hbar\omega_0 \gg kT. \tag{7.47}$$

The frequency shift will also vary exponentially with temperature if the diagonal phonon coupling is given a deltoid dependence. For $\Delta(\omega) = \delta(\omega-\omega_0)$,

$$\Delta\omega_p \simeq \exp(-\hbar\omega_0/kT) \quad \hbar\omega_0 \gg kT. \tag{7.48}$$

7.1.4 Lineshape Studies in Chemically Mixed Crystals

Studies of impurity molecule absorption lineshapes in chemically mixed crystals provide much the same information about dynamical processes in excited states as lineshape studies in pure crystals. In the case of chemically mixed crystals, lineshape studies are frequently used to extract information concerning the relative magnitude of linear and quadratic electron-phonon coupling. The temperature dependence of absorption lineshapes is related to optical T_1 and T_2 processes through (7.1, 2). In the following sections, we summarize the results of several representative experimental studies of lineshapes in chemically mixed systems.

a) Naphthalene Chemically Mixed Crystals

Figure 7.8 illustrates a naphthalene vibronic absorption band obtained in a biphenyl host lattice at 2 K [7.34]. This spectrum provides an excellent example of the effect of large linear electron-phonon coupling in a molecular crystal. The zero-phonon line at 21804 cm^{-1} is not visible in the absorption spectrum. A progression of more than fifteen members in a 12.5 cm^{-1} localized phonon mode builds to a maximum in the 10th member. The large linear electron-phonon coupling term for this transition represents a displacement of the excited-state potential hypersurface along a librational lattice coordinate. From a quantitative application of the Franck-Condon principle to the intensity distribution in the phonon progression, it is possible to obtain the amplitude of the librational displacement upon electronic excitation. Perdeuteration of the guest naphthalene molecule results in a 4.5% decrease in the phonon frequency. A completely localized harmonic guest phonon model predicts the mass loading effect for the naphthalene libration to be 6-9% depending upon the axis of libration.

The phosphorescence electronic origin of naphthalene in durene, reproduced in Fig. 7.9, provides an example of intermediate linear electron-phonon coupling involving a 17 cm^{-1} librational mode. Here, the zero-phonon line is

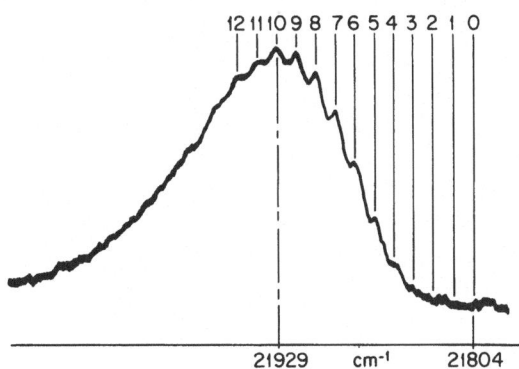

Fig. 7.8. Profile of the 433 cm^{-1} (b_{3g}) absorption vibronic origin of naphthalene in biphenyl at 2 K [7.34]

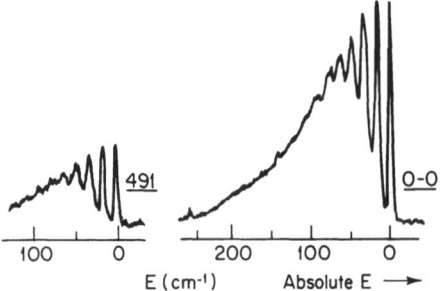

Fig. 7.9. Profile of the electronic phosphorence origin of naphthalene in durene at 2 K [7.34]

the most intense feature of the spectrum, thus the displacement of the excited-state potential is less than the zero-point amplitude of the librational motion.

b) p-Benzoquinone in p-Dibromobenzene

The lowest $T_1 \rightarrow S_0$ transition of p-benzoquinone (PBQ) is electric dipole forbidden by orbital parity as well as spin angular momentum selection rules. Therefore, the phosphorescence spectrum of PBQ in p-dichlorobenzene (DCB) provides an excellent opportunity to investigate phonon interaction with a forbidden electronic transition. The transition gains dipole character via Hertzberg-Teller vibronic coupling and consists, therefore, of vibronic origin bands. A zero-phonon vibronic origin band and associated phonon sidebands from the phosphorescence spectrum of PBQ at 5 K [7.35] are shown in detail in Fig. 7.10. The thermal modulation spectrum, which displays the change in phosphorescence intensity with temperature [7.36], is also shown in Fig. 7.10. The lowermost spectrum in Fig. 7.10 is the integrated intensity of the thermal modulation spectrum. The most intense feature of the spectrum is the zero-phonon band of the vibronic origin. Built upon this line is a single, well-defined progression in a localized lattice mode of about 21 cm^{-1} which can be followed through approximately five members. The active phonon mode represents a lattice libration and its appearance is consistent with intermediate linear electron-phonon coupling in this mode.

The thermal modulation spectrum and the integrated thermal modulation intensity illustrate the role of electron-phonon coupling in the formation of the vibronic origin band system. The effect of increasing temperature is to broaden the zero-phonon band predominantly to higher frequencies. The integrated intensity of the thermal modulation band system is identically zero. Therefore, the effect of small changes in temperature on the vibronic origin band system is to redistribute the dipole intensity within the band system and not to enhance or diminish it. This is the result anticipated in the Condon approximation for an allowed transition of an impurity site.

The phosphorescence electronic origin serves as an example of a parity-forbidden electric dipole transition. The region of the phosphorescence spectrum about the electronic origin is shown in Fig. 7.11, together with the thermal modu-

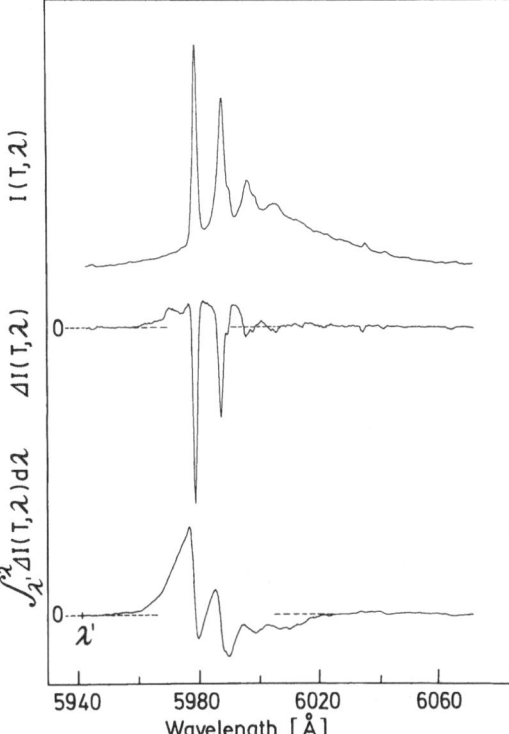

Fig. 7.10. *Upper*: Profile of the electro 1655 cm^{-1} (b_{1u}) vibronic phosphorescence origin of BQ in DCB at 5 K. *Center*: Thermally modulated phosphorescence spectrum. *Lower*: Integrated thermal modulation spectrum. [7.36]

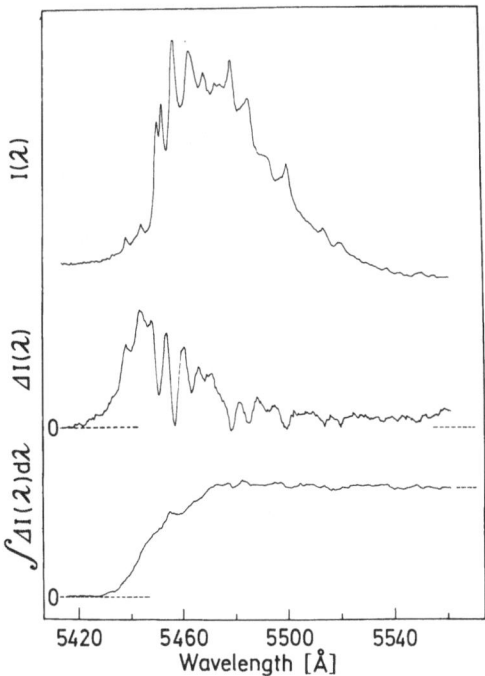

Fig. 7.11. *Upper*: Profile of the electronic origin of phosphorescence electronic origin of BQ in DCB at 5 K. *Center*: Thermally modulated phosphorence spectrum. *Lower*: Integrated thermal modulation spectrum. [7.36]

lation spectrum and the integrated thermal modulation intensity. The zero-phonon electronic origin is rigorously forbidden by electric-dipole selection rules for a PBQ molecule located substitutionally at centrosymmetric sites in the DCB. Thus, the spectrum consists of approximately fifteen phonon sidebands, but the electronic origin itself is not observed. The appearance of phonon sidebands associated with the forbidden electronic origin represents a deviation from the Condon approximation and is due to the dependence of the electric-dipole transition moment upon lattice coordinates given in (7.40). The phonon modes observed correspond to modes which destroy the inversion symmetry of the site, or ungerade localized phonon modes. The increase in intensity with increasing temperature, evident in the thermal modulation spectrum, is due to thermal population of ungerade modes in the initial state. The thermally induced intensity in the region of the forbidden electronic origin is in agreement with (7.44) and characteristic of electron-phonon interaction in an electric dipole forbidden transition.

c) Phenylmonoasaazulene (PMAA) in p-Terphenyl (TP)

The examples discussed above illustrate the relationship between the intensity of the zero-phonon line and the associated phonon sidebands as a function of the linear electron-phonon interaction. The temperature dependence of the frequency

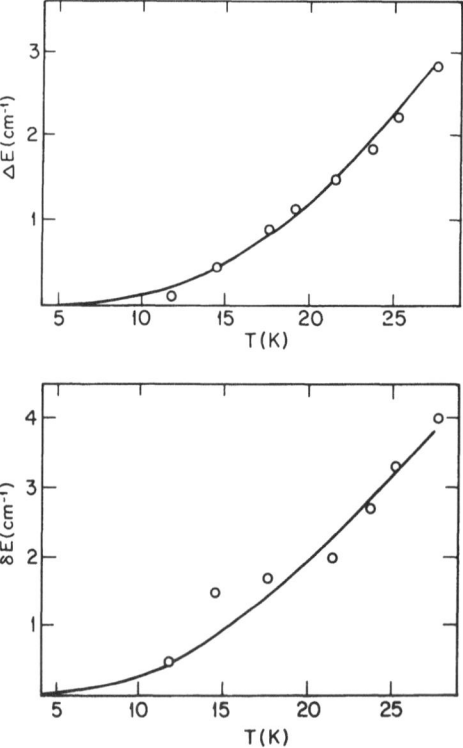

Fig. 7.12. Band broadening and band shift vs temperature for PMAA in TP. Solid lines are theoretical fit to the experimental data. [7.37]

and linewidth of the zero-phonon line in chemically mixed crystals may be illustrated using studies of PMAA in TP reported by *Burke* and *Small* [7.37].

Figure 7.12 shows the temperature dependence of $\Delta\omega_p$ and $\Delta\omega$ for the $S_1 \leftarrow S_0$ electronic origin of PMAA in TP. The solid line fits are obtained from (7.45, 46) using Δ, B, and ω_D as adjustable parameters.

d) Hole Burning and Fluorescence Line Narrowing (FLN)

Both hole burning and FLN experiments have been discussed in detail in Chap. 4 (see also Chap. 6) and the interested reader is referred to that discussion for historical background and the experimental and theoretical details of these techniques. Our purpose here is to gather some representative data from hole burning and FLN experiments on lineshapes in molecular crystals. There are, in fact, relatively few examples of either photochemical/photophysical hole burning or FLN in pure or chemically mixed crystals.

The first photochemical hole burning experiment on a chemically mixed crystal was reported by *DeVries* and *Wiersma* [7.39] for 1,2,4,5 tetrazine in benzene at 2 K. The inhomogeneous linewidth of the $S_1 \leftarrow S_0$ electronic origin at 2 K is on the order of 0.3 cm^{-1} and therefore typical of many molecular solids at this temperature. The linewidth of the photochemically burned hole in the inhomogeneously broadened electronic origin line was 700 MHz (FWHM) at 2 K, corresponding to a T_2 of about 455 ps. According to the authors, T_2 is lifetime limited at 2 K, thus, $T_2^* \gg T_1$ and $T_2 \cong T_1$. The T_2 of an excited vibrational level of the S_1 state was found to be 25 ps.

The above results may be contrasted with the hole burning experiments conducted on the $S_1 \leftarrow S_0$ electronic origin band of dimethyl tetrazine in durene [7.40]. The absorption electronic origin exhibits an inhomogeneously broadened Gaussian lineshape with a linewidth of 0.25 cm^{-1} (FWHM). A vibronic origin of the same system exhibited a Lorentzian lineshape with linewidth of 1 cm^{-1} (FWHM). Hole burning was possible only on the inhomogeneously broadened electronic origin. The hole burned had a linewidth of 120 MHz (FWHM) from which a width of 55 MHz was computed as an upper limit for the homogeneous linewidth, after correction for laser linewidth and frequency jitter. T_1 was determined to be on the order of 25 MHz, so that T_2 is about 30 MHz. Thus, T_2 is very nearly equal to its lifetime limited value.

7.2 Excitation Migration in Ordered Crystals ("Neat" and Lightly Doped)

The term migration implies a mode of excitation dynamics that is elastic or quasi-elastic, i.e., the excitation energy is approximately conserved. It implies nonoscillatory motion in both space and time. While it may refer to the motion of either particles or wave-packets, in the latter case it implies some degree of incoherence, i.e., scattering. With a certain degree of exciton scattering due to phonons, a small additional scattering due to disorder (defects or impurities) is not likely to cause drastic changes in the mode of exciton propagation. We thus include under the heading of "ordered" crystals, practical crystals that are either "neat" (undoped) or slightly doped. We justify the inclusion of the "lightly doped" crystals by the fact that often the latter are actually purer and more free of defects than some of the "neat" crystals used experimentally. Traditionally, certain impurity "traps" have been utilized (as acceptors) both in "doped" and in "neat" crystals. Our distinction between ordered and disordered crystals is thus a pragmatic one. Needless to say, our ordered crystals also have surfaces, including rugged ones.

7.2.1 Exciton Fusion Experiments

The only direct exciton migration experiments known to us that do *not* make use of impurities, defects or surfaces, are exciton fusion experiments. An example of fusion is the process of "triplet-triplet annihilation" of excitons [7.41, 42], i.e., the merging of two triplet excitons to give one exciton containing the total energy of the two. The latter may be a singlet, triplet or quintet exciton, but usually only the resulting singlet exciton has been observed, through "delayed fluorescence" [7.43]. The degree and time dependence of triplet-triplet annihilation is a measure of the triplet exciton migration and has been exploited accordingly. Ordinarily, such fusion is a second-order reaction, and is thus proportional to n^2, the *square* of the triplet exciton density, and thus to I^2, the *square* of the photon flux used for the excitation. We note that singlet-singlet and singlet-triplet annihilation are also possible, but require higher excitation fluxes, due to the shorter lifetime of singlet excitons.

a) Diffusion Picture

Avakian and *Merrifield* [7.44] measured triplet exciton diffusion coefficients, utilizing fusion to measure the spatial distribution of excitons. They generated triplet excitons in multiple regions of a crystal small enough to allow a significant fraction of these excitons to diffuse out of the illuminated region within their lifetimes. Even though $\langle n \rangle$, the average exciton density, is unchanged, $\langle n^2 \rangle$ is diminished and, as stated above, fusion is proportional to n^2.

In their experiment, *Avakian* and *Merrifield* achieved an appropriate spatial distribution by irradiating a crystal with light from a continuous He-Ne laser passed through a grating with alternating opaque and transparent strips (Ronchi ruling). The dependence of the delayed fluorescence intensity on the ruling period was measured. The exciton density n was low enough so that it was not significantly diminished by the fusion process, the latter only acting as a probe. *Ern* et al. [7.45] subsequently performed time-dependence measurements. An analogous experiment was carried out by *Levine* et al. [7.46].

b) Range and Anisotropy

The room temperature measurement mentioned above gave for the anthracene triplet excitons a diffusion length $l = (2D\tau)^{1/2}$ of about 25 μm. An important source of uncertainty is the crystal-to-crystal variation of the exciton lifetime ($\sim$ 10 ms), probably due to trapping at defects.

Careful measurements by *Ern* [7.47] have revealed the anisotropy of the anthracene crystal triplet exciton diffusion tensor. The values are $1.5 \cdot 10^{-4}$, $1.8 \cdot 10^{-4}$ and $1.2 \cdot 10^{-5}$ cm^2/s for D_{aa}, D_{bb} and $D_{c^*c^*}$, respectively, where a and b are the (monoclinic) crystal axes and c^* is perpendicular to the ab plane. *Ern* et al. [7.48] measured the temperature dependence of D_{aa} and found an increase to $4.0 \cdot 10^{-4}$ with a decrease in temperature to 118 K.

For the naphthalene crystal triplet exciton *Ern* [7.49] found $D_{aa} = (3.3 \pm 0.4) \cdot 10^{-5}$ and $D_{bb} = (2.7 \pm 0.4) \cdot 10^{-5}$ cm^{2}s^{-1}. Interestingly the ratio $D_{bb}/D_{aa} \approx 0.8$ is consistent with $b^2/a^2 = (5.99/8.22)^2 = 0.53$ (*Ern* considered it to be consistent with a theoretical value of 0.5). With a concomitant triplet lifetime of about 200 ms, this gives $l \approx 35$ μm.

c) Hopping Model

The diffusion constants (tensor components) have been successfully interpreted in terms of a hopping model [7.46, 48, 50] (see also Chap. 3). The details of the hopping model adopted may differ, but the following features seem to be generally accepted

$$D_{\mu\nu} = (1/2) \sum_i W_i R_{i\mu} R_{i\nu}, \tag{7.49}$$

where W_i is the "hopping rate" from the origin molecule to molecule i, whose position vector is R_i. The hopping rate W_i is related to β_i^2, where β_i is the exciton pairwise interaction ("transfer matrix element") between the ith molecule and the molecule at the origin. Also the "hopping time" $t_i = W^{-1}$ is often used. Beyond this point there are various treatments. In the approach of *Trlifaj* [7.51], which follows *Dexter*'s theory [7.52] of energy transfer in impurity systems,

$$W_i = \beta_i^2/\Gamma \tag{7.50}$$

where Γ is an exciton-phonon interaction ("scattering rate"). The approach of *Haken* and co-workers [7.53] gives [7.48] effectively,

$$W_i = 2\gamma_i + \beta_i^2/(\Gamma + \gamma_i), \tag{7.51}$$

where γ_i represents the "nonlocal fluctuation rate" (fluctuation of β_i), presumably due to lattice phonons, and

$$\Gamma = \gamma_0 + \sum_i \Gamma_i \tag{7.52}$$

where γ_0 is the "local fluctuation." The above formalism is expected [7.47–49] to apply only for kT large compared to the bandwidth, and does not necessarily result in diffusive motion. However, diffusive motion has always been assumed. Also, in spite of the fact that "correlated hops" (see below) may be expected, explicit or implicit utilization of random walk theory has followed [7.46, 54]. A random walk description also obviously follows from a "band model," in the limit of an extremely short mean-free path, as pointed out by *Levine* et al. [7.46]. The latter authors have argued that their experimental results were indeed consistent with both a hopping model and such a "band model" with extremely high scattering. This should not be surprising, as these two theoretical approaches should agree in this limit.

d) Relation to Spectroscopic Parameters

It has been long recognized that the pairwise exciton interactions can be derived from "Davydov splittings" [7.55] (see Chap. 1), observed in the spectra of the pure crystal, and, in more detail, by the exciton densities of states of pure crystals and the "resonance pair" or cluster spectra of isotopic mixed crystals [7.56, 57]. Similarly, the exciton-phonon scattering must surely be related to the homogeneous linewidths of these spectral transitions [7.55, 42, 58] and is usually [7.47-49] related to Γ. Thus, in principle, one can combine optical measurements and the fusion measurements to check for mutual consistency as well as to extract the nonlocal scattering rates (γ_i's). While it appears that mutual consistency has been achieved [7.48, 50], it is not clear to us whether the specific values of the γ_i's are reliable. They may, however, be considered as a measure of an *upper limit*, i.e., about 0.1 cm^{-1} for anthracene and 0.02 cm^{-1} for naphthalene [7.50]. It is not clear whether these quantities are indeed related to phonon scattering or to imperfection introduced scattering [7.48]. On the other hand, the values of Γ are reasonably well-established. They range between 15 to 65 cm^{-1} for a temperature range of 77 to 370 K, for both the anthracene and naphthalene triplet excitons [7.48, 50]. The corresponding hopping times are about 10^{-11} to $3 \cdot 10^{-11}$ s for anthracene and $5 \cdot 10^{-11}$ to 10^{-10} s for naphthalenes. At the low temperature of 4 K these values are given as $7 \cdot 10^{-13}$ and $7 \cdot 10^{-12}$ s, respectively. The most gratifying results are those of the exciton pairwise interaction (β_i) which are very consistent in their values, irrespective of experimental method or temperature. It is not yet clear to us which of the two is a greater marvel. For instance, for naphthalene these are $\beta_{1/2(a+b)} = 1.25$ cm^{-1} and $\beta_b = 0.6$ cm^{-1} for the pure electronic (0–0) triplet exciton. Any out-of-plane (perpendicular to *ab)* interactions are believed to be lower by 5 orders of magnitude [7.59, 60]. The anthracene parameters are similar (2.5 cm^{-1} and about 0.5 cm^{-1}). It is generally agreed that the triplet pairwise interactions (β) are due to *electron plus excitation exchange integrals* ("exchange" in abbreviated form), as originally suggested by *Dexter* [7.52]. These interactions are thus extremely short-ranged. While singlet exciton interactions have traditionally been assumed to be due to *transi-*

tion-dipole-transition-dipole integrals [7.61, 60] and thus long-ranged, this is not always the case. In the lowest singlet systems of naphthalene and benzene short-ranged "exchange" or "multipole" terms dominate [7.56, 57].

7.2.2. Exciton Trapping Experiments

The trapping of excitons by chemical impurities (and impurity induced physical defects) have plagued and bedevilled spectroscopists who studied the fluorescence and phosphorescence of "pure" crystals. The unexpected efficiency of this trapping, even by minute mole fractions (10^{-6} to 10^{-9}) of impurities, led to a powerful method of studying exciton migration (e.g., the fluorescence of tetracene doped anthracene [7.62] or anthracene doped naphthalene [7.63]).

a) Phosphorescence

Pure crystal phosphorescence is difficult to detect from crystals like naphthalene. It took extreme purification methods, reducing betamethylnaphthalene impurities to below 10^{-7} mole fraction [7.64, 65] to enable the observation of phosphorescence from "pure crystalline" material [7.66] and, even then, impurity effects are noticeable at higher temperatures (77–100 K) and significantly quench (trap) the phosphorescence at 4.2 K. The measured lifetime (> 150 ms) is more than an order of magnitude shorter than the expected "intrinsic" lifetime (2.5 s). As even purer samples have become available this lifetime has been "stretched," but only to 500 ms [7.67]. On the other hand, in lightly doped samples the phosphorescence is typically from impurity molecules ("traps") or impurity induced host defect sites ("X traps"). A typical description of such behavior is given in [7.50]. It should be remembered that triplet exciton mobility also leads to fusion ("triplet-triplet annihilation"), especially at higher excitation levels, and thus to *delayed fluorescence*. It turns out that the latter phenomenon is a much superior tool for the investigation of excitation transport in lightly doped (artificially or "naturally") crystals.

b) Delayed Fluorescence

As mentioned above, delayed fluorescence is a multiplicity allowed emission ("fluorescence"), typically from a singlet excited state, which is "delayed" by the much longer lifetime of the "phosphorescent state" (typically an excited triplet state), because it results from the fusion of two lower energy excitations (i.e., triplet-triplet annihilation). While a typical fluorescence lifetime is measured in nanoseconds, that of delayed fluorescence is measured in milliseconds or even seconds. It turns out that "sensitized delayed fluorescence," which is the delayed fluorescence emanating from a trap in a lightly doped crystal, is the most sensitive measure of energy transport and trapping. The delayed fluorescence from a 10^{-8} mol/mol anthracene trap in naphthalene is easily observed [7.50], and is equal in intensity to the host (naphthalene) delayed fluorescence at 77 K. Thus, adopting a simple hopping model, the excitation "visits" about 10^8 sites within a lifetime of about 10 ms. This gives a hopping time of about 10^{-10}s, in excellent agreement with the values mentioned above. We should not, however, consider such numbers as more than order of magnitude indicators. Even for a simple random walk hopping, one should correct for revisitation factors (about 4 for 2-dimensional motion [7.68]) and for "trapping efficiencies," in addition to experimental uncertainties. It is possible though that the revisitation factor is partially cancelled by a larger than unity trapping efficiency (e.g., creation of a "funnel" of 4-6 "X traps" around the anthracene impurity).

c) Fluorescence

Ordinary fluorescence ("prompt fluorescence") gives a measure of the energy transport in the singlet state, in contrast to "delayed fluorescence," which serves as a measure of energy transport in the triplet state. The former is far less efficient due to the above-mentioned much shorter lifetime of the singlet exciton (100 ns for naphthalene), even though the hopping rate is usually much faster. Based on experiments with anthracene-doped naphthalene, one observes equal host and guest emission intensities at about 10^{-5} mole fraction (compared to 10^{-8} for delayed fluorescence – see above), resulting in a singlet hopping time for naphthalene of the order of 10^{-12} s [7.50, 68].

d) Temperature Effects

Excitation trapping seems to become much more efficient at lower temperatures. This is true even for "deep traps" like the singlet state of anthracene in naphthalene [7.50], where no simple thermal (statistical) detrapping effects account for the result (however, a multistep detrapping process has been suggested [7.69]). The interesting implication is that the *energy transport in the host crystal becomes more efficient with lower temperatures.* We remember that an increase in the triplet diffusion constant with lower temperature has also been observed (Sect. 7.2.1), while in those experiments increased impurity or defect trapping would have led to the opposite result. We note that, at very low temperatures, the so-called X traps might considerably reduce the trapping by the dopant. This, however, can be avoided by extensive sample purification [7.69].

e) Hopping vs Coherence

The increased trap fluorescence yield at lower temperature has led *Hammer* and *Wolf* [7.70] to suggest a gradual transition from "hopping" to "coherent" exciton motion [7.71] for the singlet naphthalene exciton. While a detailed picture of exciton coherence will be discussed below, we note here that a satisfactory quantitative picture for both anthracene-doped naphthalene and naphthalene-doped perdeuteronaphthalene ($C_{10}D_8$) has been obtained by *Argyrakis* [7.68] simply by fitting the data with a coherency parameter *l*, where *l* is the number of "correlated hops." As T decreases, l increases. Based on a comparison of the Davydov splittings with the spectroscopic linewidth [7.72], one indeed expects l to become significantly larger than unity upon cooling to about 100 K. We return to this topic below (see Sect. 7.3.a-d).

7.3. Excitation Migration in Substitutionally Disordered Crystals

Historically the study of doped molecular crystals was aimed at studying energy migration in undoped (perfect) crystals. Eventually the study of these doped, i.e., substitutionally disordered crystals, became an end in itself.

7.3.1 Trap-to-Trap Transfer

These experiments were also designed to study the pairwise excitation interactions and energy transport in undoped crystals. However, it was eventually noted that a quasilattice of quasidegenerate traps may form an "impurity band" (see below).

a) The Nieman-Robinson Experiment

This experiment [7.73] is the first one known to us in which a tenary system has been used where all three species are intimately involved in the energy transfer. The three benzene ingredients differ only by isotopic substitution. The one with the highest excitation energy (say C_6D_6) is the "host" (about 0.99 mole fraction). One lower energy ingredient is the "trap" (say C_6H_5D) and the lowest energy ingredient is the "supertrap" (say C_6H_6). Both "trap" and "supertrap" are usually present at small concentrations ($\sim 1\%$ or less). The experiment monitors the relative emissions (phosphorescence) from "trap" and "supertrap." If these two species are present in equal amounts, and have equal trapping efficiencies, but the supertrap emission intensity is significantly higher, this is an indicator of the "trap-to-supertrap" energy transfer.

A tunneling mechanism was suggested [7.74] for the low temperature (2 K) trap-supertrap energy transfer, where the trap-depths are large compared to kT. The trap and supertrap are "coupled" via virtual states of the *intervening* host molecules. The excitation "tunnels" through an energy barrier whose height is the "trap depth" Δ (trap-host energy separation) and whose width is given by the spatial extent of the line of $(n-1)$ host sites intervening between trap and supertrap. The coupling is given [7.75] as

$$J_n = \beta^n / \Delta^{n-1} \tag{7.53}$$

where β is the (vibronic) exciton pairwise interaction (roughly equal for all three species) and n is the barrier width in lattice distances (nearest neighbor separations). n is derived from the guest concentrations (using geometrical considerations), Δ is determined spectroscopically and β is determined as mentioned earlier (7.2.1). J_n then gives the frequency (or probability) of energy transfer. Historically, *Nieman* and *Robinson* determined J_n from the intensity ratio and thus derived β (which turned out to be about 10^6 times the then theoretically accepted estimate). However, more precise methods for the determination of β have been mentioned in Sect. 7.2 (i.e., from resonance pair spectra). We emphasize here that there is no real "guest" energy band (only a "virtual band") and that the only "real" exciton band is that of the host, with typical bandwidth: $B = 2z\beta$, where z is the (nearest neighbor) coordination number. The *Sternlicht* et al. [7.74] tunneling formalism has been put on a theoretically sounder base, involving the perturbed host Green's function [7.76,77], and has been termed *exciton superexchange*. In the limit where $\Delta \gg \beta$ the simple formula(7.53) is retrieved, provided that the trap and supertrap energies are *degenerate*,but with a correction factor $\overline{\Gamma}$ [7.78, 79], accounting for the fact that in a lattice (2 or 3 dim.) there is a·"taxicab geometry" [7.80] with $\overline{\Gamma}$ shortest paths connecting the two traps, each path traversing $(n$-$1)$ host sites.

b) Update

The basic concept and the applicability of the original Nieman-Robinson experiment have been confirmed in more sophisticated recent experiments on benzene by *Tudron* and *Colson* [7.81]. They, as well as others [7.78], have emphasized the role of lattice phonons in the realistic case where the trap and supertrap excitation energies are not degenerate. *Colson* et al. [7.82] have demonstrated that the extent of the phonon assistance, expressed by a "lattice Franck-Condon factor" modifying (7.53), is determined by the relation between the trap-supertrap energy level mismatch and the phonon density-of-states of the host benzene crystal. Similar arguments have been given for naphthalene by *Ochs* [7.78]. Phonon assistance has also been discussed for trap-trap energy transfer where the energy degeneracies are removed by crystal inhomogeneities [7.82] or by guest clusterization [7.79, 83]. As long as both trap and supertrap concentrations are low, the energy transfer usually involves a single step and has been described [7.81] via a *Perrin* sphere model [7.84].

7.3.2 Cluster-to-Cluster Migration

As soon as the *presence of localized exciton clusters or aggregate states* in binary crystals became obvious [7.85, 86], the idea of cluster-to-cluster energy transfer, as well as that of an energy cascade involving a series of distinct clusters [7.87], suggested itself.

a) The Mauser-Port-Wolf Experiment

In this experiment [7.88], the singlet *fluorescence* spectra from an isotopic mixed naphthalene crystal $(C_{10}H_8/C_{10}D_8)$ were monitored at concentrations between

0.1 and 50%. A wealth of discrete cluster states was observed, as theoretically predicted by *Hong* and *Kopelman* [7.85], based on their interpretation of the experiments of *Hong* and *Robinson* [7.89]. Moreover, it was clearly shown that the relative intensities, at higher concentration, do not merely reflect the statistical cluster distribution but indicate significant energy transfer via what was termed a "dilute exciton band" of the most abundant trap (e.g., monomer). It was also shown that the energy transfer increases with temperature due to phonon-assisted cluster-to-monomer thermalization. Whether a real "monomer band" exists is still an open question (see below). However, monomer-to-monomer transfer is certainly possible as discussed above (via superexchange or "tunneling," involving the host band, or via long-range "dipole" interactions). It should also be noted that no discrete cluster emissions are observed in the heavily doped crystals (50%), as was predicted [7.85], on the basis of cluster "percolation" (see below).

Further confirmation of these ideas on the role of repeated monomer-to-monomer transport and the thermally activated excitation transfer from cluster states (i.e., exciton detrapping from dimers, trimers, etc., back into the monomers) was demonstrated by *Port* et al. [7.90].

b) Triplet Experiments

While the long-range interactions leading to trap-trap energy transfer in the singlet state could be either direct ("dipole" or exchange) or indirect ("superexchange"), only the latter mechanism is possible for triplet excitons. Experiments similar to those of *Mauser* et al. [7.88] have been performed on the same binary naphthalene system ($C_{10}H_8/C_{10}D_8$), but monitoring the triplet emission [7.91]. While less detailed, these experiments indicate behavior similar to that of the singlet, i.e., cluster-to-cluster energy transfer and cascading. Of interest are also the phosphorescence experiments performed on isotopic mixed quasi-one-dimensional crystals [7.92]. Here the "trap" concentration was kept high and the "host" molecules acted mainly as narrow barriers between clusters.

7.3.3 Critical Concentrations

The realization that the actual geometrical distribution of guest clusters in a host lattice determines the absorption and emission spectra [7.85] (see also Chap. 1) leads to an expectation that it could also strongly influence the energy transport. To monitor genuine energy transport (i.e., migration), rather than short-range transfer, one again utilizes acceptor species (supertraps), which are introduced in low enough concentrations to minimize short-range, direct donor-acceptor transfer, and to essentially measure the extent of *multistep* donor-donor transport. In other words, the excitation has to "spread" over many donor sites to be detected by the sensor (supertrap). The ternary systems, based on a dilute third component, have the considerable advantage that the sensor (supertrap) concentrations can be controlled independently of the donor concentration.

a) Naphthalene Triplet

The first such ternary system to be investigated was naphthalene ($C_{10}H_8$). The higher energy $C_{10}D_8$ system serves as "host" while the excitation carrying "guest" (donor) is $C_{10}H_8$. The dilute acceptor is betamethylnaphthalene (BMN), where one naphthalene hydrogen atom has been chemically substituted by a CH_3 group. The BMN enters the naphthalene lattice substitutionally (and randomly), and has its first triplet state 275 cm^{-1} below that of $C_{10}H_8$. This assures that thermal de-excitation of BMN supertraps is negligible at liquid helium temperatures. In the steady-state excitation experiment, the energy transport is monitored by the "quantum yield" I_s/I_{tot} where I_s is the phosphorescence yield from the supertrap (sensor) and I_{tot} is the total phosphorescence yield, properly normalized [7.93]. The early experiments [7.94] already revealed a "critical onset" of transport with concentration (Fig. 7.13) which was attributed to "dynamic percolation," meaning a kinetically controlled transport where the excitation transfer probability depends critically on cluster connectivity and its concentration dependence. The proposed mechanism (see Sect. 7.3.1) is a quantum mechanical tunneling (via exciton superexchange) in which the guest-host energy separation (Δ) plays the role of barrier height and the chain of host sites, separating the donor sites, plays the role of barrier width. As in all quantum mechanical tunnel-

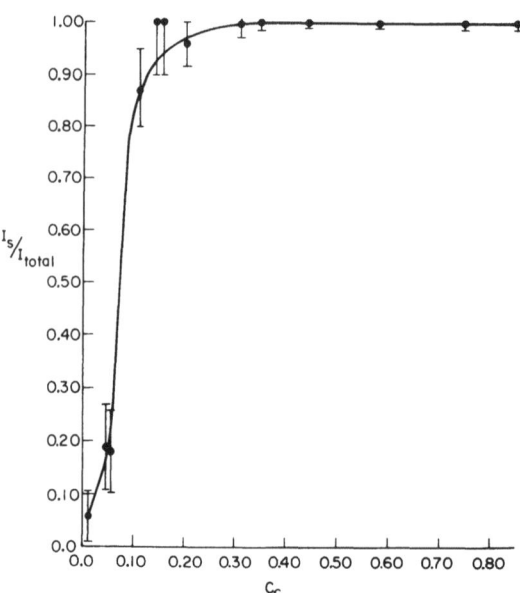

Fig. 7.13. Experimental excitation percolation: First triplet excitation of naphthalene. The experimental points are integrated phosphorescence intensity ratios, where I_s is the sensor (BMN) intensity and $I_{tot} = I_s + I_d$, where I_d is the $C_{10}H_8$ naphthalene (donor) intensity. As $SC \approx 10^{-3}$, $C_d \approx C$. [7.96]

ing processes, the penetration depends drastically on the barrier width. Because of the incremental nature of this barrier width there is an effective "cut-off" distance (in a taxicab geometry [7.80]). [If one imagines a grid of roads with gas stations open only at a fraction of the intersections ("guest" sites), then transportation depends on establishing a connectivity between open gas stations that guarantees the absence of stretches longer than the cruising range (mpg · tankcapacity).] The penetration of the barrier depends also on the barrier height (Δ). This has been established by an "energy denominator" (Δ variation) study, using partially deuterated naphthalene species as guest molecules [7.79].

b) Naphthalene Singlet

The analog experiment on the naphthalene singlet system [7.95] also showed an energy transport onset with concentration that is too sharp to be explained by mean-field-type theories in which the clusters are "smeared out." The data here fit a simple percolation model in which only nearest-neighbor guest sites are "connected" [7.83]. The overall behavior is very different than that of the triplet case, due to a combination of two mutually enforcing factors: 1) the singlet lifetime is more than *seven* orders of magnitude shorter (see Sect. 7.2.1), and 2) the typical energy separation among nearest-neighbor clusters is 15 times larger [7.96], making the respective thermalization probability (Boltzmann factor) about *nine* orders of magnitude smaller (at 1.7 K). Superexchange (tunneling), as well as longer-range direct excitation transfer, are thus negligible. [The analogous transportation problem is now one where the cruising range is reduced to the distance between nearest grid points, making thus any intersection with closed gas stations impassable.]

c) Benzene Triplet and Singlet

Analogous experiments have been conducted on benzene (C_6D_6/C_6H_6: pyrazine). Again "critical concentrations" have been observed [7.97], and again the triplet critical concentration is much smaller than that for the singlet. Qualitatively, the interpretation is the same as in naphthalene. Quantitatively, the situation is more complex as the crystal structure is orthorhombic, with four molecules per primitive unit cell, giving a three-dimensional exciton transport topology, probably based on a close-packed orthorhombic lattice, rather than the two-dimensional naphthalene topology (with an effectively square lattice grid).

d) Naphthalene Singlet − Temperature Effect

It has been shown [7.98] that an activation energy of about 10 K (7 cm^{-1}) suffices to move the critical concentration significantly (to lower values). This has been interpreted in terms of next-nearest neighbor (as well as third nearest, etc.) excitation transfer, leading to a shift of the singlet exciton critical concentration towards that of the triplet. One obviously has to guard against a very trivial temperature effect, namely, an exciton thermalization from the guest to the host band. However, this "trivial" thermalization requires a higher activation energy (70 K). The non-trivial activation energy (10 K) is related to the energy mismatches among the *primary* (nearest neighbor) guest clusters. For instance, the dimer-monomer energy mismatch is about 15 cm^{-1}, the trimer-dimer is 7 cm^{-1}, etc. [7.88]. The important overall result is that the energy transport is *enhanced* by the temperature *throughout* the whole concentration range, which is typical of a phonon-assisted ("non-metallic") process (see below).

7.3.4 Cluster Percolation

In this section we quote some basic elements of mathematical percolation theory. These are relevant to the various physical models of exciton percolation. We limit our discussion to the site percolation problem in a binary lattice [7.99].

a) Cluster Distribution and Averages

In a binary random lattice containing guest sites G and host sites H the definition of guest (or host) clusters depends on the definition of what constitutes a *bond*. The simplest example is that of nearest-neighbor-only bonds. Each given guest concentration C has its characteristic *cluster distribution*. This is given by the set of *cluster frequencies* i_m, where m is the cluster size. A common definition [7.100] of the average cluster size is

$$I_{AV} = \sum_m i_m m^2 / \sum_m i_m m. \tag{7.54}$$

This I_{AV} increases monotonically with guest concentration C and, for an infinite lattice, it diverges at C_c, the *critical concentration* (Fig. 7.14).

The critical concentration is a functional of topology. We extend the term topology to refer not only to the Bravais classification of two- and three-dimensional lattices but also to the gamut of bond definitions for each such lattice. For a given dimensionality, C_c decreases with bond order (Chap. 1, Table 1.1).

Another related definition is the *average finite cluster size*

$$I'_{AV} = \sum_m {}'i_m m^2 / \sum_m i_m m \tag{7.55}$$

where Σ' *excludes* infinite m values (there can be only one infinite cluster). Note that our definition includes the infinite cluster in the denominator (excluding it would give the alternative definition $\bar{I}'_{AV}$). We note that I'_{AV} diverges *only* at C_c, but is finite at both lower *and* higher concentrations (Fig. 7.14).

b) The Infinite Cluster and Percolation Probability

As mentioned above, for $C_c < C < 1$ there exists an infinite cluster. The probability of a *guest site* being included in the infinite guest cluster m' is obviously

$$\bar{P}_\infty = m' / \sum_m i_m m. \tag{7.56}$$

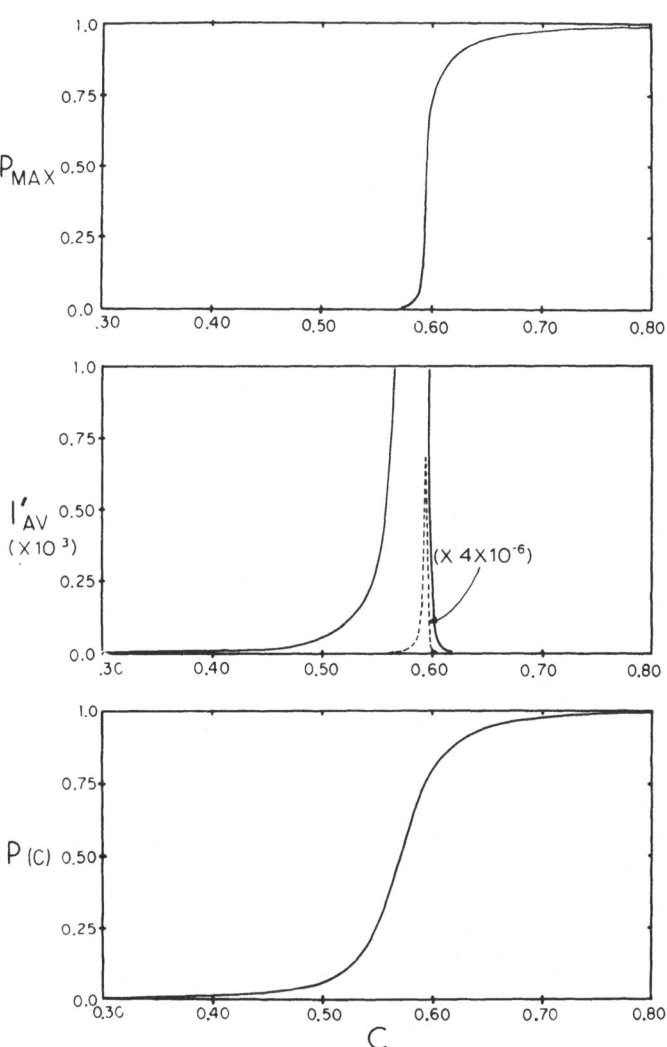

Fig. 7.14. The cluster functions: $\bar{P}_{max}$ (*top figure*), I'_{AV} (*center*) and $P(C) = P$ (*bottom*) for a square lattice with 16,000,000 sites [7.103]. For an infinitely large lattice $\bar{P}_{max}$ approaches $\bar{P}_{\infty}$ (note the P_{max} tail below $C_c = 0.5927$). The $P(C)$ function was computed for $S=10^{-3}$. (Note also that I_{AV} is practically given by I'_{AV} below C_c.) [7.93]

This is the *percolation probability*. The latter is zero below C_c and finite above it (and unity at $C=1$). For a finite crystal m' can designate the *maxicluster* [7.96], i.e., the largest cluster. The *percolation probability* P_{max} is given by the r.h.s. of the above equation (Fig. 7.14). Similarly, I'_{AV} can be defined as above (7.55), just that now the *maxicluster* has been excluded. One can also obtain a *critical concentration* C_c in a finite lattice. It is defined by the singularity in I'_{AV} [7.100].

We note that an alternate percolation probability (P_∞) definition exists in terms of the probability of any site (guest or host) being included in the infinite cluster $(P_\infty = C\bar{P}_\infty)$. As the critical concentration is reached from below, we note that many clusters (especially the larger ones) *merge* into one single cluster. This is what is meant by *cluster percolation*. Only for three lattice topologies has C_c been derived analytically (excluding the trivial answer $C_c = 1$ for a one-dimensional lattice). Even these cases are limited to two-dimensional, nearest-neighbor situations and *exclude* the simple square lattice case. No exact closed analytical solutions exist for the percolation probability $\bar{P}_\infty$ or the average cluster size I_{AV} or for the general cluster frequency i_m (except for one dimension). Approximate methods are thus utilized, especially computer simulations (Monte Carlo calculations). This explains why percolation theory has only prospered very recently.

c) Long-Range Clusters and Percolation

Long-range clusters are defined by longer-range bonds, e.g., via first-, second-, and third- nearest-neighbor bonds. We note again that C_c decreases as the bond-range increases (Fig. 7.15). Some simple approximate analytical solutions have been given for C_c as a function of n (where a succession of n nearest-neighbor lattice distances defines the maximum bond), for the case of the simple square lattice [7.101], and have been verified by extensive simulations. A graphical illustration of the effect of the interaction range (n) on the cluster distribution is given in Fig. 7.16.

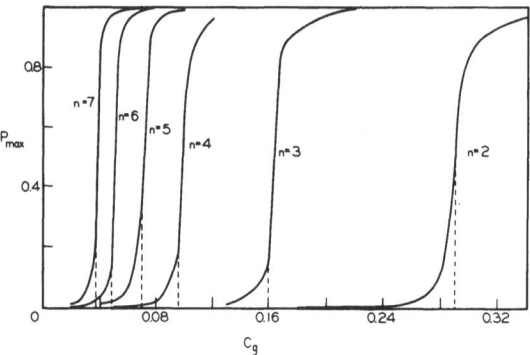

Fig. 7.15. The square lattice site percolation probability, P_{max} (probability of the guest site being part of the largest guest cluster) for different values of the long-range percolation parameter n, as a function of the guest concentration C_g. Here, n is defined as the maximum number of nearest-neighbor bonds over which an interaction (or a connection) can occur. The function $\bar{P}_\infty$ (probability of a guest being included in an "infinite" cluster) is approximated by the dashed line from the P_{max} curve down to the calculated critical site concentration C_c below which $\bar{P}_\infty$ is zero, by definition. All of the simulations for these curves were done for a square lattice of 500 · 500 sites. The effective coordination number is 12 for $n = 2$ and 112 for $n = 7$. Note that the $n = 1$ curve has been omitted as this is given in Fig. 7.14. [7.122]

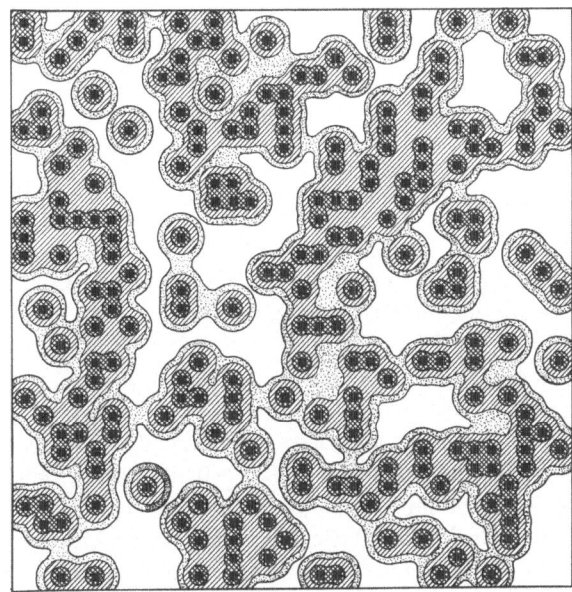

Fig. 7.16. This 32 • 32 binary square lattice with $C = 0.20$ illustrates clusters built by the definition $n = 1$ (*dark*), $n = 2$ (*dashed*) and $n = 3$ (*dotted*). Note that for $n = 3$ the percolation concentration has been exceeded. [7.93]

d) Universality, Scaling and Critical Exponents

Following the analogy shown between the Ashkin-Teller-Potts model of magnetism and percolation [7.102], the concept of scaling and the search for critical exponents started in earnest. A number of empirical results are available; for example, as $C \rightarrow C_c$,

$$\bar{P}_\infty \propto |C/C_c - 1|^\beta \qquad\qquad C > C_c \qquad\qquad\qquad (7.57)$$

$$I_{AV} \propto |C/C_c - 1|^{-\gamma} \qquad\qquad C < C_c \qquad\qquad\qquad (7.58)$$

$$I'_{AV} \propto |C/C_c - 1|^{-\gamma'} \qquad\qquad C > C_c \qquad\qquad\qquad (7.59)$$

where the *critical exponents* β, γ and γ' appear to be constant for a given dimensionality and also $\gamma = \gamma'$, as "expected." A recent test [7.103] on a square lattice with $64 \cdot 10^6$ sites shows that γ is indeed the same (2.24 ± 0.01) for both the square and triangular lattice but differs significantly from the exact two-dimensional Ising model result ($1.75000...$). Furthermore, it has been shown that (7.58) holds over a very wide range of C/C_c, and not only when $C/C_c \rightarrow 1$ [7.104,105].

Further simulations on the long-range square lattice problem have shown that the functions I_{AV} and $\bar{P}_\infty$ appear to be "universal" (superimposable) on a reduced concentration representation (plotted against C/C_c) irrespective of interaction range n [7.101]. This immediately implies that γ and β are also independent of the interaction range n. The latter has indeed been verified [7.104, 105]. For two-dimensional lattices [7.101], $\gamma = 2.2$, $\beta = 0.14$ and $\gamma/\beta + 1 \approx 17$.

7.3.5 Exciton Percolation Formalism

In a simple model, with only nearest-neighbor interactions in a random *binary* lattice, the probability P of an infinitely long-lived guest exciton registering on a detector mounted on the side opposite from the exciting radiation is

$$P = \bar{P}_\infty, \tag{7.60}$$

where the above functions only depend on the concentration and on the topology. We note that $P(C < C_c) = 0$. If we now replace the above detector with a set of randomly distributed supertraps, replacing a mole fraction S of the guest sites, then for low C (small clusters) the probability P of the same infinitely lived guest exciton registering on a supertrap is given by the probability that an average cluster contains such a supertrap:

$$P = SI_{AV} \qquad S \ll 1, C \ll C_c. \tag{7.61}$$

These intuitively obvious simple cases point out the basic connection between excitation transport and the percolation functions, which are derived from cluster distributions.

a) General Case

Assuming only a sharp interaction cut-off and a small sensor concentration ($S \ll 1$), a general expression was derived [7.106] for the probability of a guest excitation being "supertrapped," i.e., registered by the acceptor (supertrap):

$$P = G^{-1} \sum_m i_m m [1 - (1 - n_m/G)^Z], \tag{7.62}$$

where G is the total number of guest sites (including acceptors), Z is the effective number of acceptor sites (corrected for their trapping cross section), and n_m is the *average* number of distinct sites, in a cluster of size m, sampled by the exciton within the time scale of the experiment. The quantity n_m can easily be written as $n_m = \pi_m m$ where π_m is the quantum mechanical probability for a guest excitation, in a cluster of size m, to be trapped by a single supertrap (with a "cross section" of unity), within the time scale of the experiment. Obviously n_m is time dependent, and so is P. The evaluation of P demands both physical information (to obtain n_m) and the mathematical cluster distribution (i_m as a function of C for each m).

While it appears to be hopeless to evaluate P in general in analytical form, we give below a number of physical limits where this evaluation is straightforward. As the concentrations C and $S(=ZC)$ can be manipulated experimentally, one can prepare samples for which such simpler physical limits do apply.

b) Supertransfer Case

If $n_m = m$ then (7.62) simplifies considerably; this is the *supertransfer* limit in which P appears to be independent of time. Such a limit is intuitively applicable for low guest concentrations ($C \ll C_c$), and results again in (7.61). This shows that exciton percolation can give results that are extremely different from any time-dependent diffusion model, where P would always increase with t.

Actually this limit is justified [7.96] whenever $n_m S \gg 1$. Physically this means that the transfer is very efficient w.r.t. the available time. Equation (7.62) then gives [7.101]

$$P = 1 - G^{-1} \sum_m i_m m \lambda^m, \qquad \lambda \equiv e^{-s}, \qquad (7.63)$$

making S analogous to the magnetic field and P to the magnetization in cluster models of magnetism. Furthermore, one obtains [7.101]

$$P(C_c) = S^{1/\delta} \qquad S \ll 1 \qquad (7.64)$$

where

$$\delta = \gamma/\beta + 1, \qquad (7.65)$$

giving

$$\lim_{s \to 0} P(C \leqslant C_c) = 0, \qquad (7.66)$$

making our crystal an analog of a diamagnetic material below C_c. One also obtains

$$P = \bar{P}_\infty + SI'_{AV} \approx \bar{P}_\infty \qquad C \gg C_c \qquad (7.67)$$

as well as again (cf. 7.61):

$$P = SI_{AV} \qquad C \ll C_c. \qquad (7.68)$$

The excitation percolation probability can be calculated from (7.63) for the whole concentration range C and for various values of S, provided the cluster distribution (set of i_m) is available. In practice, this has been done for lattices including from 10^4 to $64 \cdot 10^6$ sites. The only region sensitive to the lattice size $(GC^{-1}$

sites) is that near C_c, where (7.63) should be used. An example of simulated P is given in Fig. 7.14. We note that our supertransfer limit is essentially the same as the "rapid transfer limit" of Chap. 3.

c) Critical Exponents and Universality

While the critical exponent δ is seen explicitly in (7.64) for $P(C_c)$, of more interest is the fact that by combining (7.57) with (7.67) one finds

$$P \propto |C/C_c - 1|^\beta, \qquad C \gg C_c \tag{7.69}$$

while combining (7.58) with (7.68) gives

$$P \propto |C/C_c - 1|^{-\gamma}, \qquad C \ll C_c \quad S = \text{const.} \tag{7.70}$$

As pointed out before, (7.57, 58) hold over a wide C/C_c range, justifying (7.69, 70). This provides a simple test of the applicability of the exciton percolation model. Once the dimensionality is specified, there are no adjustable parameters in the problem.

Alternatively, this method can be utilized to determine the dimensionality of exciton transport. We note that β, γ and δ vary drastically with dimensionality (Table 7.1). The universality of the functions $\bar{P}_\infty$ (C/C_c) and $I_{AV}(C/C_c)$ was pointed out (Sect. 7.3.4). Thus, from (7.69) one expects a universal behavior (independent of interaction range) for P at $C \gg C_c$. The same is expected from (7.70) for P at $C \ll C_c$, provided $S = \text{const.}$

Table 7.1: Percolation exponents

	Two-Dim. [7.101]	Three-Dim. [7.107]
β	0.14	0.41
γ	2.2	1.6
δ	17	5

7.3.6 Dynamic Percolation

We note that at low C we approach a "quasicontinuum" of critical concentrations and percolation curves (Fig. 7.15), caused by a quasicontinuum of maximal bond-lengths ($r = n_1 i + n_2 j$ in a square lattice), each defining a distinct interaction topology, designated by $n (\equiv n_1 + n_2)$. In a long-range excitation percolation problem we assume that there exists a maximal n for which "supertransfer" holds, implying that there is enough time to find a supertrap provided that there is a path connecting the exciton with the supertrap with no single bond larger than n. On the other hand, any $(n+1)$-bond requires a prohibitively long transfer time.

a) "Dynamic" Supertransfer

While the appropriate range n can be predicted theoretically, in principle [7.79], it is easier to determine it experimentally via (7.64), i.e., the relationship $P(C_c)= S^{1/\delta}$. If one knows S (and δ) this gives from the experimental $P(C)$ curve an experimental C_c and this given C_c defines an effective n for a given interaction topology (e.g., square lattice, see Fig. 7.15). The range n now defines the clusters, cluster distributions, the average cluster function $I_{AV}^{(n)}$ and the percolation probability $\bar{P}_\infty^{(n)}$. We now expect for $C \gg C_c$ to have $P = P_\infty^{(n)}$ [see (7.67)], for $C \gg C_c$ to have $P = SI_{AV}^{(n)}$ [see (7.68)], provided there are no "leaky clusters" (see below), and for the total curve to have P given by (7.63).

b) Primary Clusters and Dynamic Clusters

The dynamic cluster may contain primary clusters (defined by $n = 1$), secondary clusters ($n = 2$), etc. (see Fig. 7.16). The physics of energy transfer may be a composite one. For instance, the primary clusters may define coherent exciton states (totally delocalized throughout the primary cluster), while the transfer among primary clusters, but inside a secondary one, may be incoherent, though still due to direct excitation exchange, and the transfer among the secondary clusters may be due to tunneling (superexchange).

The dynamic cluster topology (n) may be a function of the acceptor concentration S, the excitation lifetime t, the excitation exchange interaction (β), the guest-host energy separation (Δ), etc. [7.79].

c) "Leaky Clusters" and Dynamic Critical Concentration

For large dynamic clusters, defined by the range n, a few additional $(n+1)$-ranged transfers will have little effect. Most of them will keep the excitation within the original clusters and the occasional jumps to another cluster will have consumed enough time to make supertrapping in the new cluster unlikely. We note that at or above C_c the excitation has to scan over roughly S^{-1} guest sites, and S^{-1} is of the order of 10^3 or 10^4. However, for small clusters (defined by the range n) an $(n+1)$-ranged transfer is likely to open a second cluster for "the exciton's search for a supertrap," roughly doubling its probability of success. Repeated $(n+1)$-type transfer will thus significantly amplify the exciton percolation probability P. It has been shown [7.93], that this amplification factor is, for the steady-state case,

$$Y = (nfS)^{-1} \qquad C \ll C_c, Y \gg 1, \tag{7.71}$$

where f is the factor

$$f = \bar{t}_{(n+1)} / \bar{t}_n \tag{7.72}$$

giving the increase in the average hopping time for the $(n+1)$-ranged transfer, relative to that of n, and η is a redundancy factor giving the total number of $(n+1)$-jumps over those $(n+1)$-jumps that move the exciton to a new cluster. Equation (7.68) is now modified to give

$$P = (\eta f)^{-1} I_{AV} = Y(SI_{AV}). \tag{7.73}$$

For time-dependent experiments, the amplification factor Y is [7.93] a monotonically increasing function of time. We thus expect the dynamic exciton percolation probability P to increase monotonically with time. On the other hand, an increase in available time will eventually make the $(n+1)$-transfers abundant over the entire C range. The critical concentration will thus move from $C_c^{(n)}$ to $C_c^{(n+1)}$, i.e., to lower C. The same effect will also be produced by significantly increasing S, i.e., reducing the average area (volume) the exciton has to sweep out. Furthermore, the same effect is also expected for an increase in temperature, which speeds up the phonon-assisted transfer (tunneling). We note that phonon assistance is necessary to overcome the energy mismatches among primary cluster states (see Sects. 7.3.2 and 3).

d) Time Studies

Time resolved studies on the triplet naphthalene system [7.93] show an increase in P with time, due to an increase in Y, in accordance with theory [7.93]. The dynamic (kinetic) nature of this system is also revealed by the drastic decrease of the naphthalene exciton decay time with concentration C [7.108] and by the rise times exhibited by the BMN acceptor emission [7.93]. However, for the naphthalene singlet system, one has $\eta fS \gg 1$, and thus no amplification but a "quasi-static" behavior occurs, i.e., the behavior prescribed by (7.68). The singlet naphthalene system indeed shows no guest lifetime decrease with increasing C [7.108] and no BMN rise time [7.109] for the very same samples that showed the simple nearest-neighbor ($n=1$) percolation topology (Sect. 7.3.3). Here the time resolution was limited to about 10 ns., i.e., $\tau_0/10$.

e) Supertrap Concentration Studies

Increasing the supertrap concentration S by an order of magnitude has a drastic effect on the critical concentration C_c, as can be seen in Fig. 7.17. This shows the essence of the "dynamic percolation" concept, as a *decrease* in *distance* (between supertraps) should be equivalent to an *increase* in excitation *time*. Actually, the results are in quantitative agreement [7.93, 110] with the relation derived from the superexchange formalism (7.53) [7.79]

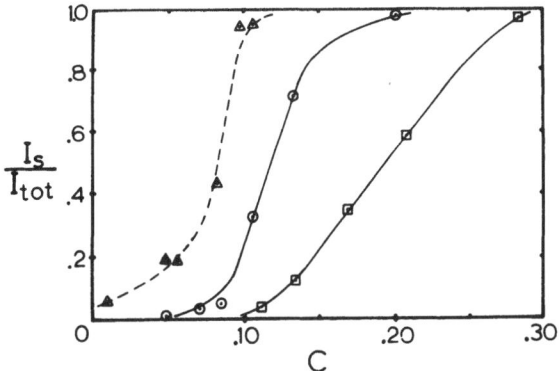

Fig. 7.17. Donor concentration (C) dependence of the energy transport measure $I_s/I_{tot} = I_s/(I_s + I_d)$, where I_s is betamethylnaphthalene phosphorescence (0–0) and I_d is that of $C_{10}H_8$, all at 1.8 K. The parameter S is 10^{-4} for the "square" data points and 10^{-3} for the "circle" points. The "triangle" points are based on Fig. 7.13 and $S \approx 10^{-2}$ at $C = 0.1$, but along this curve SC rather than S is approximately constant. The lines are merely visual aides. [7.110]

$$S_2/S_1 = (\overline{\Gamma}_2/\overline{\Gamma}_1)(\Delta/\beta)^{n_2 - n_1} = J_{n_1}/J_{n_2}, \tag{7.74}$$

where n_1 specifies C_c for the S_1 series and n_2 that for S_2. Equation (7.74) thus relates S_2/S_1 to a transfer rate ratio.

f) Temperature Studies

As mentioned above, we expect an increase in temperature to be equivalent to an increase in transfer velocity, and thus analogous to an increase in time or S (i.e., a decrease in distance). We note that for the triplet system, the primary cluster energy mismatches are about $\leqslant 1$ cm^{-1} and thus about $\leqslant kT$ (for $T = 1.7$ to 4.2 K). We thus expect for a one-phonon assistance (see Chap. 2) a *linear* relationship between transfer probability and temperature. The effect of average transfer rate on C_c has been indirectly calibrated by the effect of S (previous section). Also, we can use the effect of S to quantitatively check the temperature effect (Fig. 7.18). The experimental effect due to increasing the temperature from 1.7 to 4.2 K is equivalent to an *increase* of S by a factor of 2.5. This factor is in excellent agreement with 4.2/1.7, thus demonstrating a linear temperature effect. Comparing P for the two temperatures at constant C reveals an increase in the amplification factor Y that is also consistent with the value of 2.5. The same is true for the change in naphthalene triplet decay times with temperature [7.93].

It has been mentioned before (Sect. 7.3.3) that the effects of temperature on the singlet exciton transport are very drastic. However, the primary cluster energy mismatch is 15 times larger, which is significantly larger than kT at 1.7 K (and even at 4.2 K). We thus expect for the singlet case a multiphonon-assisted

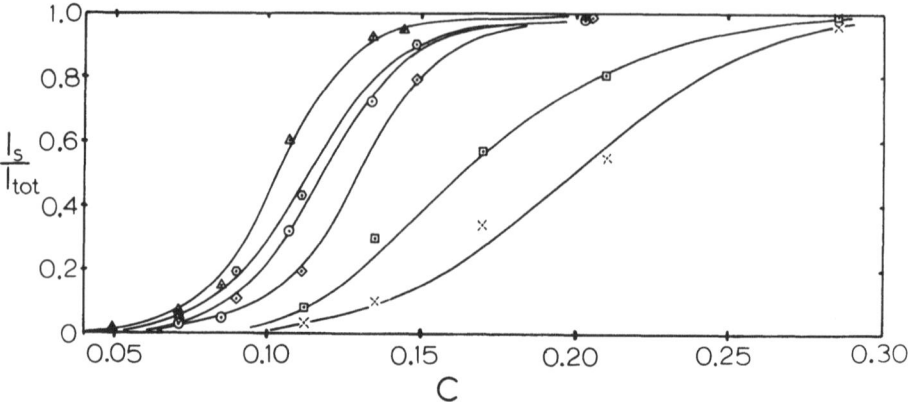

Fig. 7.18. Donor concentration dependence of the energy transport measure $I_s/I_{tot} = I_s/(I_s + I_d)$, where I_s is the acceptor ("supertrap") phosphorescence (0–0) and I_d is that of $C_{10}H_8$, for series A (X trap, $S \approx 10^{-4}$; diamonds = 1.7 K, hexagons = 4.2 K), series B (BMN, $S = 10^{-3}$; circles = 1.7 K, triangles = 4.2 K) and series C (BMN, $S = 10^{-4}$; X's = 1.7 K, squares = 4.2 K). The lines are visual guides. [7.93]

process which depends much more steeply on the temperature. However, increasing the temperature for the singlet naphthalene system causes a transition from a quasi static ($n = 1$) percolation behavior to what appears to be a dynamic percolation (about $n = 2$) case [7.98].

7.3.7 Anderson-Mott Transition

The occurrence of an Anderson-Mott transition (mobility edge) has not yet been established, either by rigorous theory or by raliable calculations [7.111, 112], and experimental verifications have proven elusive (Chap. 5). Because of the simplicity of triplet excitations in organic crystals (tightly bound Frenkel excitons, no dipole-dipole terms, well-known interactions, long lifetimes, random nature of substitutional disorder in isotopic mixed crystals, ease of spectral resolution, no complications due to backtransfer, no radiative trapping or phonon bottleneck – see Chap. 2), these have become one of the "proving grounds" for both theoretical and experimental tests of the theory.

a) The Klafter-Jortner Model

The Klafter-Jortner model [7.113-115] is based somewhat on the *Lyo-Orbach* [7.116, 117] model of an Anderson-Mott excitation mobility edge. It has the advantages of conceptual simplicity, consistency and intellectual honesty. It assumes that the local disorder in isotopic mixed crystals is little affected by the isotopic concentration and also plausibly argues that the guest exciton bandwidth is drastically affected by the concentration of the guest isotopic species (justifiably assuming separated guest and host bands [7.75]). It furthermore uses the well-accepted quantitative formulation of the guest bandwidth in terms of pairwise exciton superexchange interactions (see Sect. 7.3.1)

$$J_n \approx \beta^n/\Delta^{n-1}. \tag{7.75}$$

The basic and simple idea is that J_n varies drastically with n and n varies monotonically with C. Thus, there will be some concentration C_c defined by some *average* pairwise interaction J_c for which the criterion of an Anderson transition is met [7.113]

$$J_c \approx W/\kappa \tag{7.76}$$

where W is the local random strain (and/or inhomogeneity) energy and κ is a numerical constant of the order of 10, related to the lattice connectivity ($\approx z-1$, where z is the coordination number).

There are three major questions:

1) What is the real value of W?
2) How does one correctly calculate J_c for a given concentration C?
3) Is the Anderson-Mott model relevant to the experiments?

These questions, and related ones, are dealt with below.

b) The Role of Local Perturbation

If W is the local perturbation (local strain and/or inhomogeneity) and W' is the experimentally measured *inhomogeneous* linewidth, then obviously $W \leqslant W'$, i.e., all we obtain from the experiment is an upper limit for W. The value of W' may be determined by domain boundaries and/or differential domain strain [7.118] and may be orders of magnitude larger than W. It is a well established fact that W' can be decreased by 2 or more orders of magnitude via improved experimental techniques of crystal growth, purification, mounting, cooling, etc. We note here two extremes for singlet excitations (both at 1.5 K): *chemically* mixed crystals of naphthalene in durene have been shown [7.119] to have some lines with $W' \leqslant 0.03$ cm^{-1}, while highly purified hexamethylbenzene crystals showed an increase in linewidth from 1 to 40 cm^{-1} under *manual* strain [7.120].

c) Superlattice, Simulation and Percolation

Originally, *Klafter* and *Jortner* [7.115] used a J_c derived simply from (7.75),using $\langle n \rangle$, where the latter was obtained from a simple averaging (continuum or super-lattice) technique. In later papers *Klafter* and *Jortner* [7.114, 115] derived $\langle J_n \rangle$ from a *distribution of J* based on a *distribution of n* obtained from a simulation on a square lattice of $128 \cdot 128$ sites (at $C = 0.01$ this contains about 160 guest sites).

Monberg and *Kopelman* [7.121] argued that an extended guest band requires an extended guest cluster. For a test case of $C = 0.07$ they argued that the required connectivity for such an extended cluster ("percolation") can only be achieved at a value of $n = 5$ (or higher). The average or "most probable" value used by *Klafter* and *Jortner* in both their methods was about $n = 2$ (for $C = 0.07$). This results in a disagreement of 5-6 orders of magnitude in J_n. In the opinion of *Monberg* and *Kopelman*, the low value of $<n>$ (or high $<J_n>$) obtained in the simulation of *Klafter* and *Jortner* is heavily biased by "miniclusters" (dimers, trimers, etc.) which cannot give rise to extended ("band") states. For instance, according to the Klafter-Jortner view, nearest-neighbor ($n = 1$) interactions should be used to calculate the "band generating" J_c for $C = 0.2$. However, Fig. 7.16 shows that longer-range ($n>1$) interactions are imperative for the formation of the required extended guest cluster.

It should be noted that the Monberg-Kopelman "model" for an Anderson-Mott-like transition actually differs from the Klafter-Jortner model, as it compares calculated J_n values with W/κ and "eliminates" any $J_n < W/\kappa$. The *set of* n^*, for which the condition $J_{n*} \gtrsim W/\kappa$ is satisfied is checked for the formation of an extended cluster at the experimental C_c. We summarize the three alternatives for J_c: $J_{<n>}$, $<J_n>$ and J_{n*} and note that $<n>$ could also be obtained from a Monte Carlo simulation, rather than from a continuum model.

d) The Localization Volume

We now tackle our third question: Is the Anderson-Mott model pertinent to the energy experiments? The first subquestion is: Does the energy transport as monitored by the experiments require extended ("band") states? The experimental monitoring of energy transport is usually performed via an acceptor (supertrap) with relative concentration (w.r.t. donor) of 10^{-1} to 10^{-3}. Even a localized donor state might extend over 10 to 1000 donor sites (all included in its "localization volume"). It is this very effect of "localization volume" which has bedevilled all numerical computations performed to test out the Anderson model on a small finite lattice [7.111]. Such localization volumes are large even for one-dimensional lattices and when $J \ll W/\kappa$ [7.122]. Thus, the existence of extended band states may be irrelevant to the experimentally observed critical concentrations. We note that for the *static* percolation case [7.98], e.g., where all clusters are *primary*, the localized donor state extends over the entire cluster, with an average localization "volume" of I_{AV} sites. Thus, knowing S and measuring P one can get via (7.61) the effective localization volume (provided that time resolved measurements confirm the absence of "leaks").

e) The Time Factor

The Klafter-Jortner model, like the original Anderson criterion for "the absence of diffusion" [7.123], does *not* include *time* as a factor affecting the criterion for localization vs delocalization (7.76). In their later paper, *Klafter* and *Jortner*

[7.114] argued that above the "transition," the excitation mobility *may* be low, increasing monotonically with concentration up to a "kinetic threshold" (where the excitation mobility overcomes the natural excitation decay). Irrespective of this argument (which was forwarded to explain the high critical concentrations for singlet excitons — see Sect. 7.3.3), the main point remains that *below* the Anderson-Mott transition, for the localized states, time is certainly *not* a factor affecting energy transfer according to this model. Is this borne out by experiments? (see below).

f) The Temperature Factor

Like time, temperature is *not* included in this criteria for an Anderson-Mott transition. The theory is actually a "zero temperature" theory, neglecting the role of phonons. The implication is that, at a low enough temperature, the phonon-assisted "hops" will be considerably less efficient than the "conduction band" mobility. On the other hand, at a higher (unspecified) temperature, it is expected that the sharp mobility edge will be *eroded* [7.114] which presumably means that the critical concentration effect will be "washed out" above this temperature. Is it? (see below).

g) The Experimental Evidence

The best studied experimental case is that of naphthalene triplet transport. The original experiments showed a fairly sharp critical concentration [7.94]. Fig. 7.18 presents results from a recent study [7.93], showing the effects of temperature, acceptor concentration and acceptor species on the energy transport and the critical concentration. It is obvious that the "critical concentration" (whatever its exact definition) is drastically affected by the variation of acceptor concentration and species. This contradicts the expectation of the Klafter-Jortner (Anderson-Mott) model. Even though a slight variation of C_c with acceptor concentration has been suggested very recently [7.114, 115] as a result of the finite localization volume, this should be small, and negligible for a change in absolute acceptor concentrations from 10^{-4} to 10^{-5}, according to [7.115]. Moreover, the effect of the localization volume would be to obscure the critical concentration at high acceptor concentrations, in contrast to experimental observations. The effect of temperature is also inconsistent with the Klafter-Jortner model. The critical concentrations are drastically shifted with temperature; furthermore, the critical behavior gets *sharper*, rather than "eroded," at the higher temperature (see Fig. 7.18).

Time has also been found experimentally to affect the energy transport in the naphthalene system considerably. This was discussed in Sect. 7.3.6. We note that the time effects are for the region *below* the transition, i.e., where none are expected according to the "Anderson" model.

In summary, the naphthalene triplet excitation transport experiments, at 1.7 and 4.2 K, are inconsistent with an Anderson-Mott mobility edge model. However, it is obvious that they are consistent with a kinetic model (Sect. 7.3.6).

We claim here that the latter is also true for all the other available triplet excitation experiments, based on data available to date [7.97, 124, 126]. We also claim that the available evidence for *any* Anderson-Mott-like transition in organic crystals [7.124, 126] is ambiguous and should be tested by the methods mentioned above (use of deep-trap acceptors, variation of their concentration and species, temperature studies with deep-trap acceptors and kinetic studies).

7.3.8 Universality and Exciton Percolation

Based on the theory of Sects. 7.3.5 and 6, we can now understand some striking patterns exhibited by these experiments and utilize them to gain a rather satisfactory picture of the underlying physics. We recapitulate here some of the pertinent results of the dynamic percolation model:

$$P = \bar{P}_\infty \qquad\qquad C \gg C_c \qquad S \ll 1 \qquad\qquad (7.77)$$

$$P = S^{1/\delta} \qquad\qquad C = C_c \qquad S \ll 1 \qquad\qquad (7.78)$$

$$P = (nf)^{-1} I_{AV} \qquad C \ll C_c \qquad nfS \ll 1. \qquad\qquad (7.79)$$

We notice that the first and last equations are not explicit but implicit functions of S, because S (together with the time and the temperature) defines the topology

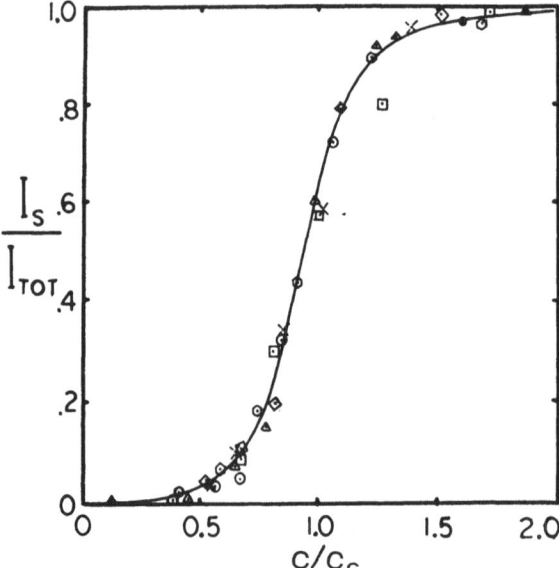

Fig. 7.19. Universal energy transport curve. The data points and designation are the same as in Fig. 7.18. For each family of data points C_c was derived from Fig. 7.18 via (7.64), using $P = I_s/I_{tot}$. [7.93]

(n), and thus $\bar{P}_{\infty}$ and I_{AV} for a given C. However, in a reduced (scaled) representation, where C is divided by C_c, this is no longer true. Thus, in a C/C_c representation P is *independent* of S (and of time and temperature!). We also note that the dependence of $P(C/C_c = 1) = S^{1/\delta}$ on S is very slight as $\delta \cong 17$ (see below).

a) Scaling Experiments

The six curves of Fig. 7.18 should now be replotted against C/C_c. We note that C_c can be determined objectively. Using $P(C=C_c) = S^{1/\delta}$, one finds for each curve the I_s/I_{tot} value equal to $S^{1/\delta}$ (for the given S) and thus determines the C_c of the given curve. The result is given in Fig. 7.19. The universal behavior is quite striking. We notice that the actual data points, rather than the curves of Fig. 7.18, have been used to produce this "universal" curve of Fig. 7.19.

b) Experimental Critical Exponents

The data of Fig. 7.19 are now replotted on a log-log basis (Fig. 7.20), using the "scaled" variable $|\, C/C_c - 1 \,|$. According to (7.69), the slope, for $C \gg C_c$, should give the critical exponent β, while according to (7.70) the slope should give, for

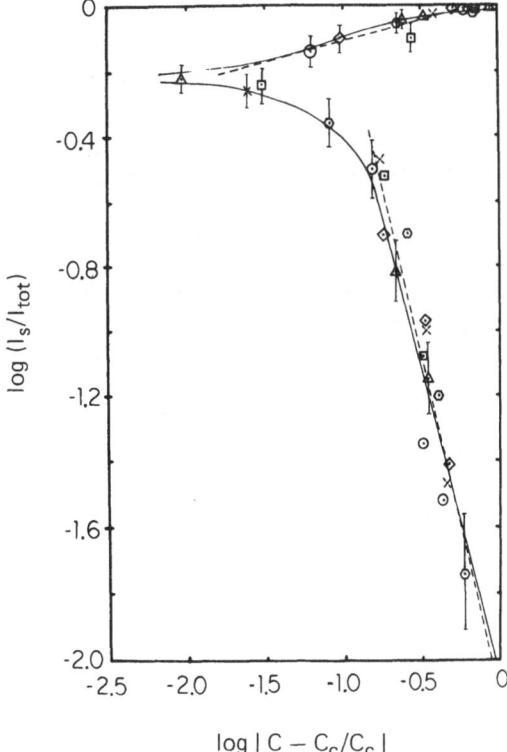

Fig. 7.20. "Scaled" energy transport curve. The data points and designations are the same as in Figs. 7.18 and 19. Error bars were added to a few points to indicate experimental uncertainties. The dashed lines are least square fits to the experimental data, giving $\gamma = 2.1 \pm 0.2$ and $\beta = 0.13 \pm 0.08$. [The full line is a theoretical curve explained in [7.93]. Using (7.73) the lower portion of the curve gives $(n f)^{-1} = 10^{-3}$. [7.93]

$C_c \ll C$, the critical exponent $-\gamma$. The values of γ and β are compared with the mathematical values [7.101] in Table 7.2.

Table 7.2. Mathematical and experimental percolation exponents

	β	γ
Theory	0.14	2.2
Experiment	0.13 ± 0.05	2.1 ± 0.2

The agreement is quite satisfactory. We note that Table 7.2 contains the mathematical values for a long-range percolation problem in *two dimensions* (see below). We also note that while (7.65) $\delta = \gamma/\beta + 1 = 17$ was used to obtain C_c, this is not crucial. An arbitrary choice of C_c at $I_s/I_{tot} = 0.5$ gives the same result as Fig. 7.19, within the experimental uncertainty, as does any choice of δ between 16 and 18.

c) Role of Dimensionality

As mentioned above, the experimental "critical" exponents β and γ substantiate two-dimensional energy transport for the triplet naphthalene exciton. This is in complete agreement with all the other evidence (Sects. 7.2.1 and 7.3.2). Percolation exponents thus give an interesting new criterion for obtaining the energy transport effective dimensionality.

d) The Cluster Picture of Energy Transport

The above demonstration of scaling and universality gives a surprisingly strong confirmation of the cluster model of energy transport ("percolation"). Furthermore, the dependence of C_c on S, T and *time* supports the dynamic cluster model and its "leaky cluster" aspects [since (7.73) is essential to the "universality"]. We now know that the inhomogeneous linewidth (W') for the naphthalene triplet case is about 10^{-1} cm^{-1} [7.125]. This will have little effect on the nearest-neighbor clusters ($J = 1.25$ cm^{-1}). Thus, the nearest-neighbor drimers, trimers, etc., shown in Fig. 7.16, must have excitations that are "coherent," i.e., delocalized over the cluster, in analogy to the spectroscopic evidence of polarized singlet dimer absorptions, etc. [7.77]. However, the next-nearest neighbor clusters ($n = 2$) are connected by superexchange (7.75) with a $J_{n=2} \approx 10^{-2}$ cm^{-1}, and the $n=3$ ones with a $J_{n=3} \approx 10^{-4}$, etc. If W is indeed given by W', and thus is significantly larger than $J_{n=3}$, etc., it is quite evident that energy transport involving $n \geq 3$ is "incoherent" in nature. This is consistent with the picture given above (e.g., Sect. 7.3.6) about the phonon-assisted nature of the transfer (tunneling). A model is thus generated of "dynamic clusters" (e.g., defined by $n = 3$) within which there is *fast* incoherent (thermally assisted) energy transport among "coherent islands" (nearest-neighbor

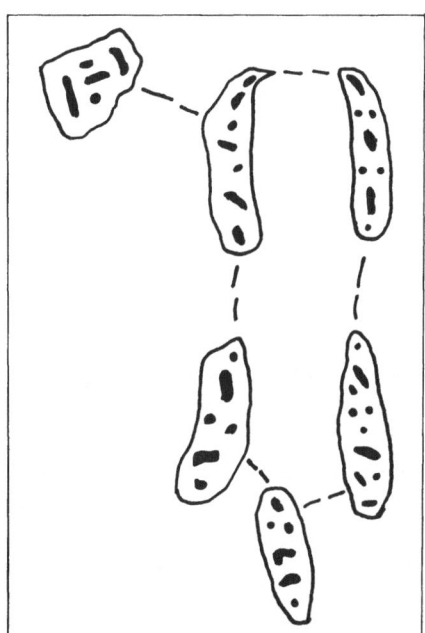

Fig. 7.21. Schematic representation of leaky dynamic clusters. The six clusters (Y = 6) include islands of coherent excitation and occasional intercluster"leaks" (dashed lines). Most of the time the energy transport is of intracluster and incoherent nature. Note that acceptors (sensors, supertraps) are *not* shown here

i.e., primary dimers, trimers, etc., as well as monomers). In addition, there are occasional "leaks" (e.g., n = 4 transfers) from one dynamic cluster to another. The picture is schematically given in Fig. 7.21.

7.3.9 Ordered vs Disordered Crystals

In the previous sections we saw that information derived from pure crystals (e.g., Davydov splitting) can be used to derive disordered crystal properties (e.g., superexchange). It is also possible to learn from disordered crystals about ordered ones. One classical example has been the utilization of pairwise interactions obtained from dilute isotopic mixed crystal spectra ("dimers") to arrive at the exciton density of states and dispersion relations of the pure crystal [7.57]. Here we discuss the utilization of heavily doped isotopic mixed crystals to study the exciton coherence problem, i.e., the exciton-phonon scattering in pure crystals. Studies on "perfect" crystals are always hampered by the fact that the contribution of scattering due to defects and impurities is hard to evaluate. The advantage of studying heavily doped isotopic mixed crystals is that the defect scattering is "saturated," i.e., it is negligible compared to the scattering from a large known amount of isotopic impurity species.

a) Random Walk

The random walk model for pure crystals is well known [7.127]. However, if next-nearest-neighbor "hops" are allowed, one must apportion relative probabilities for nearest-neighbor hops, next-nearest hops, etc. We must also emphasize that the well-known analytical formulae [7.127] are valid only for a very large number of hops, especially in the two-dimensional case. Little work has been done on the "efficiency" of the random walk as a function of time (number of steps). We present the results of such a study in Fig. 7.22. We also note here that certain statistical concepts like the "first passage time" [7.127] depend drastically on an "infinite" lifetime of the random walker; this is not appropriate for short-lived excitations. Other concepts, like "the number of distinct sites visited" [7.127] have a more transparent relationship with time and excitation lifetimes.

Random walk on a "quasilattice" (a cluster or infinite cluster) has seldom been considered. We do not expect closed analytical solutions in such a case; computer simulation seems to be the only viable alternative. Visitation efficiency can be studied as a function of time (number of steps) and relative jump probabilities. However, there is a new parameter added, C, the concentration (mole fraction) of the species that forms the quasilattice. An overview of the effects of C on the random walk is given in Fig. 7.22. We emphasize that the walker is limited to the guest quasilattice and not permitted onto the "host" quasilattice. Thus this random walk is also a special case of exciton percolation [7.96].

b) Correlated Walk

This is an old variant of the random walk problem [7.128] but has been rarely discussed [7.129]. In this model one can choose the number of correlated walks

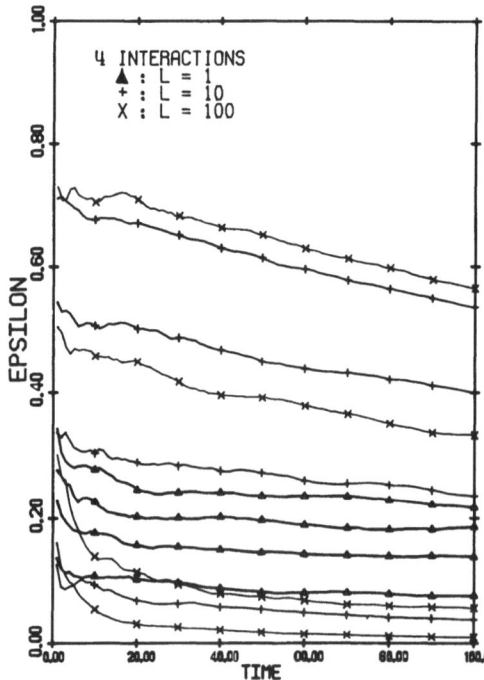

Fig. 7.22. Visitation efficiency vs time (square lattice with nearest-neighbor hops only). For each coherency parameter (l = 1, 10,100; with d = 0, 3, 30, respectively) the curves from top to bottom are, respectively, for C = 1.00, 0.90, 0.80 and 0.70 (except 0.75 for the l = 100 case). The 100 units of time correspond to 200,000 steps. [7.68]

(steps, hops) to be l, so that all l hops are in the same direction (if allowed by the structure of the quasilattice). In practice, l is replaced by a *distribution* of l values, with a standard deviation d. The number (or average number) of correlated walks simply defines our "coherency" parameter (the number of "correlated hops," Sect. 7.2.2).

The effect of walk correlation is intuitively obvious for a pure lattice — it increases the "efficiency" of visitation (see Fig. 7.22). This is not necessarily true for a correlated walk on a quasilattice. Figure 7.22 shows some specific cases, the results of which may be contrary to intuition.

c) Coherency Experiments

Both steady-state and time resolved experiments can be utilized to evaluate "coherency." We limit ourselves here to the naphthalene singlet system [7.68]. As before, a ternary system was used: $C_{10}H_8/C_{10}D_8$: BMN. However, the beta-methylnaphthalane (BMN) acceptor was the perdeuterated species, for practical reasons (availability as a "natural" impurity in the commercial host samples at the proper concentrations, and easier spectral resolution from the $C_{10}H_8$ vibronic emission).

The steady-state investigation is very similar to the one described in Sect. 7.3.3, except that much lower acceptor (supertrap) concentrations were employed, thus moving the behavior of the sample from the supertransfer limit (Sect. 7.3.5) to the "general case." Here the all important physical quantity is n_m, the number of distinct sites visited within the excitation lifetime. With fixed and known hopping time (t_h) and lifetime (τ), n_m depends on C, l and d.

The experimental results are shown in Fig. 7.23, together with a few simulation results [7.130]. There is *poor* agreement with a simple random walk model $(l = 1)$, i.e., "complete incoherence." There is *good* agreement with $20 \leqslant l \leqslant 100$, but values of $l > 100$ cannot be excluded. The latter is true because, by the introduction of between 0.01 to 0.4 of a concentration C_H of scatterers $(C_{10}D_8)$ one cannot hope to differentiate between correlation lengths once $l \gtrsim C_H^{-1}$. Thus further experiments with lower $C_H = 1 - C$ (i.e., $C > 0.99$) are desirable.

In the time resolved experiments [7.131] the same ternary system is studied, monitoring the decay of the guest $(C_{10}H_8)$ excitations, as well as the rise and decay of the acceptor (BMN) excitations in the nanosecond domain. Figure 7.24 shows the results of the experiment, again together with computer simulations for different coherency (l) values. The latter are based on a rigorous relation between the kinetics, the general percolation formalism (Sect. 7.3.5) and the computer simulations of correlated walks (Fig. 7.22), as given below.

A simple kinetic formalism rigorously applies to the ternary system, provided that there is no acceptor-donor thermalization. With the pertinent energy denominator being about 300 cm^{-1} and the temperature 2 K, this is indeed fully justified. Designating the fraction of donor (naphthalene) excitons as $N(t)$ and that of the acceptor (BMN) as $B(t)$, one obtains

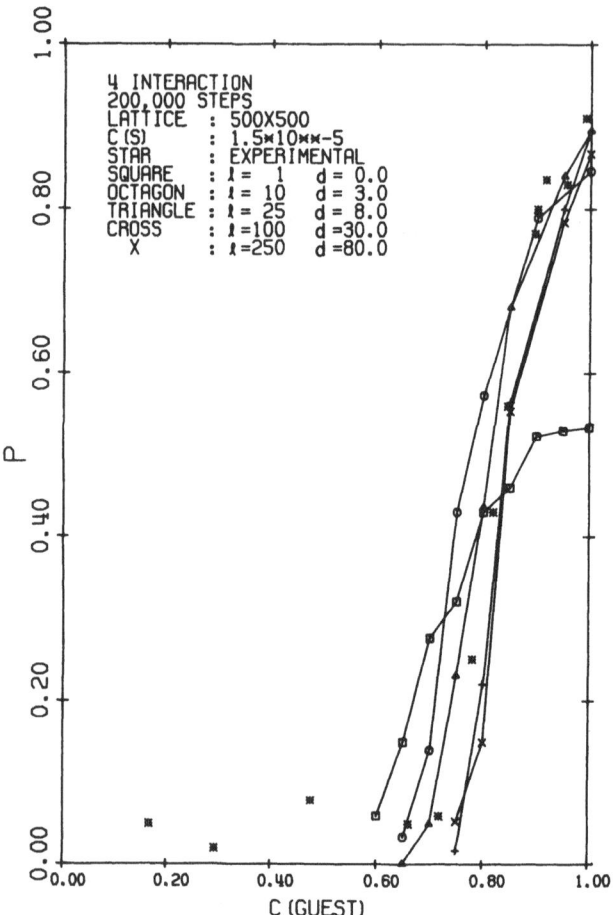

Fig. 7.23. Experimental and theoretical percolation probability P vs guest concentration C [7.130]. The experimental points (stars) are derived from the 2 K fluorescence of BMN (super-trap) in $C_{10}H_8$ naphthalene "guest" (donor), with $C_{10}D_8$ naphthalene as "host." Experimentally $P = I_s/(I_s + I_d)$

$$dN/dt = -N\tau_N^{-1} - K(t)N \tag{7.80}$$

$$dB/dt = -B\tau_B^{-1} + K(t)N \tag{7.81}$$

where K is the *time-dependent* energy transport rate (donor to acceptor) and τ_N and τ_B are the "natural" lifetimes of the donor (N) and acceptor (B). The key factor, $K(t)$ is related to the exciton percolation probability (Sect. 7.3.5) for the ternary system (note that n_m is now a function of t):

$$P(t) = \bar{P}_\infty \left\{ 1 - \exp[-n_m(t)\,\gamma\,S] \right\} + SI'_{AV}. \tag{7.82}$$

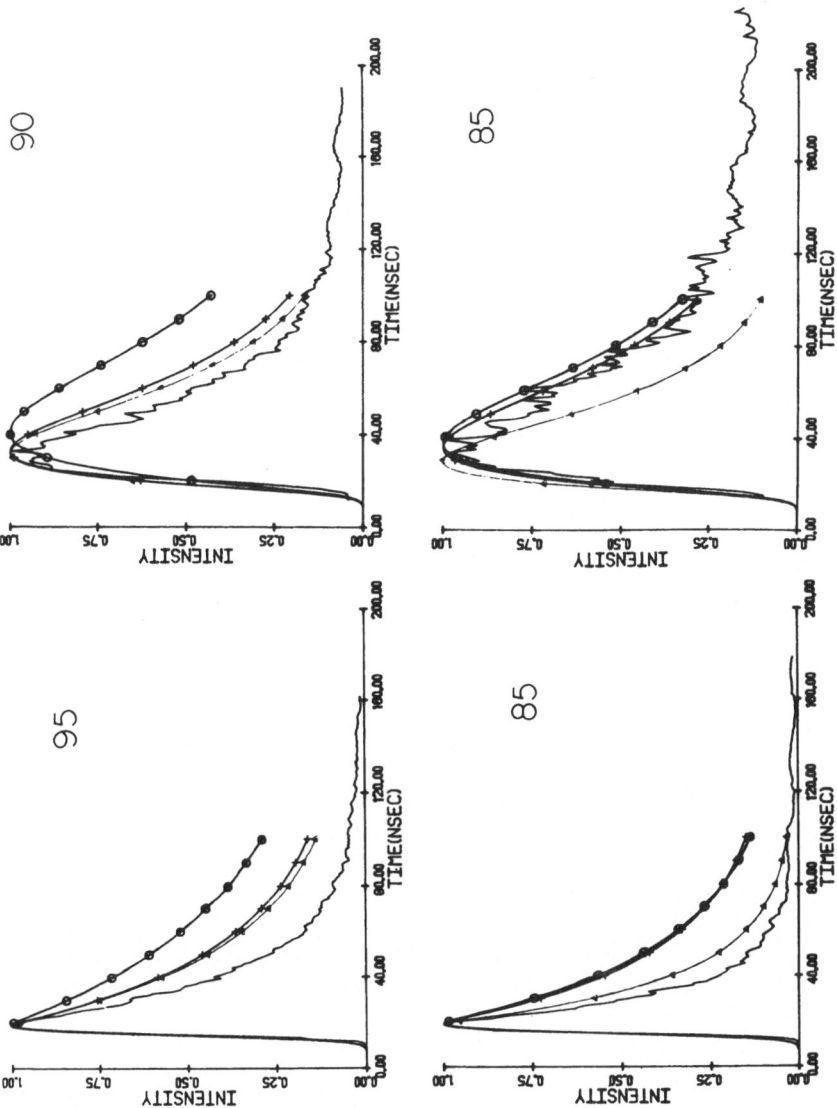

Fig. 7.24. *(Left):* Time evolution of the ternary crystal naphthalene-perdeuteronaphthalene-perdeuterobetamethylnaphthalene. The emission monitored is of the naphthalene 0–"510" vibronic band at 2 K. The intensity is normalized to unity at the maximum. The mole fractions of naphthalene are 0.95 *(top)* and 0.85 *(bottom)*. The uncertainties are about 20% (y axis). The theoretical curves (convoluted with the laser excitation line) are calculated for the 4-nearest-neighbor case with coherency values of $l = 1$ (○○○○), 10 ($\triangle\triangle\triangle\triangle$, $d=3$), and 100 (××××, $d=30$). Here nominal values were used for the lifetimes (naphthalene 100 ns, sensor 35 ns) and the sensor trapping efficiency was set to unity. Reducing the latter by about a factor of two improves significantly [7.68] the fits for $l = 100$ (but not for $l = 1$). *(Right):* Time evolution study of the ternary crystal naphthalene-perdeuteronaphthalene-perdeuterobetamethylnaphthalene. The emission monitored is the 0–0 line of betamethylnaphthalene-d_{10}. The mole fractions of naphthalene are 0.90 (top) and 0.85 *(bottom)* and those of betamethylnaphthalene are $3.4 \cdot 10^{-5}$ *(top)* and $5.1 \cdot 10^{-5}$ (bottom). All else is the same as in Fig. 7.24 *(left)*, except that the betamethylnaphthalene mole fraction is $1.7 \cdot 10^{-5}$ for the $C = 0.95$ sample. [7.131]

Defining the percolation efficiency (Fig. 7.22) as

$$\epsilon(t,c,l) = n_m/\bar{t} \equiv n_m t_h/t \tag{7.83}$$

where t_h is the hopping time and $\bar{t}$ the number of steps after time t, one obtains

$$P(t,c,l,\gamma,S) = \bar{P}_\infty \, [1-\exp(-a\epsilon t)] + SI'_{AV} \tag{7.84}$$

where some physical, time-dependent, quantities have been incorporated into the constant

$$a \equiv \gamma S/t_h. \tag{7.85}$$

It can be shown *rigorously* [7.68] that

$$K(t) = (1-P)^{-1} \, (\partial P/\partial t). \tag{7.86}$$

This gives, for $C \geqslant 0.7$, where $\bar{P}_\infty \to 1$ and $I'_{AV} \to 0$, the simple expression

$$K(t) = a(\epsilon + t \partial \epsilon/\partial t). \tag{7.87}$$

Thus, a computer simulation of $\epsilon(t,c,l)$ generates $K(t)$, enabling one to solve numerically the kinetic equations (7.80, 81). The results (Fig. 7.24) again show that $l \neq 1$ and indicate that, roughly, $l \geqslant 100$ at 4.2 K and probably higher at 1.8 K. They are also consistent with experiments on lightly doped crystals (see Sect. 7.2.2-e).

d) Role of Phonons and Inhomogeneities

The most recent value of the homogeneous linewidth for the naphthalene singlet exciton is 0.03 cm^{-1} at 1.8 K [7.132]. Using the *simple* approach (Sect. 7.2.2) to derive l, based on

$$l \cong B/\Gamma \tag{7.88}$$

where B is the exciton bandwidth and Γ the homogeneous bandwidth, one obtains $l \cong 160/0.03 = 5 \cdot 10^3$ at 1.8 K and about 10^3 at 4.2 K. This is consistent with the coherency values of the previous section (however, see Sect. 7.1.1 and 7.1.2).

The major contribution to Γ is that of exciton-phonon scattering [7.72]. This is assumed to be about the same for pure $(C_{10}H_8)$ and mixed $(C_{10}H_8/C_{10}D_8)$ crystals, as the phonons in the latter case are in the amalgamation limit [7.133] and thus are practically indistinguishable from the pure crystal phonons. Also, the addition of the BMN acceptor does *not* create localized phonons [7.134]. It is also assumed that the exciton-phonon interactions do not change drastically in the concentration range 0.7 to 1, i.e., well above percolation, where the coherent potential approximation works well for the exciton band [7.57]. Thus the coherency of heavily doped isotopic mixed crystals should in this case reflect the coherency of the pure crystal. We note that even in the "pure" crystal some of the scattering, and definitely some of the homogeneous linewidth Γ, is caused by the natural abundance of ^{13}C, which produces a 0.1 mole fraction of $^{13}CC_9H_8$ in the nominally pure $C_{10}H_8$ crystal [7.135]. On the other hand, there may be an additional inhomogeneity of a few cm^{-1} introduced in the $C_{10}H_8/C_{10}D_8$ crystals (an upper limit is given by the betamethylnaphthalene linewidths [7.136]). This may reduce the effective l to about 10–100.

7.4 Epilogue

The excitation pairwise interaction parameters of molecular solids are usually small, relative to the thermal energy (kT), the exciton-phonon interaction terms (and lineshapes) and the lattice vibrational energies. This is especially true for donor systems of low or medium concentration. As a result, the donor excited states are usually *localized* and the energy transport is essentially based on a hopping mechanism in its wider connotation. The hopping may be correlated rather than random. Incoherent hopping may be restricted to intercluster transfer while the intra-cluster transfer may be "coherent," due to the delocalized nature of the cluster states. Thus the extent of delocalization of the excited states is intimately related to the dynamics of excitation transfer. The study of excitation transport dynamics in such systems not only provides a test of our models of the solid state but is also of practical interest in itself, in relation to photosynthesis, potential solar cell concentrators and possibly organic lasers.

7.5 Addendum (to the Second Edition)

The most important recent advances concerned exciton fusion (annihilation) experiments in mixed crystals and disordered solids and their relation to some novel theoretical insights, especially concepts of fractal demensionalities in space and even time. Refined excitation relaxation experiments, involving the direct excitation of triplet states with powerful cw lasers (and modulated electro-optically) were carried out by *Gentry* [7.136] and by *Klymko* [7.137]. The first experiments led to a refined understanding of triplet exciton transport and its relation to master equation formalisms [7.138, 139]. The other experiments led to unexpected new insights involving "fractal" excitation kinetics and a totally new approach to the characterization of material heterogeneities via laser excitations. We feel that the last story deserves some amplification.

Referring to (7.85–87), relating to the excitation trapping rate coefficient $K(t)$, we rewrite (7.87) as

$$K(t) = a\partial n_m / \partial \bar{t} = K_0 \epsilon(t) \tag{7.89}$$

where we redefined

$$\epsilon \equiv \epsilon + t\partial \epsilon / \partial t. \tag{7.90}$$

Equation (7.89) gets generalized, i.e., $K(t) = K_0 \epsilon(t)$ will serve as a rate coefficient for recombination reactions such as homofusion and heterofusion. We note that $\epsilon(t)$ is the *number of distinct sites visited per unit time*. Over the time t the number of distinct sites visited by the excitation is

$$n(t) = \int_0^t \epsilon(t)dt. \tag{7.91}$$

This gives novel equations for exciton annihilation (homofusion), similar to (7.80)

$$dN/dt = -N\tau_N^{-1} - K_0 \epsilon(t)N^2. \tag{7.92}$$

The *delayed fluorescence rate* $F(t)$ is now

$$F(t) = K_0 \epsilon(t)N^2, \tag{7.93}$$

and the integrated equation relating to (7.92) is

$$N^{-1}(t) - N^{-1}(0) = K_0 n(t). \tag{7.94}$$

This relation has been tested out by simulations [7.140] for $n(t)$ being the *number of distinct sites visited by a random walker* in various percolation systems.

For a homogeneous system such as a perfect crystal $n(t)$ is well known to be linear in time (t) and thus both ϵ and K are constants in time, which is the "classical" result. However, recently it has been demonstrated [7.141] that for heterogeneous systems, the visitation efficiency ϵ is a function of time, $\epsilon(t)$, and that for those heterogeneous systems that are described by *fractal structures* one has

$$\epsilon(t) = \epsilon_0 t^{-h} \quad 0 \leqslant h \leqslant 1. \tag{7.95}$$

Moreover, the above can be related to the spectral ("fracton") dimension d_s [7.141–143],

$$h = 1 - d_s/2 \quad 0 \leqslant d_s \leqslant 2. \tag{7.96}$$

A typical value for the spectral dimension is $d_s = 4/3$, which is the case for percolation structures at criticality. This gives $h = 1/3$ for percolating clusters (the connected cluster at the critical concentration). Recent experiments on the mixed crystal system of naphthalene (described in detail above) give [7.144–146] $h = 0.36 \pm 0.03$ and thus a spectral dimension $d_s = 1.28 \pm 0.06$. The "exact" theoretical value (correcting for finite cluster effects) is [7.147, 148] $d_s = 1.26$, giving $h = 0.37$, while the simulation results [7.148] give $d_s = 1.24 \pm 0.02$ and thus $h = 0.38 \pm 0.01$. This new excellent agreement between the percolation model and the experimental results parallels the story of Sect. 7.3.8. The latter story can now be recast in terms of "fractals". The fractal dimension [7.149] d_f for a percolating cluster is simply given by [7.150]

$$d_f = d(1 - \delta^{-1}) \qquad (7.97)$$

where δ is the critical exponent $\delta = \gamma/\beta + 1$ discussed in Sect. 7.3.8. Using the *experimental* values of Table 7.2 one gets $d_f = 1.88$. Today's theoretical value [7.150] is $d_f = 1.89$ (and $\delta = 91/5$). The laser excitation experiments thus give very reasonable values for both the fractal dimension d_f and the spectral ("fracton") dimension d_s of the naphthalene percolation clusters that are defined by the triplet exciton dynamics. This seems to be, so far, the only experimental system that exhibits fractal properties down to the molecular level (5 Å). We note that the effective range of excitation transport (within the triplet lifetime) is 10–1000 Å, the effective $n(t)$ is about 500 and the effective number of exciton hops is about 10^4.

The exciton fusion (annihilation recombination) reaction,

$$triplet + triplet \rightarrow singlet \rightarrow h\nu \qquad (7.98)$$

is a *typical elementary binary reaction.* Classically one expects

$$N_s = kN_T^2 \qquad (7.99)$$

where N_s is the singlet population and N_T the triplet exciton population. Because the singlet natural decay (fluorescence) is so much faster (120 ns) than the triplet lifetime (2.7 s), one expects the delayed fluorescence (F) to follow the simple rule

$$F = KP^2 \qquad (7.100)$$

where P is the phosphorescence rate. However, for non-classical, i.e., heterogeneous media, one has a more general relation:

$$F = kP^X \qquad (7.101)$$

where X is the *reaction order.* It has been shown theoretically [7.146, 151], as well as by simulations [7.146, 105] that

$$X = 1 + 2/d_s \qquad (7.102)$$

where d_s is the effective spectral dimension. We note that for $d_s = 1.25$ one gets $X = 2.6$ (rather than 2). Indeed, based on Eq. (7.101) exciton annihilation experiments [7.146, 152] have given X values of 2.5 and 2.6 at the critical percolation concentration. Furthermore at lower concentrations one can have $0 < d_s < 1$ and thus $3 < X < \infty$. Indeed, the anomalous experimental X values [7.153] of up to 20 are in excellent agreement with very recent simulations [7.105].

The above new insights have opened up a new approach to the characterization of vapor-deposited films, polymeric and porous glasses, synthetic membranes and living tissues, using laser induced excitations [7.141]. Specifically, naphthalene (old "moth balls") can be sublimed or incorporated into a variety of media. Then these heterogeneous materials can be excited by radiation and studied in the time and energy domains.

Acknowledgement. We thank Laurel Harmon for helping to make this chapter more readable and comprehensible.

References

7.1   See for example: W.H. Flygare: *Molecular Structure and Dynamics* (Prentice-Hall, Englewood Cliffs, N.J. 1978) p. 105
7.2   F. Urbach: Phys. Rev. **92**, 1324 (1953)
7.3   Y. Toyozawa: Prog. Theor. Phys. **20**, 53 (1958)
7.4   R. Kubo in: *Fluctuation, Relaxation and Resonance in Magnetic Systems*, ed. by D. ter Haar (Oliver and Boyd, Edinburgh 1965)
7.5   H. Sumi: J. Phys. Soc. Jpn. **32**, 616 (1972)
7.6   S. Fisher, S. Rice: J. Chem. Phys. **52**, 2089 (1970)
7.7   M.K. Grover, R. Silbey: J. Chem. Phys. **52**, 2099 (1970)
7.8   D.P. Craig, L.A. Dissado: Chem. Phys. **14**, 89 (1976)
7.9   J. Klafter, J. Jortner: Chem. Phys. Lett. **50**, 202 (1977)
7.10   G.C. Morris, M.G. Sceats: Chem. Phys. **3**, 342 (1974)
7.11   C.B. Harris: Chem. Phys. Lett. **52**, 5 (1977); J. Chem. Phys. **67**, 5607 (1977)
7.12   R. Kubo, K. Tomita: J. Phys. Soc. Jpn. **9**, 888 (1954)
7.13   P.W. Anderson: J. Phys. Soc. Jpn. **9**, 316 (1954)
7.14   S.L. Robinette, G.J. Small, S.H. Stevenson: J. Chem. Phys. **68**, 4790 (1978)
7.15   J.J. Hopfield: Phys. Rev. **112**, 1555 (1958)
7.16   S.L. Robinette, G.J. Small: J. Chem. Phys. **65**, 837 (1976)
7.17   A.H. Francis, C.B. Harris: Chem. Phys. Lett. **9**, 181 (1971)
       A.H. Zewail, C.B. Harris: Chem. Phys. Lett. **28**, 8 (1974)
7.18   D.M. Burland, D.E. Cooper, M.D. Fayer, C.R. Gochanour: Chem. Phys. Lett. **52**, 279 (1977)
7.19   A.H. Francis, C.B. Harris: Chem. Phys. Lett. **9**, 188 (1971)
       D.D. Dlott, M.D. Fayer: Chem. Phys. Lett. **41**, 305 (1976)
7.20   D.M. Burland: J. Chem. Phys. **59**, 4283 (1973)
       D.M. Burland, R.M. Macfarlane: J. Lumin. **12/13**, 213 (1976)
       D.M. Burland, U. Konzelmann, R.M. Macfarlane: J. Chem. Phys. **67**, 1926 (1977)
7.21   C.B. Harris: Chem. Phys. Lett. **52**, 5 (1977)
7.22   M.R. Philpott: J. Chem. Phys. **52**, 5842 (1970)
7.23   G.C. Morris, M.G. Sceats: Chem. Phys. **3**, 332 (1974)
7.24   T. Muto: Prog. Theor. Phys. (Kyoto) **4**, 181 (1949)

7.25   K. Huang: Phys. Proc. R. Soc. (London) **A204**, 406 (1950)
7.26   S. Pekar: Zh. Eksp. Theo. Fiz. **20**, 510 (1950)
7.27   M. Lax: J. Chem. Phys. **20**, 1752 (1952)
7.28   R.C. O'Rourke: Phys. Rev. **91**, 265 (1953)
7.29   T.H. Keil: Phys. Rev. **A140**, 601 (1965)
7.30   D.L. Dexter: Solid State Phys. **6**, 353 (1958)
7.31   K.K. Rebane: *Impurity Spectra of Solids* (Plenum Press, New York 1970)
7.32   G.J. Small: J. Chem. Phys. **58**, 2015 (1973)
7.33   D.E. McCumber, M.D. Sturge: J. Appl. Phys. **34**, 1682 (1963)
7.34   P.H. Chereson, P.S. Friedman, R. Kopelman: J. Chem. Phys. **56**, 3716 (1972)
7.35   B.H. Loo, A.H. Francis: J. Chem. Phys. **65**, 5076 (1977)
7.36   B.H. Loo, A.H. Francis, K.W. Hipps: J. Chem. Phys. **65**, 5068 (1976)
         K.W. Hipps, A.H. Francis: J. Phys. Chem. **83**, 1879 (1979)
7.37   F.P. Burke, G.J. Small: Chem. Phys. **5**, 198 (1974)
7.38   G.J. Small: Chem. Phys. Lett. **57**, 501 (1978)
7.39   H. de Vries, D.A. Wiersma: Chem. Phys. Lett. **51**, 565 (1977)
7.40   H. de Vries, D.A. Wiersma: Phys. Rev. Lett. **36**, 91 (1975)
7.41   H. Sternlicht, G.C. Nieman, G.W. Robinson: J. Chem. Phys. **38**, 1326; **39**, 1610 (1963)
7.42   R.E. Merrifield: Pure Appl. Chem. **27**, 481 (1971)
7.43   R.G. Kepler, J.C. Caris, P. Avakian, E. Abramson: Phys. Rev. Lett. **10**, 400 (1963)
7.44   P. Avakian, R.E. Merrifield: Phys. Rev. Lett. **13**, 541 (1964)
7.45   V. Ern, P. Avakian, R.E. Merrifield: Phys. Rev. **148**, 862 (1966)
7.46   M. Levine, J. Jortner, A. Szoke: J. Chem. Phys. **45**, 1591 (1966)
7.47   V. Ern: Phys. Rev. Lett. **22**, 343 (1969)
7.48   V. Ern, A. Suna, Y. Tomkiewicz, P. Avakian, R.P. Groff: Phys. Rev. B **5**, 3222 (1972)
7.49   V. Ern: J. Chem. Phys. **56**, 6259 (1972)
7.50   H.C. Wolf, H. Port: J. Lumin. **12/13**, 33 (1976)
7.51   M. Trlifaj: Czech. J. Phys. **8**, 510 (1958)
7.52   D.L. Dexter: J. Chem. Phys. **21**, 836 (1953)
7.53   H. Haken, P. Reinecker: Z. Phys. **249**, 253 (1972)
         E. Schwarzer, H. Haken: Opt. Commun. **9**, 64 (1973)
7.54   R.C. Powell, Z.G. Soos: J. Lumin. **11**, 1 (1975)
7.55   A.S. Davydov: *Theory of Molecular Excitons* (Plenum Press, New York 1971)
7.56   G.W. Robinson: Ann. Rev. Phys. Chem. **21**, 429 (1970)
7.57   R. Kopelman: In *Excited States II*, ed. by E.C. Lim (Academic Press, New York 1975)
7.58   R.M. Hochstrasser, P.N. Prasad: In *Excited States I*, ed. by E.C. Lim (Academic Press, New York 1974)
7.59   L. Altwegg, M.A. Davidovich, J. Funfschilling, I. Zschokke-Granacher: Phys. Rev. B **18**, 4444 (1978)
7.60   L. Altwegg, M. Chabr, I. Zschokke-Granacher: Phys. Rev. B. **14**, 1963 (1976)
7.61   Th. Forster: Ann. Phys. **6**, 55 (1948)
7.62   F. Lipsett, A. Dekker: Can. J. Phys. **30**, 165 (1951)
7.63   H.C. Wolf: Z. Phys. **143**, 266 (1955); **145**, 116 (1956)
7.64   D.M. Hanson, G.W. Robinson: J. Chem. Phys. **43**, 4174 (1965)
7.65   G. Castro, G.W. Robinson: J. Chem. Phys. **50**, 1159 (1969)
7.66   E.B. Priestley, A. Haug: J. Chem. Phys. **49**, 622 (1968)
7.67   K.W. Benz: Ph.D. Thesis, University of Stuttgart (1970)
7.68   P. Argyrakis: Ph.D. Thesis, University of Michigan (1979); P. Argyrakis, R. Kopelman: Chem. Phys. **51**, 9 (1980)
7.69   H. Auweter: Ph.D. Thesis, University of Stuttgart (1978)
7.70   A. Hammer, H.C. Wolf: Mol. Cryst. **4**, 191 (1968)
7.71   V.M. Agranovich, Yu. V. Konobeev: Phys. Status Solidi **27**, 435 (1968)
7.72   S.L. Robinette, S.H. Stevenson, G.J. Small: J. Chem. Phys. **69**, 5231 (1978)

7.73   G.C. Nieman, G.W. Robinson: J. Chem. Phys. **37**, 2150 (1962)
7.74   H. Sternlicht, G.C. Nieman, G.W. Robinson: J. Chem. Phys. **38**, 1326 (1963)
7.75   G.W. Robinson, R.P. Frosch: J. Chem. Phys. **38**, 1187 (1963)
7.76   H-K. Hong, R. Kopelman: Phys. Rev. Lett. **25**, 1030 (1970)
7.77   H-K. Hong, R. Kopelman: J. Chem. Phys. **55**, 724 (1971)
7.78   F.W. Ochs: Ph.D. Thesis, University of Michigan (1974)
7.79   R. Kopelman, E.M. Monberg, F.W. Ochs: Chem. Phys. **19**, 413 (1977)
7.80   E.F. Krause: *Taxicab Geometry* (Addison-Wesley, Menlo Park, Calif. 1975)
7.81   F.B. Tudron, S.D. Colson: J. Chem. Phys. **65**, 4184 (1976)
7.82   S.D. Colson, R.E. Turner, V. Vaida: J. Chem. Phys. **66**, 2187 (1977)
7.83   R. Kopelman, E.M. Monberg, F.W. Ochs: Chem. Phys. **21**, 373 (1977)
7.84   F. Perrin: C.R. Acad. Sci. **178**, 1978 (1924)
7.85   H-K. Hong, R. Kopelman: J. Chem. Phys. **55**, 5380 (1971)
7.86   V.L. Broude, A.V. Leiderman, T.G. Tratas: Fiz. Tverd. Tela **13**, 3624 (1971)
7.87   R. Kopelman: Cambridge Conf. Molecular Energy Transfer (1971)
7.88   K.E. Mauser, H. Port, H.C. Wolf: Chem. Phys. **1**, 74 (1973)
7.89   H-K. Hong, G.W. Robinson: J. Chem. Phys. **54**, 1369 (1971)
7.90   H. Port, D. Vogel, H.C. Wolf: Chem. Phys. Lett. **34**, 23 (1975)
7.91   C.L. Braun, H.C. Wolf: Chem. Phys. Lett. **9**, 260 (1971)
7.92   R.M. Hochstrasser, J.D. Whiteman: J. Chem. Phys. **56**, 5945 (1972)
7.93   D.C. Ahlgren: Ph.D. Thesis, University of Michigan (1979)
7.94   R. Kopelman, E.M. Monberg, F.W. Ochs, P.N. Prasad: J. Chem. Phys. **62**, 292 (1975)
7.95   R. Kopelman, E.M. Monberg, F.W. Ochs, P.N. Prasad: Phys. Rev. Lett. **34**, 1506 (1975)
7.96   R. Kopelman: In *Radiationless Processes in Molecules and Condensed Phases*, ed. by F.K. Fong, Topics in Applied Physics, Vol. 15 (Springer, Berlin, Heidelberg, New York 1976) p. 297
7.97   S.D. Colson, S.M. George, T. Keyes, V. Vaida: J. Chem. Phys. **67**, 4941 (1977)
7.98   E.M. Monberg, R. Kopelman: V.L. Broude Memorial Issue, Mol. Cryst. Liq. Cryst. **57**, 271 (1980)
7.99   V.K.S. Shante, S. Kirkpatrick: Adv. Phys. **20**, 325 (1971)
7.100 J. Hoshen, R. Kopelman: Phys. Rev. B. **14**, 3438 (1976)
7.101 J. Hoshen, R. Kopelman, E.M. Monberg: J. Stat. Phys. **19**, 219 (1978)
7.102 P.W. Kasteleyn, C.M. Fortuin: J. Phys. Soc. Jpn. **26**, 11 (1969)
7.103 J.Hoshen, P. Klymko, R. Kopelman: J. Stat. Phys. **21**, 583 (1979)
7.104 R. Kopelman, E.M. Monberg, J.S. Newhouse, F.W. Ochs: J. Lumin. **18/19**, 41 (1979)
7.105 J.S. Newhouse: Ph.D. Thesis, University of Michigan (1985)
7.106 J. Hoshen, R. Kopelman: J. Chem. Phys. **65**, 2817 (1976)
7.107 A. Sur, J.L. Lebovitz, J. Marro, M.H. Kalos, S. Kirkpatrick: J. Stat. Phys. **15**, 345 (1976)
7.108 D.C. Ahlgren, E.M. Monberg, R. Kopelman: Chem. Phys. Lett. **64**, 122 (1979)
7.109 E.M. Monberg: Ph.D. Thesis, University of Michigan (1977)
7.110 D.C. Ahlgren, R. Kopelman: J. Chem. Phys. **70**, 3133 (1979)
7.111 D. Weaire, B. Kramer: J. Non-Cryst. Solids **32**, 131 (1979)
7.112 D.J. Thouless: In *Les Houches Ill-Condensed Matter*, ed. by R. Balian (North-Holland, Amsterdam 1979) p. 1
7.113 J. Klafter, J. Jortner: Chem. Phys. Lett. **49**, 410 (1977)
7.114 J. Klafter, J. Jortner: Chem. Phys. Lett. **60**, 5 (1978)
7.115 J. Klafter, J. Jortner: J. Chem. Phys. **71**, 1961 (1979)
7.116 S.K. Lyo: Phys. Rev. **B3**, 3331 (1971)
7.117 R. Orbach: Phys. Lett. **A48**, 417 (1974)
7.118 S.D. Colson: Private communication (1977)

7.119   R. Kopelman: Symp. Molecular Structure and Spectroscopy, Columbus, Ohio (1967)
        Paper R7
7.120   S.D. Woodruff: Ph.D. Thesis, University of Michigan (1976)
7.121   E.M. Monberg, R. Kopelman, Chem. Phys. Lett. **58**, 492, 497 (1978)
7.122   L. Harmon, R. Kopelman: Unpublished
7.123   P.W. Anderson: Phys. Rev. **109**, 1492 (1958)
7.124   D.D. Smith, R.D. Mead, A.H. Zewail: Chem. Phys. Lett. **50**, 358 (1977)
7.125   H. Port: Private communication (June 1979)
7.126   D. Burland, A.H. Zewail: Adv. Chem. Phys. **40**, 369 (1979)
7.127   E.W. Montroll: J. Math. Phys. **10**, 753 (1969)
7.128   S. Chandrasekhar: Rev. Mod. Phys. **15**, 1 (1943)
7.129   K. Lakatos-Lindenberg, K.E. Shuler: J. Math. Phys. **12**, 633 (1971)
7.130   P. Argyrakis, R. Kopelman: J. Chem. Phys. **66**, 3301 (1977)
7.131   P. Argyrakis, R. Kopelman. Chem. Phys. Lett. **61**, 187 (1979)
7.132   G.J. Small: Private communication (June 1979). See also ref. [7.72]
7.133   P.N. Prasad, R. Kopelman: J. Chem. Phys. **57**, 863 (1972)
7.134   R. Kopelman: J. Phys. Chem. **80**, 2191 (1976)
7.135   J.C. Laufer, R. Kopelman: J. Chem. Phys. **57**, 3202 (1972)
7.136   S.T. Gentry, R. Kopelman: J. Phys. Chem. **88**, 3170 (1984); J. Chem. Phys. **81**, 3014
        (1984);
        S.T. Gentry: Ph.D. Thesis, University of Michigan (1983)
7.137   P.W. Klymko: Ph.D. Thesis, The University of Michigan (1984)
7.138   S.T. Gentry, R. Kopelman: J. Chem. Phys. **81**, 3014, 3022 (1984)
7.139   R.P. Parson, R. Kopelman: J. Chem. Phys. **82**, 3692 (1985)
7.140   P. Argyrakis, R. Kopelman: In *Advances in Chemical Reaction Dynamics*, ed. by P.M.
        Rentzepis, C. Capellos (Reidel, Dordrecht, Holland 1985)
7.141   R. Kopelman: J. Stat. Phys. **42**, 185 (1985)
7.142   S. Alexander, R. Orbach: J. Phys. (Paris) Lett. **43**, 625 (1982)
7.143   R. Rammal, G. Toulouse: J. Phys. Lett. **44**, 13 (1983)
7.144   R. Kopelman, P.W. Klymko, J.S. Newhouse, L.W. Anacker: Phys. Rev. B **29**, 3747
        (1984)
7.145   L.W. Anacker, P.W. Klymko, R. Kopelman: J. Lumin. **31/32**, 648 (1984)
7.146   L.W. Anacker, R. Kopelman: J. Chem. Phys. **81**, 6402 (1984)
7.147   I. Webman: J. Stat. Phys. **36**, 603 (1984)
7.148   P. Argyrakis, R. Kopelman: Phys. Rev. B **29**, 511 (1984)
7.149   B.B. Mandelbrot: *The Fractal Geometry of Nature* (Freeman, San Francisco 1983)
7.150   D. Stauffer: *Introduction to Percolation Theory* (Taylor and Francis, London 1985)
7.151   L.W. Anacker, R.P. Parson, R. Kopelman: J. Phys. Chem. **89**, 4758 (1985)
7.152   P.W. Klymko, R. Kopelman: J. Phys. Chem. **86**, 3686 (1982)
7.153   P.W. Klymko, R. Kopelman: J. Lumin. **24/25**, 457 (1981)

8. Addendum (to the Second Edition)

W. M. Yen

Since the first publication of this volume some five years ago much progress has been achieved in all areas of laser spectroscopy as applied to ordered and disordered solids. The volume of original literature on this subject has experienced a nearly exponential growth in this period attesting to the widening interest in this field. As one of the consequences, many more excellent reviews of individual aspects or problems in the laser spectroscopy of solids have appeared in the recent past. Thus in this addendum, which is intended to be brief, we will rely on citations of a sampling of this reference material in order to provide the reader with detailed updates on the status of this field.

One of the aims of the first edition was to illustrate that various physical and chemical disciplines shared a common interest in the study of solid-state optical properties though much of the terminology used has tended, because of tradition, to vary from field to field. It is gratifying to see that our effort was a preliminary step in the increased awareness of these shared interests between physicists and chemists and that this has, in turn, resulted in a substantial growth in interdisciplinary forums focussing on laser spectroscopic studies of solids. An excellent example of such a merging of interests can be found in the recent conference proceedings edited by *Trommsdorf* and *Jacquier* [8.1]; this meeting which dealt with dynamical processes also provides a sampling of the status and current interests in the field as of 1985.

As to specific chapters in the present quasi-monograph, the first chapter was intended to serve as a brief introduction to the general terminology of solid-state spectroscopy, consequently this chapter remains useful as such and requires no update. Other than the reference literature already cited in the various chapters, an additional comprehensive introduction to optical properties of solids by *Imbusch* and *Henderson* [8.2] will become available shortly. The volume provides a thorough overview of the general spectroscopic properties of active centers in solids.

Chapters 2 and 3 require not major alteration at this time, as experimental results obtained in the past few years have generally been in agreement with these theoretical developments. For example, the phonon mediated interactions believed to be responsible for the inter-ionic transfer of optical energy are solidly established. Recent experimental studies have shown the existence of several multipolar processes as well as having demonstrated the importance of exchange mechanisms in pair-wise transfer [8.3]. The controversy which existed in the interpretation of transfer in ruby, particularly *viz* the problem of Anderson localization, has mostly been resolved [5.156, 162]. It would appear that the initial observations of *Koo* et al. [5.123] were spurious, that the transfer in the Cr^{3+} system is dipolar and that resonant transfer among Cr^{3+} ions is present but is very slow. This problem and its resolution was discussed in detail by *Gibbs* and his co-workers in a recent review [8.4]. Thus, localization phenomena still await confirmation through an appropriate optical experiment and the transition from incoherent to coherent optical energy transport remains an important problem to be investigated.

Considerable progress has also been achieved in the understanding of the macroscopic aspects of the excitation transfer problem. Laser-spectroscopic techniques such as TRFLN have provided us with a simple and convenient way with which to measure donor-donor dynamics directly. The availability of this additional input has made it possible to experimentally quantify the ranges over which various macroscopic transfer models are valid. A number of comprehensive studies have been carried out in model rare-earth systems which have relied explicitly on our understanding of the parametric dependences of the microscopic interionic interaction [3.29, 8.5]. The status of these studies has also been reviewed by *Yen* and others [8.6, 7]. More recently a theoretical update of macroscopic aspects of optical excitation transfer processes have been provided by *Burshtein* [8.8]; in this review the author also furnished a complete list of Russian references which had not been generally available to the Western readership.

In reference to Chap. 4, the general methodology for conducting laser spectroscopy in the solid state has not changed a great deal over the past few years. This is not to say, however, that the specific tools and techniques have not improved in the meantime. The interim since the first edition has witnessed impressive growth in the commercialization of diverse laser sources such as high power Nd^{3+}-YAG and excimer lasers. The availability of F-center lasers have also provided us with the means to expand spectroscopic investigations into the near ir [8.9]. It is to be noted that the continuing interest in the development of optical materials for solid-state tunable lasers has provided considerable stimulus to advances in this field and to the study of optical properties in general [8.10]. Improvements have been realized not only in increased ranges of spectral coverage but also in increases in both temporal and frequency resolution. Picosecond time-resolved spectroscopy, though a relatively expensive endeavour, is now being conducted routinely in many laboratories even as the frontiers in this technology have already been pushed well into the femtosecond region. Many of the problems arising from laser jitter and instabilities which plagued coherent transient and other measurements of this type have been solved or cleverly bypassed so that truly intrinsic spectra has been obtained in many cases. These techniques coupled with other forms of resonant spectroscopy have helped to reveal a host of additional interactions which affect the electronic states of active centers in solids. Coherent techniques as well as recent results obtained through the use of them have been reviewed by by a number of authors [8.11, 12]. These reviews, as well as some more general ones [8.14] serve to supplement material appearing in Chaps. 4 and 5.

The improvement of our general experimental capabilities as well as the widespread increase in research interest and activities in this field have had considerable impact on the material described in the experimental overview chapters, i.e., Chaps. 5–7.

For the case of crystal or ordered materials, the advances achieved are noticeable. The whole methodology of laser spectroscopy is now routinely applied to characterize materials and dynamical optical properties which are of interest. For example, a recent paper reports the successful indirect extraction of an intrinsic width of 9.8 Hz for the ^2E level of Cr^{3+} in $YAlO_3$ through the use of a variation of degenerate four-wave mixing [8.15]. As has been noted above, various controversies alluded to in Chap. 5 have found resolution inclusive of the problem of transfer in ruby and of certain properties of stoichiometric materials [8.16].

The availability of high-power tunable devices has also contributed to the expansion of studies which rely on nonlinear optical properties of solids. Again these methods have been reviewed in detail by *Levenson* [8.17] with the theory being reviewed by *Shen* [8.18]. We have also witnessed during this period a revival in interest in multiphoton and multicenter processes in solids. Results in the former have been reviewed by *Bloembergen* [8.19]; studies of two-photon absorption processes have proven to be extremely interesting, as they have led to a re-examination of the Judd-Ofelt approximations [8.20]. In the case of multi-ion processes, studies in this area using upconversion have helped ellucidate exchange processes which produce pairing in rare-earth systems and have yielded information on the microscopic origins of broadening in these systems [8.3, 4.21].

Several reviews updating the status of laser-spectroscopic studies in insulating glasses have also appeared recently [8.22–24]. Notable advances have been achieved in the study of structural details of glasses through the use of simulation and through the application of site-selective spectroscopies. There has, in addition, been considerable theoretical and experimental interest in the relaxation processes which affect the optical line widths of transitions in glasses. Though at this writing the exact role of various excitations of the disordered system, i.e., TLS, fractals, is not completely clear, the likelihood that this problem which is of interest in organic and inorganic compounds alike will be resolved in the near future [8.25]. As is noted in Sect. 7.9, fractals have proven to be a useful concept in dealing with the kinetics of organic materials, and this concept has been invoked in an experiment dealing with relaxation of centers in an etched inorganic glass [8.26]. Again it is not entirely clear at this time whether these concepts are universally translatable to all insulating glasses.

Finally, a brief update of developments in the spectroscopy of organic materials has been provided by Sect. 7.9. The organic field has benefitted specially from the availability of subnanosecond sources and a variety of fast relaxation and transfer processes are being actively investigated in a number of laboratories around the world; various reviews of these advances have also become available [8.27–30].

As we continue to emphasize, this field of endeavors has enjoyed a period of considerable activity and interest. The source of the impetus in the field originates not only from the interest in fundamental-physical processes but also from the demand from various techniques for well quantified and understood efficient optical materials. For example, the requirement for compact and powerful laser sources which are tunable have naturally led to a re-examination of a number of solid-state systems. The importance of this area of studies is then likely to be maintained so that the usefulness of this volume as an introduction and as a historical, if nothing else, reference will not be limited by the passage of time. Our present intention is to prepare sequel volumes to this one for detailed updates and to provide coverage in areas which for various reasons we have had to omit in the present work.

References

8.1 *Proceedings of the Dynamical Processes Conference, DPC '85*, ed. by H.P. Trommsdorf and B. Jacquier, J. Physique **46** (C7) (1985)

8.2 G.F. Imbusch, B. Henderson: *Optical Properties of Inorganic Solids* (Oxford U. Press, Oxford 1986)

8.3 J.C. Vial, R. Buisson: J. Physique Lett. **43**, L745 (1982)

8.4 H.M. Gibbs, S. Chu, S.L. McCall, A. Passner: In *Coherence and Energy Transfer in Glasses*, ed. by P.A. Fleury and Brage Golding (Plenum, New York 1984) p. 373

8.5 J. Hegarty, D.L. Huber, W.M. Yen: Phys. Rev. **B25**, 5638 (1982)
G.P. Morgan, D.L. Huber, W.M. Yen: J. Physique **46** (C7), 25 (1985)

8.6 W.M. Yen: In *Spectroscopy of Rare Earth Ions in Crystals*, ed. by R.M. Macfarlane and A.A. Kaplyanskii (North Holland, Amsterdam 1986);
also in *Rare Earth Spectroscopy*, ed. by B. Jezowska-Trzebiatowska, J. Legendziewicz and W. Strek (World Scientific, Singapore 1985) p. 484

8.7 V.M. Agranovich, M.D. Galanin: *Electronic Excitation Transfer in Condensed Matter* (North Holland, Amsterdam 1982)

8.8 A.I. Burshtein: J. Lumin. **34**, 167 (1985)

8.9 L.F. Mollenauer: In *Laser Handbook*, ed. by M.L. Stich and M. Bass (North Holland, Amsterdam 1985) Vol. 4, Chap. 2

8.10 B. DiBartolo: *Spectroscopy of Solid-State Laser Type Materials* (Plenum, New York 1986)
P. Hammerling, A.B. Budgar, A. Pinto (eds.): *Tunable Solid State Lasers*, Springer Ser. Opt. Sci., Vol. 47 (Springer, Berlin, Heidelberg 1985)
R.L. Byer, E.K. Gustafson, R. Trebino (eds.): *Tunable Solid State Lasers for Remerte Sensing*, Springer Ser. Opt. Sci., Vol. 51 (Springer, Berlin, Heidelberg 1985)

8.11 Richard C. Brewer, Ralph G. DeVoe: In *Coherence and Energy Transfer in Glasses*, ed. by P.A. Fleury and Brage Golding (Plenum, New York 1984) p. 171

8.12 R.M. Macfarlane, R.M. Shelby: *Ibid.* p. 189

8.13 R.M. Macfarlane, R.M. Shelby: In *Spectroscopy of Rare earth Ions in Crystals*, ed. by R.M. Macfarlane and A.A. Kaplyanskii (North Holland, Amsterdam 1986)

8.14 David S. Kliger: *Ultrasensitive Laser Spectroscopy* (Academic, New York 1983)

8.15 D.G. Steel, S.C. Rand: Phys. Rev. Lett. **55**, 2285 (1985)
B. Zeldovich, N. Philipetsky, V. Shkunov: *Optical Phase Conjugation*, Springer Ser. Opt. Sci., Vol. 42 (Springer, Berlin, Heidelberg 1985)

8.16 M.M. Broer, D. L. Huber, W.M. Yen, W.K. Zwicker: Phys. Rev. **B29**, 2382 (1984)

8.17 M.D. Levenson: *Introduction to Nonlinear Laser Spectroscopy* (Academic, New York 1982)

8.18 Y.R. Shen: *Principles of Nonlinear Optics* (Wiley, New York 1984)

8.19   N. Bloembergen: J. Lumin. **31/32,** 23 (1984) and references therein
8.20   B.R. Judd: In *Rare Earth Spectroscopy,* ed. by B. Jezowska-Trzebiatowka, J. Legend-ziewcz and W. Strek (World Scientific, Singapore 1985) p. 575
8.21   C.G. Levy, S. Huang, S.T. Lai, W.M. Yen: J. Lumin. **24/25,** 659 (1981)
8.22   M.J. Weber: Ceramics Bull. **64,** 1439 (1985) and references therein
8.23   W.M. Yen: In *Optical Spectroscopy of Glasses,* ed. by I. Zschokhe-Gräncher and D. Haarer (Reidel, Amsterdam 1986)
8.24   W.M. Yen: In *Coherence and Energy Transfer in Glasses,* ed. by P.A. Fleury and Brage Golding (Plenum, New York 1984) p. 145
8.25   S.K. Lyo: In *Optical Spectroscopy of Glasses,* ed. by I. Zschokhe-Gräncher and D. Haarer (Reidel, Amsterdam 1986)
8.26   U. Even, K. Rademann, J. Jortner, N. Manor, R. Reisfeld: Phys. Rev. Lett. **52,** 2164 (1984)
8.27   A.H. Zewail, D.D. Smith, J. Lemaistre: In *Excitons,* ed. by E.I. Rashba and M.D. Sturge (North Holland, Amsterdam 1982)
8.28   V.M. Agranovich, R.M. Hochstrasser: *Spectroscopy and Excitation Dynamics of Condensed Molecular Systems* (North Holland, Amsterdam 1983)
8.29   V.L. Broude, E.I. Rashba, E.F. Sheka: *Spectroscopy of Molecular Excitons,* Springer Ser. Chem. Phys., Vol. 16 (Springer, Berlin, Heidelberg 1985)
8.30   M. Ueta, H. Kanzaki, K. Kobayashi, Y. Toyozawa, E. Hanamura: *Excitonic Processes in Solids,* Springer Ser. Solid-State Sci., Vol. 60 (Springer, Berlin, Heidelberg 1986)

Subject Index

Topics in Applied Physics Founded by Helmut K. V. Lotsch